非凡之光　LKL 建筑照明设计（第 4 卷）

Licht Kunst Licht 4
Innovative Lighting Design for Extraordinary Spaces

[德国] 安德烈亚斯·舒尔茨　著
江怡辰　译

江苏凤凰科学技术出版社 · 南京

图书在版编目（CIP）数据

非凡之光 ：LKL 建筑照明设计 . 第 4 卷 ：汉文、英文 /（德）安德烈亚斯 · 舒尔茨著 ；江怡辰译 . -- 南京 ：江苏凤凰科学技术出版社，2022.1

ISBN 978-7-5713-2440-7

Ⅰ . ①非… Ⅱ . ①安… ②江… Ⅲ . ①建筑照明－照明设计－汉、英 Ⅳ . ① TU113.6

中国版本图书馆 CIP 数据核字 (2021) 第 200727 号

非凡之光　LKL 建筑照明设计（第 4 卷）

著　　者	［德国］安德烈亚斯 · 舒尔茨
译　　者	江怡辰
项目策划	刘立颖　庞　冬
责任编辑	赵　研　刘屹立
特约编辑	庞　冬
出版发行	江苏凤凰科学技术出版社
出版社地址	南京市湖南路 1 号 A 楼，邮编：210009
出版社网址	http://www.pspress.cn
总　经　销	天津凤凰空间文化传媒有限公司
总经销网址	http://www.ifengspace.cn
印　　刷	凸版艺彩（东莞）印刷有限公司
开　　本	787 mm × 1092 mm　1/8
印　　张	40.5
字　　数	259 200
版　　次	2022 年 1 月第 1 版
印　　次	2022 年 1 月第 1 次印刷
标准书号	ISBN 978-7-5713-2440-7
定　　价	368.00 元

图书如有印装质量问题，可随时向销售部调换（电话：022-87893668）。

目录

CONTENTS

序言

建筑照明设计——理论及背景介绍

没有光就没有生命

光是生命本身的先决条件，也是我们所熟知的开展人类社会活动的重要前提。随着人们对光线不断深入地理解，逐渐意识到自然光和人工光照明不仅仅为我们提供了视觉感知活动的条件，而且还影响着人类身体和情感的各个方面。正因如此，社会对专业照明设计的需求才全所未有的迫切。

起源——“大爆炸”

这个始于137亿年前的事件，被天文学家称为“大爆炸”。这是一个有趣的术语，因为在我们的固有认知中，爆炸总是在现有空间中发生的，但这个“大爆炸”同时创造了空间和时间，也就创造了宇宙本身。

在最开始的时候，辐射能量非常集中。光子的密度非常之高，因此产生了不透明的雾。因为当时还没有出现生命，所以爆炸声不曾被听见，雾气也不曾被看见。虽然“大爆炸”产生了对视觉和听觉感知的物理刺激，但是当时没有眼睛可以去看，也没有耳朵可以去听。

即便这样，我们仍然从“大爆炸”的那一瞬间起测量本底辐射。这其中包含的信息构成了“大爆炸”理论的基础。我们生活的星系，即银河系，诞生于大约45亿年前。根据计算，银河系由2000亿～4000亿颗恒星组成，其中之一就是太阳，它被8个行星环绕。

太阳是生命起源的先决条件。作为长期而复杂的生命进化过程中的一部分，现代人（智人）出现在大约20万年前的非洲大草原上。随后，现代人从那里迁徙至全球各地，并在约25 000年前出现在欧洲大陆。

人类在生理结构上适应了当初在非洲大草原的生活。从本质上看，我们与当时的人类完全相同，只是如今我们的生活环境和当初大不一样，当代人主要生活在城市环境中。从生理结构上看，人类习惯于户外生活，但当代社会环境造成人类将更多的时间花在室内空间。这种差异给我们的身体机能带来了挑战，进而影响着当代社会与建筑环境。这也是现代照明设计中最需要考虑的因素之一。

建筑照明设计的理论基础

建筑照明设计是一门交叉学科，包括理论知识和基于经验的实践探索。人体生理与感知系统是建筑照明设计的两大理论依据。

光的物理定义是一种“电磁辐射”。实际上，人眼只能感知到整个光谱中相对较窄的一段波段，因此通常被我们称之为“光”的东西其实只是被我们感知到的部分光谱。但问题在于，对光的物理定义无法很好地描述我们所看到的光。

在20世纪很长的一段时期内，我们遵循着一种误导性的信念，即我们可以通过对物理刺激（即电磁辐射）量化的、详细的描述来表达视觉印象。如今，我们才意识到通过物理定义提供的信息来理解日光和人工光照明是非常有限的。

视觉感知理论（V/P- theory）将人类看见了什么以及人类视觉感知系统是如何工作的列入其研究范畴。从通过大脑自主处理和分类功能所形成的眼球刺激，到最终大脑有意识地以及潜意识地对刺激信号的反应，该理论研究人类如何通过一系列感知活动来获取可用的视觉信息。

关于获取可用视觉信息的过程，需要注意的第一点是，它与环境相关，并且随着环境的变化而变化。因此，关于它的测量和计算方法尚不完善。在人类视域范围内的视觉对比会显著影响我们的视觉感受。总体而言，以下7个基于人类视觉的照明因素影响着所需光线的品质：①光照水平，②光线分布，③眩光，④阴影，⑤反射，⑥光色，⑦显色性。

对照明设计师经典的系统化训练包含如何在实践中运用基于视觉的照明因素进行设计，进而积累设计经验，为做出差异化的和与环境紧密相连的照明设计奠定基础。

历史背景——维特鲁威理论和随时间发展的理论趋势

古罗马建筑师、工程师马可·维特鲁威大约在基督出生的时代就撰写了《建筑十书》。

维特鲁威认为，建筑设计师一定不能仅仅是建筑物的建造者。他必须具有与解剖学、几何学、地质学、地理学相关的知识，并且需要具备音乐和艺术方面的知识储备，才能从事相关工作并取得成功。由此推理，对现代建筑照明设计师的要求也应与此相似。

根据维特鲁威的观点，建筑应满足坚固性、实用性和美观性。遵循当代照明设计的出发点，我们将它们称之为可持续性、功能性和美学性。

维特鲁威的理论成为1500年后文艺复兴时期建筑设计的指导原则。他启发并影响了达·芬奇、布拉曼特、布鲁内莱斯基和米开朗琪罗等人。1490年，达·芬奇根据其建筑理论创作了著名的手稿作品《维特鲁威人》。

时过境迁，几个世纪以来，关于优质照明的概念已经发生了变化。在文艺复兴时期，理想的照明是合理的环境照明，旨在真实地反映空间和细节。因此当时的照明解决方案是根据该时期所奉行的哲学理念来进行设计的。

16世纪和17世纪是巴洛克时期，这个时期的风格与文艺复兴时期的风格主张形成了强烈的对比。当时的照明充满戏剧性，多用彩色，具有剧场感，并较多地采用聚光效果来突出环境中的主体部分，而将其余部分隐藏在暗处。这种风格得到了名门望族的认可，突出了皇室的辉煌和国王、君主的权威感。

这些照明趋势的转变是文明发展过程中一个自然形成的现象。如果跳过其中部分历史，直接进入20世纪，那么我们就来到了现代主义时期。现代主义的理想照明构成与文艺复兴时期的照明相似，但其根本理念却不同。在现代主义时期，建筑和空间的广泛使用是其本质理念的一个典型特征。采用通用性设计方法的另一层意义在于社会价值观倾向于为普通老百姓及其生活打造建筑空间。

随着社会的发展，现代主义逐渐转变为后现代主义，它将经典的历史元素与当代的设计和构造概念融合在一起。因此在后现代主义时期，理想的照明解决方案也代表着一种具有不同分布特性的自然光和人工光照明的融合。

事实上，当代建筑照明设计师所面临的挑战之一是人们所接触到的往往是多种建筑风格并存以及多种建筑形式混合的状态。譬如现代主义、后现代主义、国际主义、历史折中主义以及受地区与乡村田园灵感启发的建筑风格。除此之外，对于绿色建筑的追求将越来越深入地影响着未来建筑环境的塑造，同时影响着照明的解决方案。如今的这些设计演变发生得非常之快，有时只需要几年时间，整体理念就会改变。而在过去，设计原则与理念的演变往往需要跨越几个世纪。当下，我们注意到设计方法的加速发展与日新月异的变化，与此同时，我们也可以观察到更多关于文化、时尚和群体偏好发展过程中的异曲同工之处。

技术上的转变——白炽和电致发光

我们生活在一个瞬息万变的时代，在这个时代中已经有一些专业拥有了使用新技术的端口，而且这些专业已从根本上改变了人们开展工作的条件、工具和方法。

其中受技术研发进步影响最深远的专业之一就是照明设计。在过去的15年中，照明行业发生了巨大的变化。

技术的变革意味着产生光照的方式发生了根本性的变化。在以前的很长一段时间，我们一直通过火、火焰和热制造人工光。大约150年前，人类在真空环境中通过电力加热灯丝，发明了白炽灯泡。白炽是热辐射的一种特殊情况，它在可见光谱波段也可以发出电磁辐射，其辐射特性与灯丝的温度直接相关。（拉丁语中的Incandescere意思为白热化）

不同于热辐射光源，LED技术是通过另一种原理产生光，我们将其发光过程称之为电致发光。它是冷体辐射的一种形式，可以通过化学反应或电能产生。发光二极管（LED）由半导体晶体层和用于颜色转变的荧光粉层组成。其中一头较低的电压会导致电子在层间移动，当电子回到其平衡位置时，它将发出光辐射。这从根本上区分了冷发光与白炽辐射。

LED光源最重要的优势之一就是其出色的效能，它的发光效率正在飞速提升，且潜力尚未被充分发掘出来。LED光源另一个显著的优势是使用寿命超长，它的使用寿命约为50 000小时。

LED是一种通过直流电（DC）工作的电光源，这意味着LED灯具需要配置一个镇流器，将日常使用的交流电（AC）转换为直流电之后才能正常使用。直流电的应用特性使LED光源能够较容易地利用太阳能板或电池发光。LED的电子特性也有助于使其控制方式与现代信息技术（IT）相融合。

记录、宣传和发展

建筑照明设计仍然是一个相当年轻的专业，因此关于这个专业的重要性的宣传还是十分有必要的。毋庸置疑，要达到此目的最为有效的方法就是提供并记录高品质的参考案例。多角度的照明解决方案展示了目前照明设计的复杂性以及基于当下照明技术发展所带来的无比丰富的设计可能性。

持续发展照明设计专业的重要性显而易见。对于高等教育而言，也始终需要专业的实践案例和科学性的研究与发展作为参考。

基于照明设计的科学研究尚未达到较高的高度。目前占主导地位的研究课题主要是在照明技术、医学、心理学等相关交叉领域里进行的。这些研究对照明设计专业来说是相当有价值的知识扩展。但是，为了更好地理解照明设计，我们的目标应该是将这些研究成果与发现结合到一个跨学科、整体的知识库中，从而更深入地剖析照明设计的复杂之处，也更好地展现照明设计的魅力。

基于这些不同方面的研究与发现，我们有必要制定一种新的、普适的、跨学科的照明设计理论。当我们将其与建筑设计放在一起比较时，就会清楚地发现其中蕴含了许多不同的理论基础和知识体系。例如空间的、美学的、技术的、功能性的、社会性的理论与知识，等等，而这些知识体系在目前的照明设计领域中是极度缺乏的。

新维度的灯光

Licht Kunst Licht照明设计事务所（以下简称“LKL”）撰写的这本《非凡之光 LKL建筑照明设计（第4卷）》是其对所参与的专业照明设计项目的最新文献记录。这本书也是他们以往书籍的自然延续。他们长期以来连续的、多样化的项目经历使本书中的内容极具启发性。这些内容也适合相关专业的人士参考，成为广受欢迎的阅读材料。

从这本书中，读者可以体会到两个趋势。一个是尽管近年来相关技术发生了巨大的变化和发展，但对根本性的照明设计品质的追求和把控是始终保持不变的。另一个趋势是随着建筑设计方向的改变，照明解决方案也随之变化。

照明在建筑中的角色已不仅仅聚焦于其功能性，越来越多的人开始重视照明能够带来的舒适性与愉悦感。这意味着照明设计在建筑概念的生成和实施过程中扮演着无可替代的角色。在当代照明设计中，设计师不仅要根据具体的空间活动特性提供照明方案，而且还应将使用者的需求纳入每一个设计方案的考虑范畴。照明需要满足人类的视觉需求，满足其生理和昼夜节律的需要，并营造出令人心情舒畅的空间氛围。每一个具体的项目对照明任务和需求都有先后顺序，尽管如此，但人们总是对整体的环境做出反应。一个良好的照明方案可以潜移默化地或直接地，甚至十分精准地向使用者传递信息。这有助于使用者获得良好的空间引导感和幸福感。这本书通过很多优秀案例来说明这种整体性的设计方法。

简·埃希德，终身荣誉教授

ARCHITECTURAL LIGHTING DESIGN – THEORY AND BACKGROUND

WITHOUT LIGHT NO LIFE

Light is the prerequisite for life itself and thus also for social human life as we know it. As we now know, light and lighting do not only provide the conditions for our visual perception, they affect the human being physically and emotionally. Therefore, the need of professional lighting design has never been more pronounced than it is today.

THE BEGINNING – THE BIG BANG

It started 13.7 billion years ago with what the astronomers call the Big Bang. It is a curious term because in our conception, an explosion always takes place in an existing space, but the big bang simultaneously created space and time, and therefore the universe itself.

In the very beginning there was an enormous concentration of radiation energy. The density of photons created a non-transparent fog. But no bang was heard and no fog was seen, because no life existed. The physical stimuli for the visual and acoustic perception were created but there were no eyes to see and no ears to hear.

Still we measure background radiation from the moment of the big bang. The information contained therein forms the basis for the big bang theory. Our galaxy, 'The Milky Way' was born about 4.5 billion years ago. It has been calculated that the galaxy consists of approximately 200 – 400 billion stars and one of these is our sun, which is orbited by eight planets.

The sun is the condition for life. As part of a long and intricate evolution of complex life forms, the modern man (Homo Sapiens) emerged at the African Savannah approximately 200,000 years ago. From there he emigrated and appeared in Europe 25,000 years ago.

The humans are biologically adapted to life in the African Savannah. Essentially, we are exactly the same, except nowadays we live in a totally different, mostly urban environment. The discrepancy between our biological need to be outdoors and our contemporary tendency towards indoor life creates challenges for our organism and therefore for our society and built environment. This is one of the most relevant tasks of modern lighting design.

ARCHITECTURAL LIGHTING DESIGN – THE THEORY

Architectural lighting design is a multidisciplinary topic and it includes both theoretical evidence and experience-based knowledge. The two different theoretical bases are physical and perceptual.

The physical definition of light is 'electromagnetic radiation'. Only a relatively narrow band from the entire spectrum can actually be perceived by the human eye and is therefore commonly referred to as 'light'. The problem is that physical definition of light is poorly describing what we see.

For a long period in the 20th century, we followed the misleading belief that we can describe the visual impression by a quantity and detailed description of the physical stimuli, the electromagnetic radiation. Nowadays, we are aware of the limited information the physical definition provides for the understanding of light and lighting.

The visual perceptual theory (V / P-theory) takes into consideration what we see and how the human visual perception is working; from the stimuli of the eye via the autonomous processing and classification function in the brain, to the final conscious and subconscious interpretation of the processed stimuli for getting useable visual information.

The first statement of this process is that it is relative and dynamically related to the environmental context. Its methods of measurement and calculation are therefore rather inadequate. The contrast in the visual field significantly influence our visual sensation. At large, the following seven visually based factors are describing the required lighting quality: 1) Light level, 2) Light distribution, 3) Glare, 4) Shadows, 5) Reflections, 6) Light colour and 7) Colour rendering.

A classically structured training of lighting designers encompasses learning how to use the visual based factors in practice, building up a bank of experiences and thus forming the basis for a differentiated and context related lighting design.

HISTORICAL BACKGROUND – VITRUVIUS' THEORY AND TREND SHIFTS OVER TIME

The Roman architect and engineer Marcus Vitruvius Pollio wrote De architectura libri decem (Ten books about architecture) around the time of the birth of Christ.

Vitruvius argued that an architect must be much more than a building constructor. He must have knowledge about anatomy, geometry, geology, geography, but also music and art for doing a relevant job and being successful. In a transferred meaning, the conditions are similar for a modern architectural lighting designer.

According to Vitruvius, architecture should satisfy the three principles of firmitas, utilitas and venustas or translated: durability, utility and beauty. Following the aspects of contemporary lighting design we would call them sustainability, functionality and aesthetic.

Vitruvius thesis formulated the guiding principles for the ideal Renaissance architecture 1,500 years later. He inspired and influenced among others Leonardo da Vinci, Bramante, Brunelleschi and also Michelangelo. In 1490, Leonardo da Vinci even drafted the world famous sketch of the Vitruvian Man.

The conception of good lighting quality has changed over the centuries. During the age of Renaissance, the ideal light was a rational ambient lighting, which realistically rendered the space and the details. The lighting solutions were according to the valid conception of the philosophy of the period.

The following era of the 16th and 17th centuries is the Baroque period and was a strong reaction to the Renaissance ideals. The lighting was dramatic, colourful, theatrical and worked with spot effects, to emphasize or screen parts of the environment while leaving other parts in deep shadow. This style was endorsed by the powerful and wealthy and staged the power and splendour of kings and emperors.

These trend shifts are a natural part of the cultural evolution. If we make a historical leap forward to the 20th century we reach Modernism. The ideal lighting of Modernism is in its conformation similar to Renaissance lighting but the underlying philosophy is different. The universal use of the buildings and space is a characteristic aspect. Another meaning of the universal approach is a social ambition to create buildings also for common people and their whole life.

The reaction of the Modernism is the Postmodernism which mixes classical historical elements with current design and constructions. The ideal lighting solution also represented a blend of natural and artificial light with different distribution characteristics.

One of the challenges for the contemporary architectural lighting designers lies in the fact that nowadays we have a number of simultaneously existing architectural styles and also a lot of hybrids. Modernism, Postmodernism, International Style, historical eclectic style, regionally and vernacularly inspired architecture. Beyond that, the green building demands will more and more shape the built environment of the future and affect the lighting solution. These changes happen very fast today, sometimes taking only a few years, whereas formerly the evolution of historical stiles often stretched across centuries. Nowadays we notice an accelerated evolution of design approaches and simultaneously, a greater overlap of cultures, fashions and preferences can be observed.

TECHNOLOGICAL SHIFT – INCANDESCENCE VS ELECTRO-LUMINESCENCE

We are living in an ever-changing world and there are a few professions which have got access to new technologies, which have essentially changed the conditions, tools and methods for conducting their work.

One of the professions most profoundly affected by this development is lighting design. Here, a great transformation has occurred during the past 15 years.

The technological shift implies a fundamentally different way to produce light. Up until recently, we have produced artificial light by fire, flames and heat. By heating up a filament in a vacuum by electricity, the incandescent bulb was born. Incandescence is a special case of thermal radiation, which emits electromagnetic radiation also in the visible spectrum. The character of the radiation is directly related to the temperature of the filament. (Latin incandescere, to glow white)

The LED technology emits the light by another principle without heating and this process is named electro-luminescence. It is a form of cold body radiation. It can be caused by e.g. chemical reactions or electrical energy. The light emitting diode (LED) consists of layers of semiconductor crystals and phosphors for colour conversion. A low electric voltage causes an electron movement between the layers and when the electron is coming back to a rest position it emits light radiation. This fundamentally distinguishes luminescence from incandescence.

Among the LED lamp's greatest assets is its outstanding energy efficiency, which is strongly evolving and has not reached its full potential, yet. Another great advantage is its phenomenal life span of roughly 50,000 hours.

The LED is an electronic light source and uses direct current (DC) which means that a driver is necessary for transforming the alternating current (AC) to DC. The DC makes it is easy to run the LED light source e.g. on solar panels or batteries. The electronic nature of the LED also facilitates its control and integration into IT technologies.

DOCUMENTATION, DISSEMINATION AND DEVELOPMENT

Architectural lighting design is still a young profession and the dissemination of the importance of the topic is a necessity. The most efficient way to achieve this is for sure to provide and document relevant and qualified references. The multi-faceted solutions show the complexity of current lighting designs and the big variation of possibilities provided by the current lighting technology.

The importance of continuous development of the lighting design topic is obvious. For an advanced education there is a constant need for professional references and scientific research and development.

The lighting design based research has yet to attain a stronger position. Today, the dominating amount of studies are performed in adjoining topics such as technology, medicine, psychology etc. This is a very valuable expansion of knowledge. However, for a better understanding of lighting design, the goal is to combine these research findings into an interdisciplinary, holistic pool of knowledge, allowing to better unfold the intricacies and effects of lighting design.

Particularly the non-existing theory part of the lighting design topic has to be formulated. A comparison with the realm of architecture shows that here exist a great number of different established theories; spatial, aesthetic, technical, functional, social etc. as a foundation of knowledge.

LIGHT IN NEW DIMENSIONS

The fourth book of Architectural Lighting Design, edited by Licht Kunst Licht is the latest contribution to the documentation of professional lighting design projects. It is a natural continuation of their presentations in the earlier books. The long-term continuity and the variation of projects makes the material inspiring and applicable for a wide group of professionals as well as a more popular reading.

It is possible to gather two different tendencies of the presentations. One is the basic lighting design quality consistent and unchangeable over time despite later year's dramatic technological changes and development. Another general observation is the application changes related to the architectural shifting trends over time.

The role of lighting in architecture has expanded from a focus of functionality to also considering comfort and pleasantness. This means that the lighting design has a natural part in to completing and fulfilling the architectural concept. In contemporary lighting design, it is not enough to provide lighting for specific activities; the lighting has to take human requirements into consideration in every application. The lighting has to supply the human's visual demands, satisfy physical and circadian needs and create an emotional comfortable and pleasant atmosphere. In each specific project there is of course a hierarchy of demands but despite this, humans are always reacting to the whole environment. Lighting unconscious as well as intentional, even sophisticated messages to the observer. This supports the user in his sense of orientation and well-being. A lot of good examples showing this holistic approach are presented in this book.

数十载的建筑灯光设计

安德烈亚·雷尔勒女士与安德烈亚斯·舒尔茨先生的访谈对话

自2005年以来，LKL就以作品集的形式向外界发布其竣工的设计项目。本书作为该公司作品集的第四卷，延续了公司一直以来的优良传统。它进一步证明了这家照明设计事务所具备的高超的创意设计品质和技术能力。同时，由于这本书2016年发行的时候恰逢LKL成立25周年，因此它也代表这家事务所自创始以来的一座特殊的里程碑。为了纪念这个事件，安德烈亚·雷尔勒女士采访了安德烈亚斯·舒尔茨先生，聊了聊他在过去25年中所经历的事情，包括事务所是如何持续运营的，以及建筑与照明设计领域之间相互关系的变化与观念的发展。

安德烈亚·雷尔勒：事务所的一切是如何开始的，您是出于什么机缘进入了照明领域？

安德烈亚斯·舒尔茨：我大学所学的专业让我成了一名电气工程师，但是我对建筑和艺术情有独钟，因此从未满足于将自己的发展局限在电气工程领域。要知道，在德国当时并没有专门针对照明设计的学术教育体系，因此无法为对此感兴趣的人们提供深入研究的机会。在完成学业后，我被一家照明设计事务所录用了，这家事务所的工作让我很快地意识到自己在这方面有着极大的兴趣，这份工作也带给了我很大的满足感。由于之前的电工培训经历以及之后在电气工程领域的专业学习，我掌握了照明设计所需的技术知识，这能帮助我实现技术上可行的照明设计方案。事实证明，这对我来说是一个很大的优势，因为如果你对照明技术了如指掌、应用自如，那么你将拥有更大的设计自由度。

但你若希望充分探索这些设计的自由空间，则需要大量美学方面的知识支持，在这一点上我认为当时的自己还不具备这一能力。由此，我萌生了与充满创造力的伙伴们展开合作的想法。1991年，我与两位同事在德国波恩和柏林两地同时创立了LKL，他们其中一位是工业设计师，另一位是建筑设计师。

安德烈亚·雷尔勒：贵公司的名字第一眼看的时候会让人感觉很陌生，后来发现它其实是一个有趣的文字游戏。它背后有什么故事吗？

安德烈亚斯·舒尔茨：其实对这个名字的解释比人们想象的要简单得多。我们最初的几份设计合同都是关于人工光照明的设计项目。我们希望通过公司的名称来表达我们的业务范畴，并不仅仅是电气照明工程，还包含着大胆的创意性工作。我们并不认为自己是一群灯光艺术家，但我们有着很高的艺术抱负。正因如此，我们认为“Kunst”（艺术）一词应该成为我们品牌的一部分。由此我们的事务所名称演变成了这样一个双关语“Licht Kunst Licht”，其中“Kunstlicht”（人造光）构成了一个闭环，这个双关语也构成了我们的徽标。坦白说，在刚开始的时候，我们自己对这个名字也适应了一段时间。

安德烈亚·雷尔勒：如今，这个名字已经被成功打造成了一个品牌，并且贵司已先后参与了800多个国内外项目。请问，你们是如何将项目工作分配给两个办公室的呢？

安德烈亚斯·舒尔茨：起初，柏林的办公室负责当地和周边区域的项目，而波恩的办公室则负责德国境内其他地区和国际性的项目。但是如今，由于许多柏林的建筑设计团队也开始承接国际性项目，所以我们柏林的办公室也变得越来越国际化。值得一提的是，两个办公室的团队架构是相同的。除了建筑设计师、照明设计师和照明技术人员外，我们还聘请了室内设计师和电气工程师。我们还有一个专门致力于日光设计与研究的团队。此外，我们还设有灯具与工业设计部门。根据不同的项目要求，我们会将来自不同办公室、不同部门且拥有不同设计背景的同事们组成一个团队开展工作。同时，在不断的实践中，我们也逐步建立了相对固定的设计团队，他们经常在多个项目中协同工作。

安德烈亚·雷尔勒：从一个仅由3人组成的创业团队到如今一家拥有26名员工的事务所，创业初期的同事们现在还是事务所的一分子吗？

安德烈亚斯·舒尔茨：托马斯·莫瑞兹是我们的创始人之一，我们共同经历了行业的风风雨雨。我们事务所的许多成名作就出自他之手。多年来，我们形成了一种类似共生关系的协作与沟通模式。在我们这里，言语上的交流和审美直觉层面的融合形成了一种鼓舞人心并且富有成效的互动方式。我希望我们这种独特的创意交流方式与密切合作的关系能够长长久久地持续下去。

当然，还有其他一些同事也已经在这里工作很多年了。例如斯蒂芬妮·格罗斯-布罗克霍夫、埃德温·斯米达和我们的行政主管巴伊亚·鲁特菲，他们都是在事务所创立的第二阶段加入的，并已经在这里工作超过25年了。

安德烈亚·雷尔勒：在浏览贵司第四部作品集时，我发现由施塔布建筑设计事务所设计的项目占据了本书很大的一部分内容。

安德烈亚斯·舒尔茨：是的，弗克尔·施塔布和他的团队从公司成立之初就与我们为伴，共同成长。他们以充分的自信和极强的美学追求对待建筑设计，并且经常邀请我们作为灯光设计者在项目早期的概念草图阶段就介入到创作过程中。例如，对于阿伦斯霍普艺术博物馆，施塔布提出了一种本质上属于照明设计的建筑设计方案——因为他知道可以信任、依赖我们，并将我们视为他的合作伙伴，共同实现这样的设计概念。

安德烈亚·雷尔勒：作为LKL的“门面”担当，您肯定十分忙碌，并且需要经常出差。您是如何管理两地的办公室的？

安德烈亚斯·舒尔茨：在与我们规模相似的事务所中，也许会有设立从事市场营销、项目获取或制定服务费用建议等工作的部门，但我们并没有单独设立这样的部门。这是因为我试图保持公司内部组织较高的灵活度和工作效率，尽可能地减少工作中的行政等级与复杂的组织架构。当然，我也承认这样做有可能造成工作上的漏洞——但是，这样固执地、有意识地进行“无行政等级”状态的引导对我而言是神圣的。我宁愿接受工作中因此出现的“小弯路”，也不愿意为每一个工作新情景逐一制定新的规则。

安德烈亚·雷尔勒：您相信“创造性混沌”（creative chaos）的概念吗？

安德烈亚斯·舒尔茨：当然相信，我们的办公环境是充满创造力的。据我观察，我们团队的关键性优势在于丰富的设计经验以及打破常规的创新方法。我们乐于以积极的方式应对具有挑战性的任务。因此我们的办公环境永远不会无聊，这点对我们来说很重要。保证我们高效工作的另外两个因素是公司结构的特色和团队成员勤奋的工作态度。这些因素在团队需要创意构思及项目实施时就会显得至关重要。可以说，有时候我们的工作方式实际上很"普鲁士"。

安德烈亚·雷尔勒：您提到贵司的"创造性混沌"和"普鲁士"组织结构，在日常工作中这两个方面是否可以很好地融合在一起？还是在其中的一些环节需要大量的调解工作？

安德烈亚斯·舒尔茨：从我的角度看，这里没有冲突。我的父母都来自柏林，但我在莱茵河畔长大。我的成长过程理所当然地、深深地受到普鲁士思想的影响，但是我也同样吸收了莱茵地区人们的思维方式。我觉得自己可能同时拥有两种不同的性格——莱茵地区轻松活泼的精神和普鲁士精准谨慎的态度。我只是希望这样的混合是自己可以驾驭的……（微笑）

安德烈亚·雷尔勒：据我所知，毕业后您获得了向一位照明设计先驱学习的机会。在汉斯·冯·马洛奇事务所的工作经历对您的事业有影响吗？

安德烈亚斯·舒尔茨：他确实是在建筑和美学方面都有着很高追求的照明设计领路人之一。在那个年代，让照明设计师参与建筑设计是非常罕见的，但是他仍然十分自信地冒险涉足于建筑照明领域。很巧的是，马洛奇与我的专业背景一样，他也曾是一名电气工程师。这点很特殊，因为在当时，人们通常认为电气工程师没有受过专业设计方面的训练。但是和许多成功的照明设计师一样，马洛奇强有力地证明了事实并非一定如此。在他的事务所的学习经验对我十分受用，我对此十分感激。

从艺术角度上讲，我们曾经处于一个宏大叙事的时代，例如埃托尔·索特萨斯和孟菲斯的后现代主义，汉斯·霍林和他的阿伯特贝格博物馆等案例都是其中的代表案例。这些作品都倾向于将建筑项目作为宏大的叙事进行表达。我饶有兴致地对此进行了观察与思考，但从未与这些概念产生过共鸣。从设计早期开始，我就认定了非装饰性的、更符合建筑美学的照明设计方向。事实证明，运用这种理念进行设计的时机已经成熟，并且在当时还没有人专门去拓展这一领域。追求风格化创新的勇气和恰逢的好时机是我们事务所成功的两个关键因素。

安德烈亚·雷尔勒：即使这样，我发现 LKL的作品不仅仅是纯粹的偏理性的建筑照明设计项目。贵司也创作了一些抽象的，甚至是极具雕塑感的灯具设计作品，同时还有多媒体幕墙照明设计作品。

安德烈亚斯·舒尔茨：对，是这样的。我们也会设计与研发灯具产品，同时也完成了一些将灯光本身作为空间焦点的设计项目。但无论灯光的表现形式是什么，我们思考照明方案的出发点始终都是建筑与建筑环境需要什么样的灯光。随着经验的不断积累，我们承接的项目类型也越来越丰富。例如在近期，我们承接了不少高端酒店项目，其中包括多个超六星级度假村。在这些项目中，除了对照明效果的思考外，灯具本身的造型设计也成为整体项目特色的组成元素。此外，近期我们也承接了一些偏理性的、较为严谨的博物馆照明设计项目。在这些项目中，自然光和人工光都被视为一种内敛的介质，以烘托空间气氛。除了这些规模较小的项目外，我们也有着城市空间尺度的城市光环境规划项目经历。大部分的设计类型在过去是不存在的，例如多媒体幕墙照明设计。对我们而言，像多媒体幕墙照明这样的形式可以强有力地表达出整体项目的设计意图。

安德烈亚·雷尔勒：作为希尔德斯海姆应用科学与艺术大学照明设计专业的教授，您会与您的学生分享知识和经验。您给下一代照明设计从业人员的最重要的建议是什么？

安德烈亚斯·舒尔茨：作为本专业所有与设计相关事务的负责人，除了传授行业基本设计应用工具的使用方法外，我也非常重视对设计方法的讲解。如何系统性地开展并深化设计研究，深入研究建筑架构，做到真正理解建筑，这对我来说至关重要。此外，我会和学生们分享必要的实践经验，并经常以我们事务所当前所承接的项目作为案例进行实践教学。

至于我给下一代的建议，我想除了创意的驱动力外，扎实的技术能力对照明设计师的成功也至关重要。目前，照明控制技术正扮演着越来越重要的角色。此外，我会鼓励每一位年轻的设计师在自己创业之前，先在已经成立的照明设计事务所或者企业之中以团队成员的身份进行锻炼，从而积累丰富的设计实践经验。

安德烈亚·雷尔勒：是什么让您着迷于光这种媒介？

安德烈亚斯·舒尔茨：光为我们提供了无限的可能性。精妙的照明设计可以为建筑和空间塑形。光可以强调建筑与空间的意图、氛围和功能，或者（在最佳情况下）丰富它们。与建筑、景观和城市设计的总投资成本相比，对光的投入成本可以说是很低了。换句话说，以较低的预算，光可以对空间产生微妙且深邃持久的影响，并大大有益于建筑主体的表达。每当我们成功实现一个满意的总体效果时，我都会为之兴奋。在我们的系列作品集中，我们展示了从博物馆到行政大楼等各个领域和类型的项目。它们都在表达着光线与建筑空间的融合能力，同时也展现出这种媒介的强大美学力量。

ANDREA RAYHRER'S DIALOGUE WITH ANDREAS SCHULZ

Since 2005, Licht Kunst Licht has published its realized projects in the form of work reports. This fourth volume continues the fine tradition. It further documents the great creative quality and technical competence of the lighting consultancy office's designs. At the same time, this book represents a special milestone, as its release in 2016 coincided with Licht Kunst Licht's 25th anniversary. To mark this occasion, Andrea Rayhrer spoke with Andreas Schulz about the early years a quarter century ago, the constants in the office's work, as well as the changes and perspectives in the dialogue between architecture and lighting design.

ANDREA RAYHRER: How exactly did it all start and how did you find your way into lighting?

ANDREAS SCHULZ: Earning a degree as an electrical engineer yet having a fondness of architecture and the arts, I never felt quite at home in electrical engineering. At that time, there were no academic programs in Germany that offered a greater insight into lighting design. After my studies, I was offered a position at a lighting design office, and I quickly realized that this type of work gave me great satisfaction. Through my training as an electrician and my subsequent studies in electrical engineering, I had acquired the necessary know-how to implement technically competent lighting concepts. This proved to be a great advantage, because if you master the technical aspects, new design freedoms begin to emerge.

In order to fully exploit these freedoms, a great aesthetic input is required that I did not feel sufficient in at the time. This triggered the desire to collaborate with creative partners from the start. In 1991, I founded the office Licht Kunst Licht in Bonn and Berlin with two colleagues – an industrial designer and an architect.

ANDREA RAYHRER: The company's name is an intriguing play on words that was rather unfamiliar at first. What is the story behind it?

ANDREAS SCHULZ: The explanation is much simpler than one might think. Our first design contracts were artificial lighting projects. With our firm name we wanted to express that we not only design electric lighting but that we are ambitiously creative. We do not see ourselves as lighting artists, but we have high artistic aspirations. Therefore, we felt the word "Kunst" (art) should form part of our brand. This evolved into the pun Licht Kunst Licht, which also relates to our logo, where the word Kunstlicht (artificial light) forms a ring. To be perfectly honest, at first, the name took a little getting used to for us as well.

ANDREA RAYHRER: Meanwhile, the name has been established as a brand, and the office has participated in more than 800 domestic and international projects. How are the assignments allotted to each of the two office locations?

ANDREAS SCHULZ: Originally, the Berlin office was responsible for local and regional projects, while Bonn undertook both domestic and international work. This has now changed, as Berlin has become more international with many Berlin based architects working on projects around the world. Also, the work structure is identical in both offices. In addition to architects, lighting designers, and lighting technicians, we also employ interior designers and electrical engineers. One team is exclusively devoted to daylight. Furthermore, we have a department for industrial and luminaire design. Depending on a project's requirements, we join colleagues from various disciplines and from both offices. However, there are also constellations that have become established and often work together on various projects.

ANDREA RAYHRER: From a three-member founding team to a company 26 employees strong: Are any colleagues from that time still part of the office today?

ANDREAS SCHULZ: Thomas Möritz has participated from the very beginning. Since then, we have jointly experienced the ups and downs of the planning trade. Many designs that have made us known and renowned where penned by him. Throughout the years, we have developed a kind of symbiotic communication. Here, verbal and aesthetic intuition blend into an inspiring and productive mutuality. I hope that this particularly creative exchange and our close collaboration will continue to last for a long time.

There are of course other colleagues who have been working here for many years. Stephanie Grosse-Brockhoff, Edwin Smida and our office manager Bahia Loutfi joined us in our second step, so-to-say, and have been with us more than 25 years.

ANDREA RAYHRER: When browsing this fourth volume of your work, the projects designed by Staab Architekten take up a large portion of the book.

ANDREAS SCHULZ: Yes, Volker Staab and his team have been companions from the company's inception. He designs architecture with great self-confidence and aesthetic certitude and often involves us in the early sketches. For the Ahrenshoop Museum of Art, for example, he presented an architectural design that is essentially a lighting design – knowing that he could count on us as a partner for the implementation of the concept.

ANDREA RAYHRER: As the "front man" of Licht Kunst Licht, you are very sought-after and constantly travelling. How do you succeed at managing both offices?

ANDREAS SCHULZ: We have no separate departments engaging in marketing, project acquisition, or fee proposals, as one might expect in an office of this size. I try to keep the organization flexible and efficient and tackle assignments with as little hierarchy and superstructure as possible. Obviously, this has the potential to create gaps – but a wilfully piloted anarchy is sacred to me. I will much rather accept a small detour than create a rule for every new scenario.

ANDREA RAYHRER: Do you believe in the concept of creative chaos?

ANDREAS SCHULZ: Of course, our office is highly creative. Design-experience and unconventional approaches leading toward innovations are among the virtues I observe in our team. We enjoy tackling challenging tasks in a spirited manner. It is never boring in our office, that is important. Two other components of our efficiency are structure and diligence. They are imperative when implementing creative ideas in reality. I would even state that we are virtually Prussian in our approach, at times.

ANDREA RAYHRER: Do creative chaos and Prussian structure mesh well in the daily work routine or does it require a lot of mediation at the interface?

ANDREAS SCHULZ: As far as I am concerned, there is no conflict. I am a child of parents from Berlin and raised in the Rhineland. Certainly, Prussian virtues formed part of my upbringing, but I soaked up an equal amount of Rhineland mentality. I probably carry both – the Rhineland lightness and Prussian precision – I just hope that the mixture is sufferable ... (smiling)

ANDREA RAYHRER: After graduation, you had the opportunity to learn from one of the lighting design pioneers. Did working at Hans T. von Malotki's office influence your career?

ANDREAS SCHULZ: He certainly was one of the pioneers of architecturally and aesthetically ambitious lighting design. Considering that it used to be entirely unusual to involve a lighting designer, he ventured into architecture quite confidently. Strangely, Malotki was also an electrical engineer. This is notable because engineers are often assumed to have not had appropriate design training. Like many other successful lighting designers, Hans T. von Malotki has impressively proven that this does not have to be the case. I am thankful for the lessons learned at his office.

Stylistically speaking, it was the era of grand gestures: Ettore Sottsass and Memphis, Postmodernism, Hans Hollein and the Abteiberg Museum. It was all about making big statements. I observed this with great interest but never quite felt at home in these concepts. From an early stage I identified with the non-decorative, more rational architecture-committed lighting design. Evidently, the time was ripe for this approach and until then, no one had claimed this territory. The courage to pursue stylistic innovations and good timing were two key success factors for our office.

ANDREA RAYHRER: Yet, Licht Kunst Licht does not only cater to rational, purely architectural lighting design. Your office has produced luminaire designs with an almost sculptural, or at the very least, object-like characters as well as media facades.

ANDREAS SCHULZ: Yes, that's correct. We have developed luminaire designs and lighting solutions that have quite an assertive visual presence within a space. The question is always what the architecture needs. Furthermore, the diversity of projects that we work on has grown over the years. Presently, for instance, we design ambitious hospitality projects, such as 6-star-plus resorts around the world. Apart from the lighting effect, they require luminaires that form design elements in their own right. Furthermore, we develop austere museum projects, where light is a subservient medium – both as daylight and electric light. And we work on large urban designs for entire city quarters. Much of this simply did not exist before – for example, media facades; and working on a media facade entails making a strong design statement.

ANDREA RAYHRER: As a professor for Lighting Design at the HAWK University for Applied Science and Arts in Hildesheim, you share your knowledge and experience with your students. What is the most important advice you can offer to the next generation of professionals?

ANDREAS SCHULZ: As the person in charge of all matters pertaining to design, I attach great importance to both the mediation of fundamental tools of the trade and methods. Systematically developing a design, delving into the architecture, truly understanding it – all of this is crucial to me. Furthermore, I provide the necessary practical experience and often use projects that we are currently working on in our office as examples.

My advice to the next generation ...? – Well, apart from creative drive, profound technical competence is imperative to a lighting designer's success, and control technology is playing an increasingly important part. Also, I would encourage every young professional to acquire sufficient experience as a team member in an established lighting design practice or the industry before venturing into self-employment.

ANDREA RAYHRER: What fascinates you about the medium of light?

ANDREAS SCHULZ: Light offers an infinite wealth of possibilities. A dedicated lighting design actively configures architecture and space. It underscores its intentions, atmospheres, and functions or – in the best of cases – enriches them. The financial cost is – compared to the total investment costs for architecture, landscape and urban design – rather low. With a relatively low budget, light can make a profound difference and greatly benefit the architecture. I am delighted every time we succeed at achieving a good overall result. In our series of books, we present projects from a variety of realms, ranging from museums to administration buildings. They all illustrate how integrative light can be, but also the enormous aesthetic power this medium unfolds.

LKL团队

Team Licht Kunst Licht

KÄMPFEN
BERLIN
Voo
BROOKLYN
PARIS

AWARD
WINNER

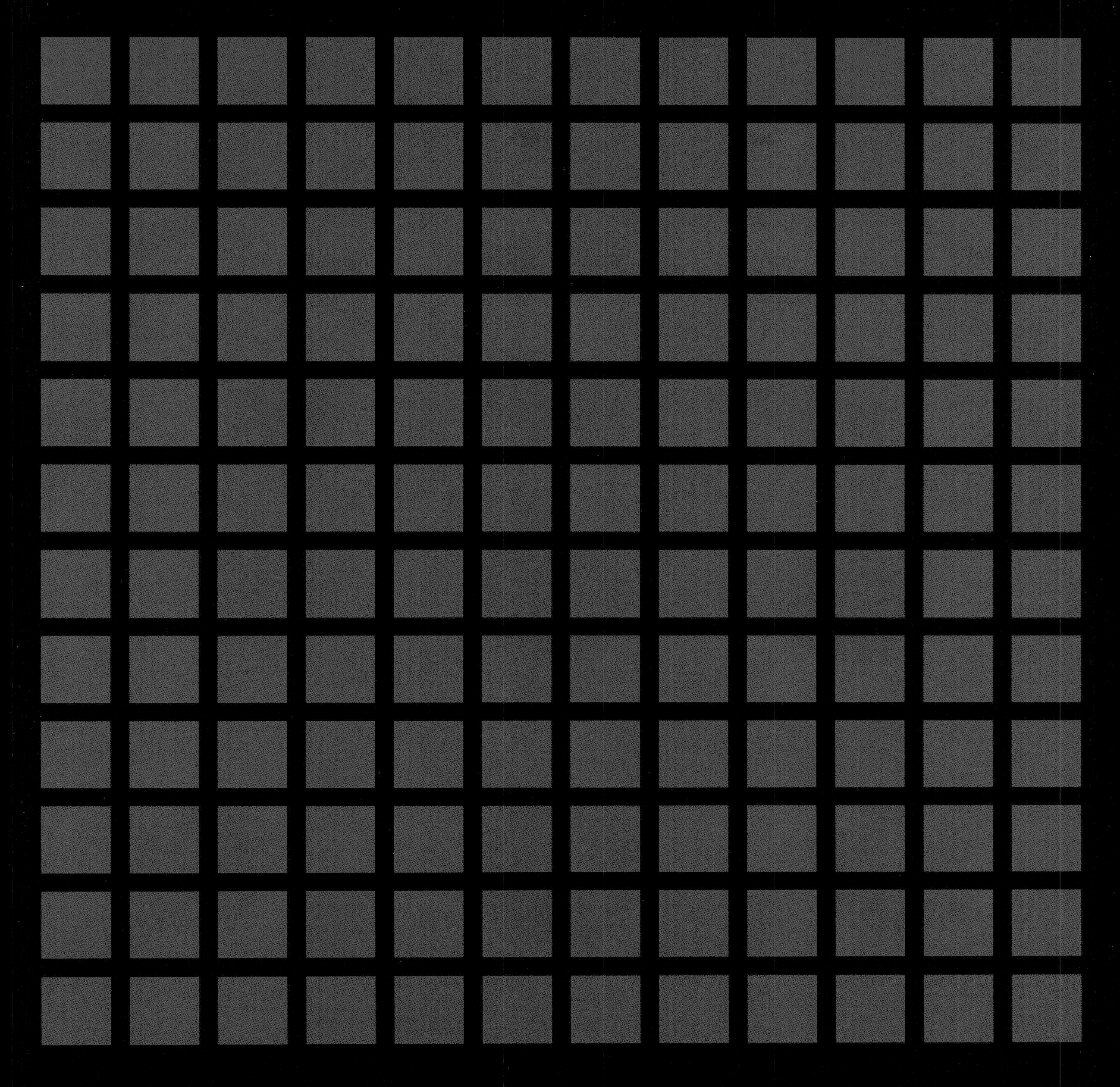

LICHT KUNST LICHT 4

施泰德博物馆 法兰克福

STÄDELSCHES KUNSTINSTITUT UND
STÄDTISCHE GALERIE FRANKFURT AM MAIN

法兰克福施泰德博物馆扩建项目

为了扩大当代艺术作品的展览空间，法兰克福施泰德博物馆经历了在其两百多年发展历史中规模最大的建筑扩建和与展览内容相关的展品补充。为此，法兰克福的施耐德+舒马赫建筑设计事务所在原施泰德花园的地下空间设计并扩建了大约3000平方米的展厅空间。在地下展厅柔和的拱顶天花上，建筑师和灯光设计师将195个圆形天窗镶嵌其中。展厅的照明理念是以用户友好型、环保节能的方式平衡室内入射日光与人工光源，同时将照明作为一个整体，融入整个建筑环境中。

日光建筑的艺术

施泰德博物馆新花园大厅建筑设计竞赛的评审团这样评价施耐德+舒马赫建筑设计事务所的设计，“白天像一颗颗璀璨的宝石，夜间像一片发光的地毯”。由于新建设施缺少可用的地上空间，建筑设计师设计了一个宽敞的、无立柱的地下画廊，并在其天花结构中加入了直径1.5米至2.7米不等的圆形天窗。

白天，这些天花开孔为展厅提供了自然光线，而在夜晚人工光源开启后，它们则变成了展厅上方花园草坪中的发光岛屿。

天窗开口部分包含了一个具有多层屏风结构的系统。它允许经过精准测量和控制的日光进入展厅空间，从而提供照明，也可以在需要时打开完全不透光的窗罩。同时，其中集成的LED照明系统可以为地下展厅提供人工光照明。通过这样的照明控制系统，无论是自动还是手动控制，地下展厅中的自然光和人工光照明都可以通过系统进行联动或独立的调节。天窗开口和人工光的结合设计满足了现代临时展览空间的所有照明需求，无须在室内空间设置传统的照明设备。

日光：极致的细节处理

通过对展厅日照情况的模拟和分析，照明设计师得出了该空间多种不同的室内光线状态。对展厅入射光线情况的分析，不仅要考虑一天之中以及全年太阳的高度和天光的变化情况，还要考虑周围建筑物对其产生的遮挡，以及室内光线分布的均匀程度。在这个展厅空间中，越靠近大厅中央的天窗，其直径越大，加之展厅本身的拱形天花结构，展厅的自然光线分布呈现不均匀的状态。

通过对入射光线的分析，照明设计师最后决定在每个天窗结构中创造4种遮光方案（全开、两种不同程度的减弱以及完全遮挡），以此来应对白天可能出现的多种日照情况。理论上来讲，每个天窗开口都可以单独控制。同时，考虑到用户操作的便捷性，天窗开口也可以通过分组进行统一控制。

根据给定空间段展品的保护要求，以及目前展览移动分隔墙的空间划分，博物馆的技术人员可以自行调整天窗的分组设置。在同一个分组之中，对日光的限制要求最高的那个天窗开口将成为整组系统控制的标准。此外，操作人员如果设置了一个固定的日光遮挡数值，系统中的自动控制模式则可以选择为禁止状态。

室内日光系统的遮挡级别是根据室外实时的光照水平来决定的，这些数据由一个户外感光元件来收集。照明设计师通过预置好的照度阈值，以及对遮阳组件开启和闭合时间的延迟设置，可以避免整套系统对短暂变化的天空条件做出响应。经计算得出的惯量设定可有效防止遮阳系统的无间断调整变化，以及不必要的磨损。使用者还可以根据自己的实际需求来设定阈值。

这样一来，展厅日光方案可以弱化自然光进入室内的动态变化，并在需要时减轻发光强度，同时也不会让室内光线氛围显得单调。对室内日光摄入的控制和干预，可以满足全时间段展品对光线的防护要求，同时具有高显色性的日光光谱确保了参观者享受纯正的艺术。大厅中央区域，在较大的天窗下，可以在80%的展览时间内实现完全的日光照明，从而节省了大量能源。

大厅的整体照明：灵活、高效、维护成本低

在日光入射量不足的情况下，展厅所需的照度可以通过开启一个能够发出均匀光线的人工光照明系统进行补足，该系统的光源部分由LED光学元件组成。在与天花板齐平的天窗薄膜层背面，可调白光色温的LED灯带被巧妙地安装在天窗开口的内侧。每个天窗的灯带色温可以通过照明控制系统进行单独调光，展厅工作人员也可以为每个展区设定环境照明整体的光线色温。对LED灯具的甄选主要是为了确保其在完整的调光范围内可以保持光色稳定性，同时保持照明的高显色性（相当于R_a≥90）。

除了为展厅提供整体照明之外，LED照明系统还承担了其他两项任务。第一个任务是为其扩建空间的夜间室外环境营造特色的光线韵律感。傍晚时分，天窗上的多层屏风会被打开，使LED光线从天窗玻璃中散发开来。对于博物馆的参观者而言，白天的天窗开口在夜间化身为一个可以在其上漫步的环形发光岛屿，在庭院中营造出了一种特别的氛围。

在紧急情况下，逃生通道的照明由部分天窗内指定的LED展陈照明提供。这些LED灯具将由应急电路供电，确保在紧急情况下仍能正常运行。

重点照明：为珍品展示提供强有力的灯光

就与展览相关的重点照明而言，设计师将聚光灯巧妙安装在混凝土裙板和天花薄膜周围的接缝点位置，并根据展品的特色配备不同的透镜配件。这些聚光灯统一采用LED技术，将光线聚焦在部分选定的展品和展示墙上。

效率和品质的完美结合

施泰德博物馆扩建项目的设计使之成为一个绿色建筑。例如通过回收地热能进行供暖，以及对空调进行废热回收等，这些措施都可以为其创造非常节能的运行条件。这一经济环保的理念也始终贯穿在展厅的照明设计之中。

白天，天窗的设计可以大大减少人工光照明的使用。这是一个尤为突出的节能措施，且不会给展馆的正常运营带来任何限制和影响。同时，通过对自然光的有效控制和使用，可以营造出高品质的视觉体验空间。由于展厅与室外环境始终保持着视觉联系，因此参观者可以在地下空间中轻松自在地欣赏展品，并感到身心愉悦。当展览需要时，灯具出色的色彩还原表现力和可调节的色温控制能力，可以营造出极佳的观看环境，并为展陈设计提供极大的设计灵活度。

业主：
施泰德艺术学院
使用人：
施泰德艺术学院
建筑设计：
施耐德+舒马赫建筑设计事务所，法兰克福
LKL项目经理：
坦贾·鲍姆
竣工时间：
2012年
项目规模：
4100平方米
整体项目预算：
5200万欧元
照明预算：
180万欧元

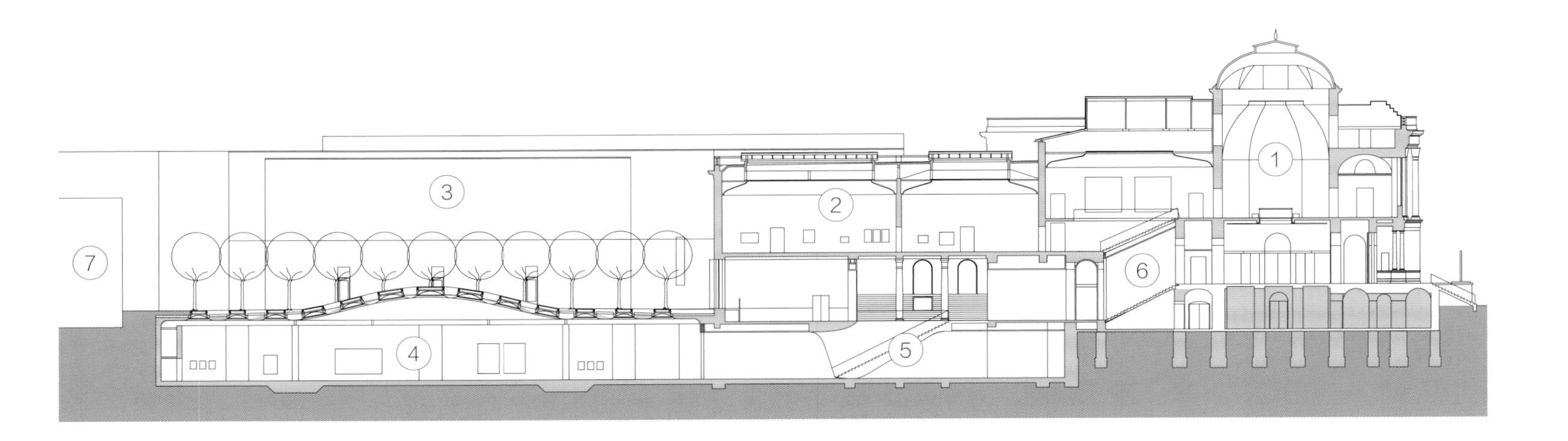

现存建筑和花园大厅的剖面图

Section of existing building and Garden Hall

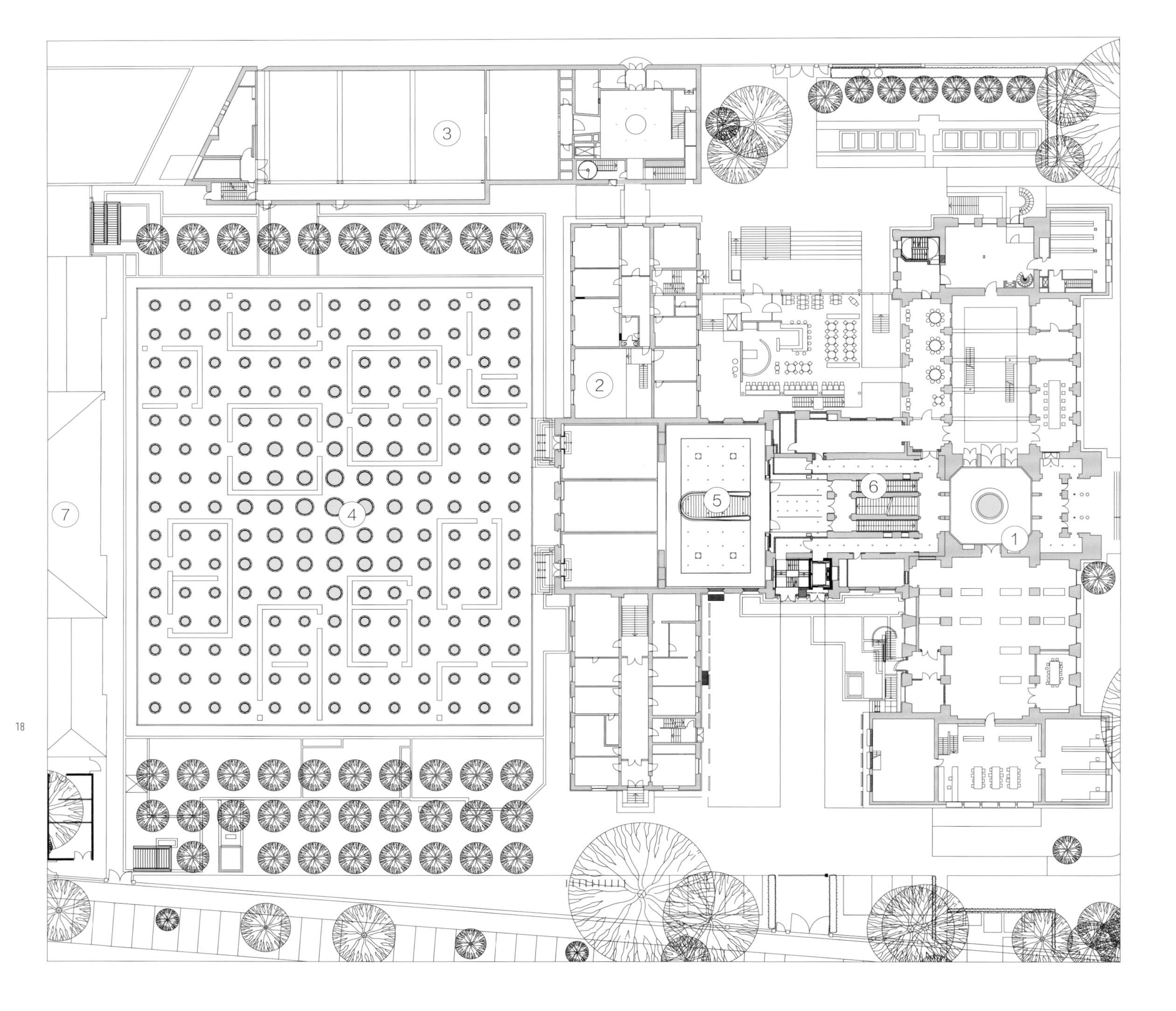

平面图

Floor plan

1. 主楼
2. 花园翼楼
3. 西配楼
4. 花园大厅扩建部分
5. 花园大厅楼梯
6. 从旧楼延伸出的新楼梯
7. 施泰德学院

1. Main building
2. Garden wing
3. West wing
4. Garden Hall annex
5. Garden Hall staircase
6. New stair from existing building
7. Städel School

With its extension for the display of contemporary art, Frankfurt's Städel Museum has received its largest architectural and content-related expansion in its nearly 200-year history. For this purpose, the Frankfurt based architectural office schneider+schumacher has created an additional 3,000 m² of exhibition space underneath the Städel Garden. 195 circular skylights perforate the softly vaulted tension ceiling of the underground hall. For the project, a lighting concept was developed that harmonizes the use of daylight and electric light in an energy efficient and user-friendly way, while simultaneously integrating itself fully into the architectural context.

DAYLIGHT ARCHITECTURE FOR THE ARTS

"A bright jewel during the day, a luminous carpet at night" was the praise from the architectural competition jury regarding schneider+schumacher's design for the Städel Museum's new Garden Hall. Given the lack of available space for the new construction, they designed a spacious, column-free underground gallery with skylights between 1.5 and 2.7 meters in diameter integrated into its roof structure.

While these overhead apertures provide natural light for the exhibition during the day, at night when the electric light is switched on, they turn into luminous islands in the park lawn above.

The skylight openings contain a multi-layered screen system. It permits a well-measured daylight intake in the exhibition rooms, but can also generate a complete black-out, if required. Meanwhile, an integral LED system generates a tailor-made electric lighting solution. Both independently or intelligently linked, either manually or automatically, the two components can be addressed via a lighting control system. This combination of luminaires and daylight openings accounts for all the requirements of a modern temporary exhibition space, without exposing the typical lighting equipment.

DAYLIGHT: TAILORED TO THE SMALLEST DETAIL

The simulation and analysis of daylight intake in the exhibition hall yielded a wide variety of lighting situations. Not only did the different combinations of solar altitudes and sky conditions throughout the course of a day and year need to be considered, but also the shadowing from surrounding buildings and uneven light distribution within the space. The unevenness results from the increase in skylight diameters towards the centre of the hall and the concave ceiling.

The multitude of possible lighting situations through daylight is tackled by a four-layered shading solution in each skylight (open, two levels of reduction, and black-out). Essentially, every daylight opening is individually controllable. In order to keep the system operation user-friendly, groups can be defined that can be controlled in unison.

According to conservational requirements for a given spatial segment and depending on its current sub-division by mobile partition walls, the technical museum staff can adjust these grouping assignments themselves. Within a group, the skylight with the highest restrictions becomes the standard to which all other group members are controlled. Optionally, the automatic control can be deactivated when specifying a fixed reduction value.

The level of reduction is selected according to the exterior illuminance values, which are gathered by an outdoor sensor. By setting defined threshold values and delay times for the opening and closing of the shading elements, they do not respond to momentary changes in the sky condition. This calculated inertia prevents the abrasion and disruption of constant readjustment of the shading screens. The tolerance values can be defined by the occupant according to their requirements.

The result is a daylight solution that dampens intensity and dynamics of natural light intake, yet does not generate a monotone atmosphere. The conservation requirements can be met at any time, while the daylight spectrum simultaneously renders an unadultered appreciation of the artwork with its excellent colour rendering properties. At the hall's center, under the larger skylights, complete daylight autonomy can be achieved 80 % of the time, thus yielding substantial energy savings.

THE GENERAL ILLUMINATION IN THE HALL: FLEXIBLE, EFFICIENT, LOW-MAINTENANCE

In case of insufficient daylight intake, the required light levels are achieved by switching on a diffuse electric light component, realized with LED light sources. Warm and cool white coloured temperature LEDs were skillfully blended and installed behind a skylight membrane layer that is flush with the ceiling. The colour temperature components can be individually dimmed via the lighting control system, and the museum can establish its preferred light colour for the ambient illumination of each space. The LED assortment was made to ensure colour stability across the entire dimming range while retaining a high colour rendering index (comparable to an CRI of 90 +).

In addition to the general lighting, the LEDs cover two additional tasks: The first is the nocturnal orchestration of the expansion's exterior image. At dusk, the skylight screens are opened, allowing the LED light to emanate through the fritted glazing. For the visitors of the Städel Park, the daylight openings are transformed into circular, walkable luminous islands that generate a special atmosphere in the museum's courtyard.

In case of an emergency, the egress lighting is provided by a designated selection of exhibition LEDs within the skylights, which are switched on by the emergency power circuit.

ACCENT LIGHTING: POWERFUL LIGHT FOR SPECIAL PRESENTATIONS

For exhibition related accent lighting, spotlights with various optical accessories can be attached to monopoints in the circumferential joint between the concrete skirting and the membrane ceiling. These spotlights implement LED technology and add directional, focused light on selected exhibits or display walls.

SUCCESSFUL COMBINATION OF EFFICIENCY AND QUALITY

The Städel expansion was designed as a Green Building. Various measures, such as the use of geothermal energy for heating and air-condition with heat waste recovery, provide particularly efficient building operation. The lighting design for the exhibition hall consistently joins into this economic and ecological concept.

During the daylight hours the skylights allow for a great reduction of electric lighting. This is especially remarkable given the fact that the energy savings do not entail limitations. The bespoke use of daylight constructs a space with sublime visual qualities, where visitors are at ease and enjoy comfort due to a constant visual relation with the outdoor environment. When needed, the excellent colour rendition and adjustable colour temperature of the electric light concept create superb viewing conditions and allow great exhibition design flexibility.

CLIENT:
Städelsches Kunstinstitut

OCCUPANT:
Städelsches Kunstinstitut

ARCHITECTS:
schneider + schumacher Planungsgesellschaft mbH, Frankfurt am Main

TEAM LEADER LICHT KUNST LICHT:
Tanja Baum

COMPLETION:
2012

PROJECT SIZE:
4,100 m²

OVERALL BUILDING BUDGET:
52 million euros

LIGHTING BUDGET:
1.8 million euros

博物馆扩建区域位于现有建筑花园的地下空间。这个预张构造的天花略微向上弯曲，横跨整个大厅，最高处可达8.2米。

The extension is located underneath the garden of the existing building. Its pretensioned, softly bulging ceiling spans the up to 8.20 meter tall room.

天花板的曲率和圆形天窗的直径变化多端，这对施工工艺提出了十分严格的要求。

The vaulted ceiling and varying diameters of the circular skylight apertures demanded the highest levels of craftsmanship from the contractors.

为了保护法兰克福绍曼凯区域的花园景观，施泰德博物馆的扩建区域被设计在了地下空间。

In order to preserve the garden on Frankfurt's Schaumainkai, the Städel Museum's extension has been located in the basement.

尽管施泰德博物馆的扩建区域被设计在了地下空间，但参观者仍然可以从地面上看见这个新构筑物。当夜幕降临时，195个圆形天窗就像光之岛屿一般，散布在施泰德花园的草坪上。

Although situated underground, the new construction is visible from above grade. Like luminous islands, the 195 circular skylights are scattered across the lawn of the nocturnal Städel Park.

从具有雕塑感的楼梯向下走，参观者可以到达花园大厅的地下二层。可调节式LED轨道聚光灯位于天花凹槽之中，为空间提供柔和的可调白光照明。

Via a sculptural staircase, the visitor reaches the Garden Hall in the second basement. Adjustable, track-mounted LED accent lights with seamless tunable white control are located in ceiling slots.

这些引人注目且设计精良的天花开口为整个花园大厅营造出生动活泼的氛围，同时给地下空间带来了意想不到的高品质体验。

The ever noticeable, yet discreet ceiling apertures create a lively environment throughout the entire Garden Hall. The result is a quality of experience quite unexpected for an underground setting.

大厅天花为向中央凸起的拱形结构，越靠近大厅中央区域的天窗，其直径越大，由此造成展厅核心区域的照度水平随之增强。通过天窗底部的柔光层，进入展厅空间的自然光和人工光都是较为均匀的光线，这有利于展品展示。

The increased skylight diameters near the middle of the space and the bulging ceiling lead to enhanced lighting levels at the building core. With their diffusing bottom layer, the skylights uniformly distribute both the incident daylight and electric light, which is beneficial for the artwork.

天窗细节剖面图
1.玻璃
2.日光屏风（遮光屏风）
3.LED基础照明
4.拉伸箔
5.重点照明

Detail section of skylight
1. Glass
2. Sunscreen / black-out screen
3. LED basic illumination
4. Stretch foil
5. Accent lighting

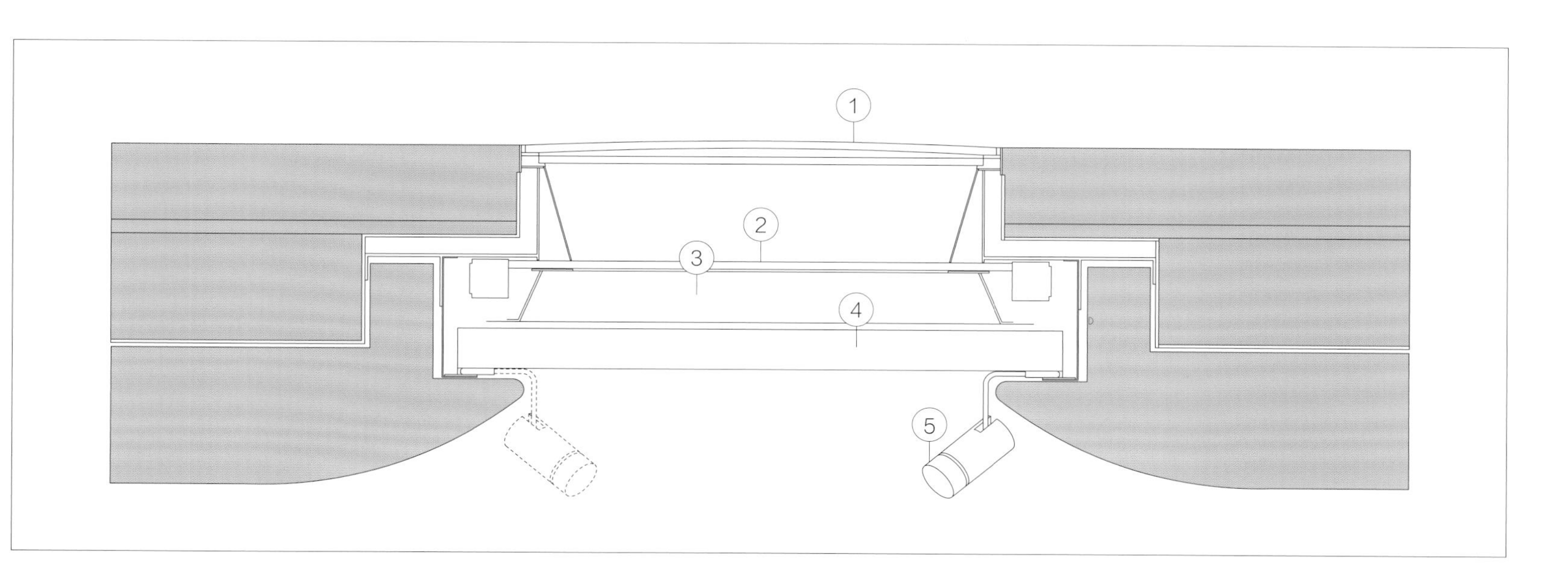

应对不同天气状况的入射自然光控制方案。通过在天窗内实施100%、60%和30%遮光率三种遮光方案的操作控制，确保室内展厅空间的照度保持在一个相对稳定的水平。

Schematic representation of the responsive daylight intake control. The 100 % / 60 % and 30 % transmittance reduction options from three operable screens within the skylights keep the lighting in the exhibition halls at a constant level.

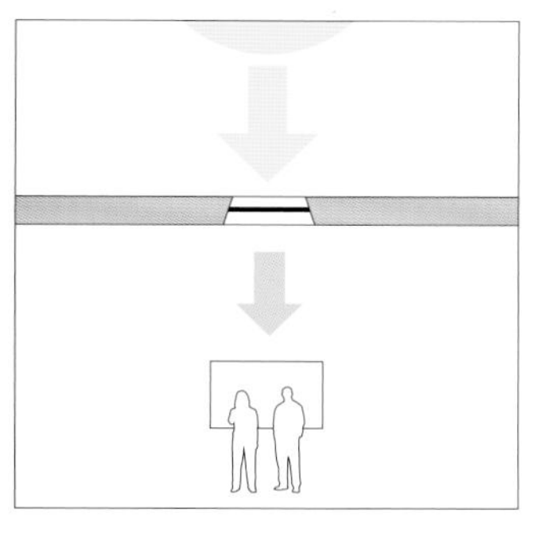

强日光情况=
展区100% 遮光率，
完全开启LED人工光

Strong sunlight =
100 % closed black-out layer
100 % LED lighting in the exhibition space

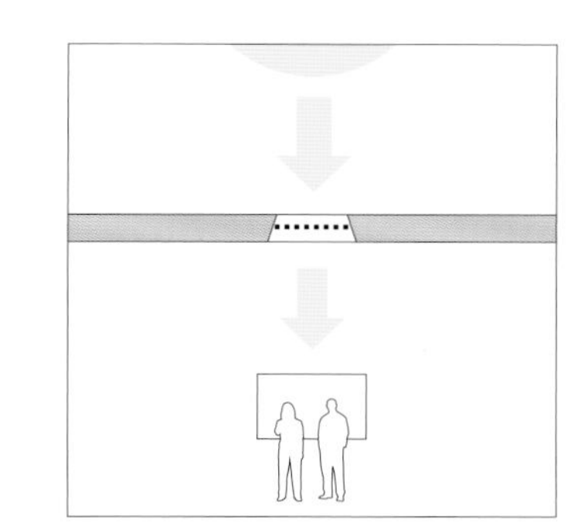

中等强度的日光情况=
展区60%遮光率，
入射自然光

Moderate sunlight =
60 % light transmission shades
Daylight in the exhibition space

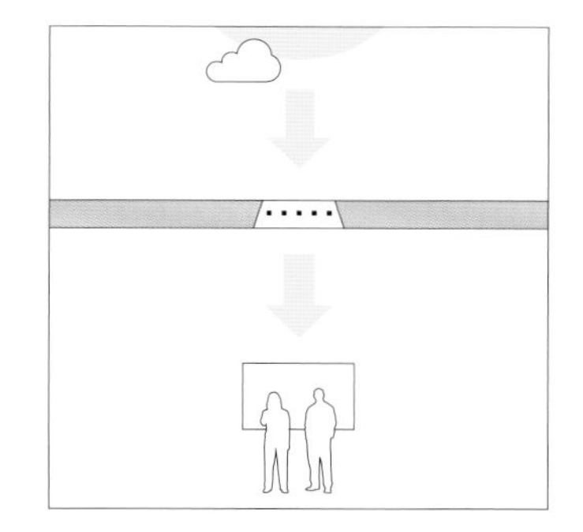

多云天=
展区30%遮光率，
入射自然光

Cloudy sky =
30 % light transmission shades
Daylight in the exhibition space

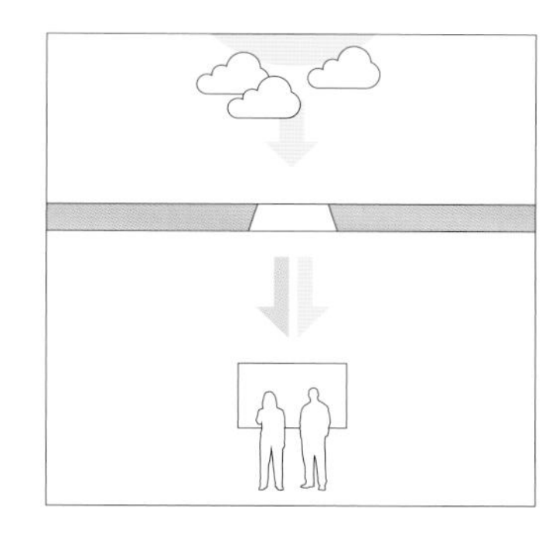

阴天=
打开展区遮光屏风，
以LED人工光补足自然光

Overcast sky =
Open shades
Daylight supplemented by
LED lighting in the exhibition space

自然光
Daylight

LED人工光
LED artificial light

自然光照度等级朝着空间边缘的方向递减。这些区域陈列着光敏度较高的展品。如果展陈需要，可以在天窗结构中增加重点照明的灯具，以补足整体照明。

Towards the perimeter, the daylight levels decrease. Here, light-sensitive exhibits are displayed. Should they be needed, accent lights can be attached in the skylights to complement the general lighting.

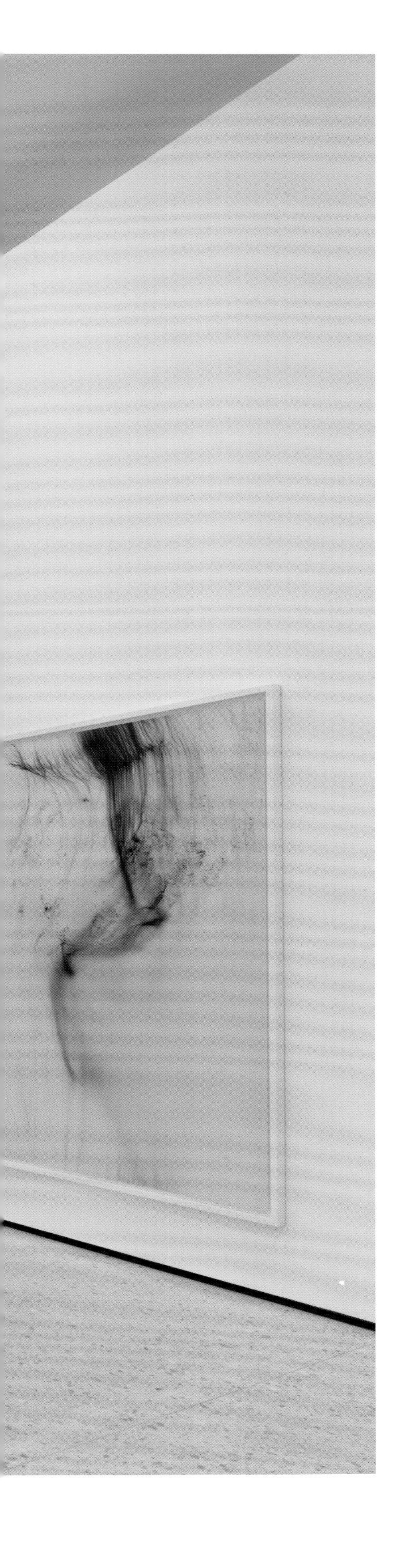

从天窗位置发出的光线为展品提供了主要照明，其中包括自然光和人工光。通过天窗的控制系统可以调节上述两种照明的入射光强度。

The skylights provide the primary illumination of the artwork, utilizing daylight and electric light. Both components have controllable light intensities.

AWARD
WINNER

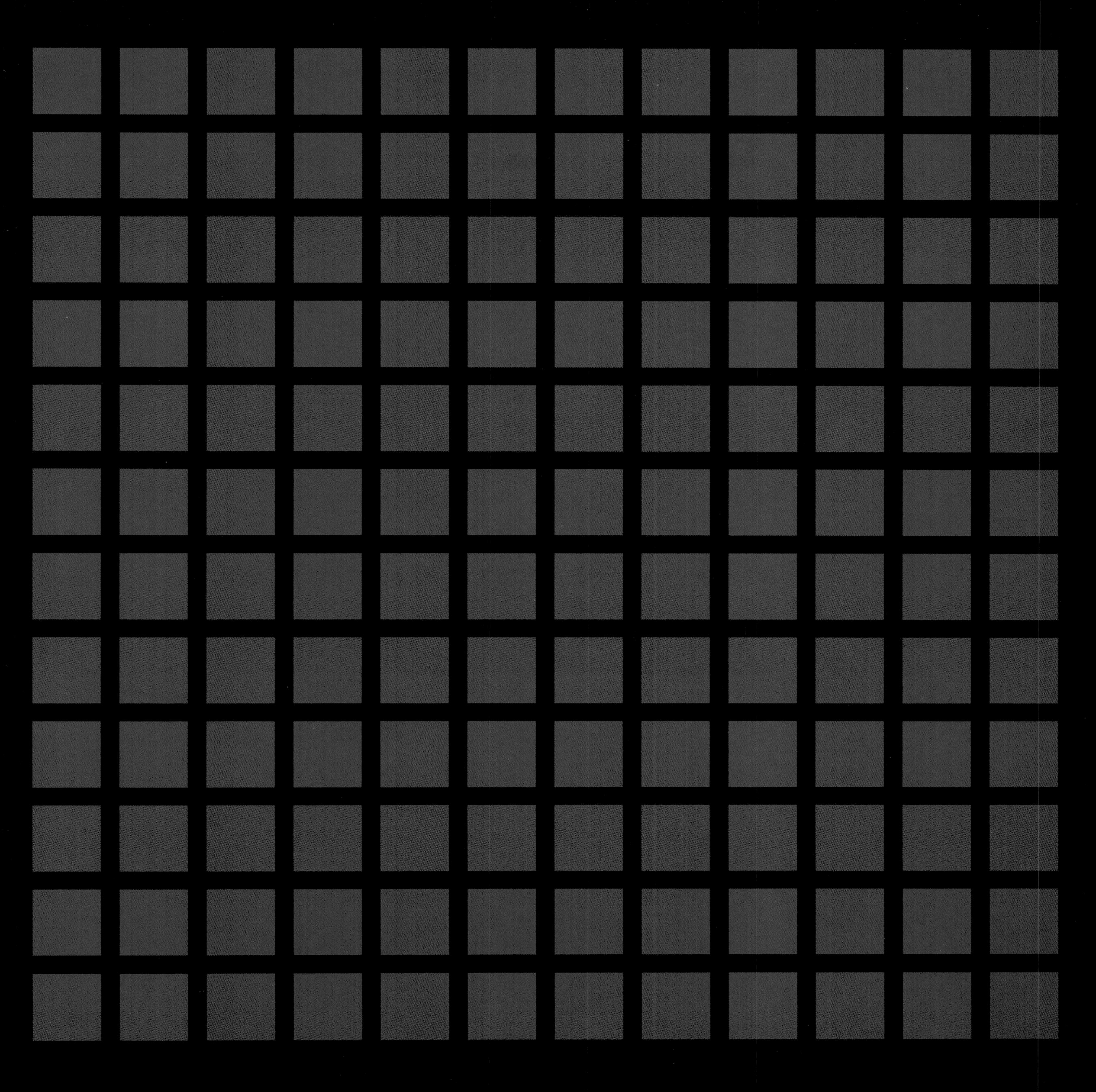

LICHT KUNST LICHT 4

蒂森克虏伯新总部园区 埃森

THYSSENKRUPP QUARTER ESSEN

归家，启程，沉思——蒂森克虏伯新总部园区

蒂森克虏伯集团的新总部设立在德国埃森市克虏伯地带的一个历史悠久的区域。用于这个项目的建筑材料和产品几乎有一半是来自集团本身的自有资源。新总部建筑群在一个公园式的园区中分布组合，建筑整体所呈现出的轻盈、通透的视觉效果很大程度上是依赖于其建筑立面的设计。玻璃立面、金属表面和骨板结构的巧妙设计为该建筑带来了丰富的想象空间。项目的照明设计方案也强有力地凸显出建筑多种材料组合的特点。这个项目整体的园区规划是由JSWD建筑设计事务所与柴克斯和莫雷尔建筑设计事务所联合完成的。围绕在园区中央水轴四周的分别是Q1总部大楼、Q2会议楼、Q5和Q7行政楼以及一个停车场。

建筑物外立面的视觉印象源自建筑学中“核”与“壳”之间相互关联的原理。金属网、穿孔金属板、遮阳板和大型有色金属板组成了建筑物内部和外部的多层饰面特征。从某些建筑剖面图来看，这些饰面与建筑玻璃立面是相互重叠的。变化丰富的日光透射进建筑空间，在透明立面、镂空立面以及实心立面之间进行调节与中和。精心设计与调适的人工光照明方案确保了极具特色的造型在夜间也能完全显现出来。夜幕降临后，建筑的魅力很大程度上是由室内空间所发出的光线以及经过精准调节的室外照明所表达出来的。

水轴、大道和连接小径

园区中轴线上布有一条长达235米的水轴。水轴两侧是主要的通行大道：左侧是一条宽阔的人行道；右侧是“世界大道”，其中种植了来自五大洲的15种树木。两条大道的周围分布着一些连接小径，部分小径还允许人们穿梭在水轴之中。

为了不干扰项目整体的室内照明效果，照明设计师对室外照明部分进行了严谨地规划。通过巧妙选择照明设备，设计师很好地规避了令人反感的光幕反射现象，以及影响人们视觉舒适度的眩光。设计师沿着两条主干道依次布置了防眩光效果极好的高杆灯，这些高杆灯的光源配光曲线是特别为该项目定制的；连接小径则由短柱灯提供照明。此外，水轴内壁上的LED灯带作为水面与“河岸”的过渡元素，能在夜间凸显出水轴的几何结构。

Q1总部大楼

矗立在水轴尽头的是50米高的新集团总部大楼，其令人印象深刻的夜间视觉效果要归功于精心营造的室内光环境。它凸显了室内建筑表面所反射出的光芒，在雕刻建筑形态的同时，营造出迷人的空间纵深感。

遵循建筑学中“核与壳”的设计原理，该建筑物的造型结构以一个玻璃立方体的形式呈现，其中一部分被穿孔金属板覆盖。傍晚时分，在三楼的室内空间中，地埋式上照灯所发出的微妙的光给阳极氧化金属表面带来了柔和的光芒。安装在大楼入口处天花退进空间的下照灯向地面发出定向照明。由此，从地面反射的光线则照亮了天花层板，使建筑体块呈现出轻盈的漂浮感。

在这个项目中，设计师并未对外立面采用传统的幕墙照明，取而代之的是，挑选了部分室内建筑饰面进行重点突出，从而在夜间将建筑的核心结构呈现在室外参观者的眼中。总部大楼的中庭区域则透过巨大的全景窗向外界呈现出引人注目的视觉形象。Q1总部大楼南北立面上宽26米、高28米的全景窗使建筑结构与周围的景观相通，并为中庭和相邻空间引入了日光照明。从室外向内看，参观者透过全景窗仿佛可以看到总部大楼透明的“心脏”。设计师在贯穿中庭的步行桥扶手中内嵌了条形LED灯带，使步行桥的造型格外醒目。这些横向的发光扶手与两侧竖向的发光直梯通道相互参照，仿佛在建筑空间中开启了一段有趣的“视觉对话”。位于中庭两侧玻璃屋顶下方的聚光灯为这个11层楼高的玻璃体建筑空间提供照明。值得一提的是，四个楼梯井空间竖向贯穿整栋建筑，以此凸显建筑的整体高度。在办公楼层中，直接和间接的地面照明元素柔和地照亮天花层板。

Q2会议楼

蒂森克虏伯集团通常在Q2会议楼中招待贵宾和客户。大楼二层全部为会议功能空间，其中设有多个招待室，以及一个可以容纳1000人的会议大厅。大楼三层设有访客餐厅和监事会的会议室。员工食堂和咖啡厅位于首层。与总部大楼相同的是，人们也可以从室外清楚地看到会议楼的室内建筑结构，这要归功于会议楼底层全透明的以及其余楼层半透明的建筑立面。这栋大楼的室内照明同样在夜间展现出了建筑的迷人风采。

这样的光环境设计原则在食堂空间的照明方案中尤为明显：强劲的擦墙光突出了室内后壁的钛色镶板，每张餐桌上方带有磨砂玻璃遮光罩的定制圆柱形吊灯对桌面进行重点照明，界定了餐厅的功能区。与此同时，走道区域也被照亮，就餐区域在水平方向和垂直方向的照度水平也保持在一个平衡的状态。

访客餐厅采用了相似的照明分区策略，通过对灯光等级的设置强调了餐厅的特色。兼具直接照明和间接照明功能的大型环形吊灯悬挂在餐桌上方，与落地窗相呼应，形成和谐的视觉节奏。齐胸高的装饰分割墙与天花上悬挂的金属细丝帘将餐桌两两分组，成为这个细长空间的特色布置。RGB彩光灯具集成在餐厅家具中，在需要时为闪亮的金属窗帘提供动态的彩光照明。与此同时，天花上的凹槽灯也勾勒出餐厅不同区域的轮廓。

设计师专门为该项目研发的LED发光天花被安装在监事会的会议室内。在这个发光天花中，除了漫射光组件外，还包含能够对会议桌进行直接重点照明的光学系统。同样的定制灯具也被应用在了Q1总部大楼的会议室中，呈现出令人信服的光线效果。

“沉静之屋”

占地135平方米的“沉静之屋”是大楼中供员工冥想、小憩和静修的地方。在这个空旷的房间中央，悬挂着一个巨型立方体。这个立方体打破了整个空间的一致性，其内径表面由30～40厘米长的钛板无缝拼接覆盖。轻质的发光天花凸显了其丰富的材料组成和造型感。可单独控制的RGB-LED光源构成了一个像素矩阵，可以呈现出或静态或动态的照明场景。

照明品质和能效控制

蒂森克虏伯集团通过集中式的可持续发展管理计划，将能源运行流程系统地整合到商业运营中。因此，总部大楼项目的设计任务中特别提出了谨慎处理环境及自然资源的要求。例如，该项目的基本能耗量比其地方标准的要求低58%。此外，整个项目场地的照明设施都实现了能源的高效利用：项目中广泛使用了LED光源，并充分利用日光系统进行照明，对空间的人工光照度根据入射日光情况和人流情况进行控制和调节。2011年5月，这个充分考虑了环境因素的设计项目获得了德国可持续建筑协会颁发的金牌证书。

业主：
蒂森克虏伯集团
使用人：
蒂森克虏伯集团
建筑设计：
JSWD建筑设计事务所，科隆
柴克斯和莫雷尔建筑设计事务所，巴黎
LKL项目经理：
亚历山大 · 罗奇
坦贾 · 鲍姆
竣工时间：
2010年
项目规模：
170 000平方米
整体项目预算：
3亿欧元
照明预算：
980万欧元

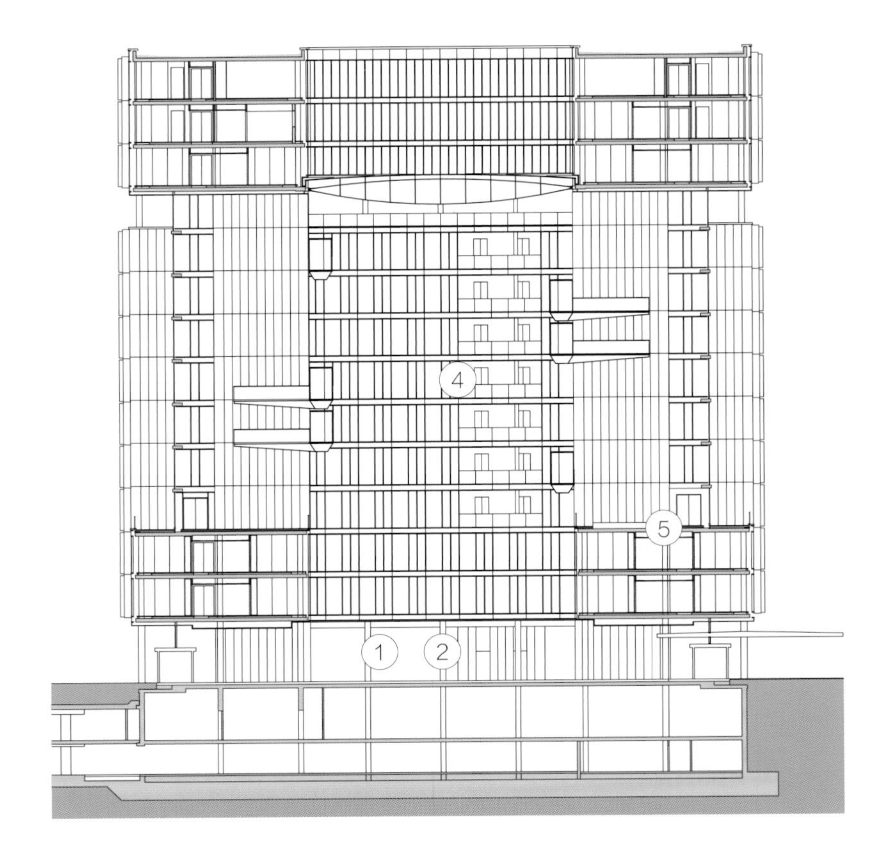

剖面图

Section

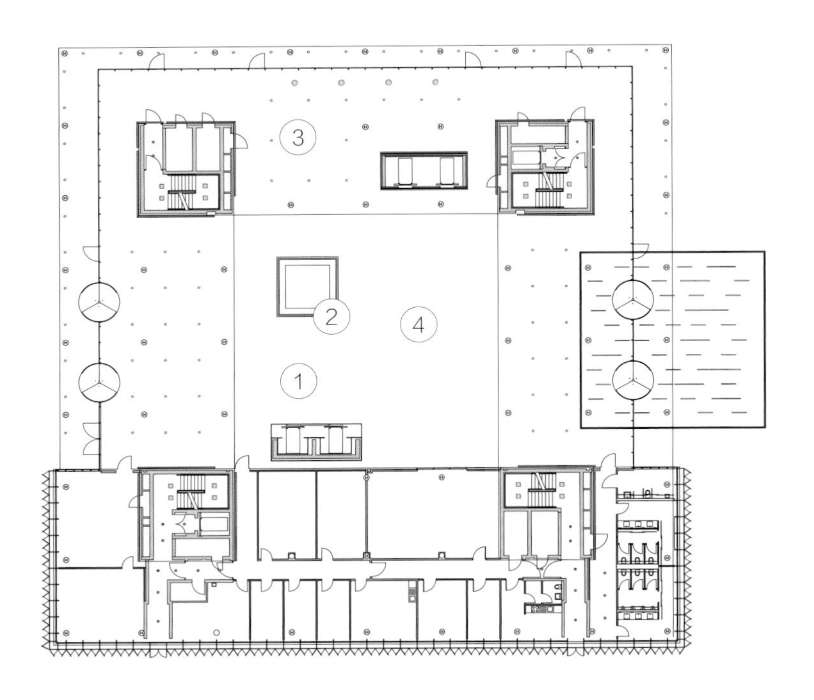

首层平面图

Floor plan ground floor

1.前厅
2.接待处
3.休息区
4.带步行桥的中庭
5.绿植区（全景花园）

1. Foyer
2. Reception
3. Lounge
4. Atrium with pedestrian bridges
5. Green zone / panoramic garden

The ThyssenKrupp group has built its new headquarters on a historic location within Essen's Krupp belt. Nearly half of the materials and products used for its construction originate from the client itself. The airiness of the buildings, grouped in a parkland, is largely based on their façade concept. The architecture plays imaginatively with glass, metal surfaces, and lamellae. The lighting concept supports this vital and multilayered mix of materials. The design from the architectural joint venture ARGE Architekten TKQ between JSWD architects, and Chaix & Morel et Associés for the ThyssenKrupp Group headquarters follows the concept of a campus. Grouped around a central linear water basin are the main building Q1, the forum called Q2, the administration buildings Q5 and Q7, as well as a parking garage.

The exterior impression of the building results from the principle of interaction between core and shell. Metal mesh, perforated sheet metal, sun shielding lamellae, and large coloured sheets of metal characterize many surfaces both in and outside the buildings. In section, the glass facade overlap. The variable nature of daylight modulates this mix of transparent, perforated and solid layers. A finely tuned lighting concept ensures that this stylistic tool is also fully functional in the evening and at night. The building's allure largely originates from the illumination emanating from within, as well as the sensitively adjusted exterior lighting solution.

WATER BASIN, AVENUES, AND CONNECTING PATHS

The central axis of the quarter is marked by a 235 meter long water basin. Flanking the basin are the main access paths: a generous footpath to the left and the 'Allee der Welten' (Avenue of the Worlds) to the right with 15 tree species from five continents. These are supplemented by smaller paths, some of which cross the basin.

In their exterior lighting design, the designers took great care not to interfere with the buildings' illumination emanating from within. Disturbing reflections and distracting glare sources have been avoided by a smart selection of luminaires: Very well shielded pole mounted fixtures with specially developed light distribution follow the two main paths, while the smaller trails are illuminated from bollards with low light positions. Marking the transition between the water surface and the 'shore' are LED-strips following the outline of the pool. As an added result, they emphasize the pool's geometric structuring of the park at night.

Q1 - THE HEADQUARTERS

Located at the front end of the pool is the new 50 meter tall headquarters building. Its impressive nocturnal visual impact is owed to a carefully orchestrated lighting scenario featuring reflective interior surfaces, carving out its volume and creating an attractive sense of spatial depth.

Following the principle of core and shell, the building's structure appears to consist of a glass cube that is partly encased by a layer of perforated sheet metal. Floor recessed uplights have been installed in the interior of the third floor. At nightfall their subtle grazing light lends a gentle glow to the anodized metal. Mounted in the ceiling of the entry level's building setback are downlights delivering directional light to the floor. The reflection from the ground illuminates the bottom of the ceiling slab, making the architecture appear weightless and floating.

No additional typical facade illumination methods are employed. Instead, selected interior surfaces are highlighted and convey the building structure to the observer. The atrium displays its striking exterior image through immense panorama windows. These 26 meter wide and 28 meter tall glass surfaces on the Q1 building's north and south facades open up the structure to the surrounding landscape and provide the atrium and adjacent rooms with daylight. From outside, the panorama windows allow a view into the transparent heart of the headquarters. The footbridges stretching across the atrium stand out spectacularly with LED light lines integrated within their handrails. These luminous horizontal references enter an enthralling visual dialogue with the vertical emphasis found within the illuminated lift portals. The illumination of the eleven-storey glass volume is provided by spotlights located underneath the glass roof on either side of the atrium. Strikingly emphasized, the four stair wells distinctively reveal the building's full height. On the office floors, the ceiling slabs are gently revealed by direct / indirect floor luminaires.

Q2 - THE FORUM

In the Q2 Forum building, ThyssenKrupp welcomes its guests and customers. The conference floor stretches across the entire second storey, with meeting rooms and a conference and event hall for up to 1,000 people. The third floor accommodates the guest canteen as well as the supervisory board room. Located on the ground floor are the staff canteen and cafeteria. Also in this building, the structure is perceptible from outside through the transparent ground floor and semitransparent facade of the remaining floors. Here, the interior lighting also dictates the building's attractive exterior image.

This is particularly evident in the lighting solution for the canteen: While powerful grazing light stages the titanium-coloured panelling of the room's rear wall, the cylindrical custom luminaires above the tables, each with a frosted glass shade, provide a zoning order to the restaurant. At the same time, all circulation areas are brightly lit and the tables are provided with a balanced combination of horizontal and vertical illuminance levels.

A lighting concept with a similar zoning order underlines the representative character of the guest canteen. Suspended above the tables, which have been arranged orthogonally to the windows, is a large ring-shaped chandelier with direct and indirect lighting components. Chest-high partition walls and ceiling suspended filigree metal curtains always group two tables together and lend structure to the elongated space. An RGB-luminaire is integrated into the furniture, offering dynamic coloured light to the shiny curtains. At the ceiling, a light cove traces the outline of the individual areas.

A luminous LED ceiling, specifically developed for this project, has been installed at the supervisory board room. In addition to a diffuse component, it also implements an optical system that creates direct, succinct light on the conference table. The convincing lighting effect and design of the custom fixture is also used to enhance the meeting rooms at the Q1 Headquarters.

ROOM OF SILENCE

The 135 m^2 'Room of Silence' serves as a place of meditation, reflection, and retreat. In the middle of this bare room, a large cube is suspended. The consistently homogenous spatial impression is broken only by the interior lining of the cube. Its inner surface is fully covered with 30 to 40 centimeter large titanium shingles. Their rich materiality and play of shapes are emphasized through a luminous ceiling. The individually controllable RGB LED diodes form a pixel matrix that can display an unlimited variety of static and dynamic lighting scenes.

LIGHT QUALITY AND ENERGY EFFICIENCY

Through a centralized sustainability management program, the ThyssenKrupp group systematically integrates ecological actions into its business operations. Thus, a careful handling of the environment and its natural resources was an important requirement in the design of the Headquarters; for example, the base energy use of the project is 58 % below its local permitted allowance. Furthermore, an efficient use of energy is implemented in all lighting elements throughout the site: LED light sources are used extensively, daylight is consistently utilized, and electric lighting is dimmed and controlled via daylight and presence detectors. This environmentally sensitive designed project was awarded with a Gold Certificate by the German Society for Sustainable Building in May 2011.

CLIENT:
ThyssenKrupp AG
OCCUPANT:
ThyssenKrupp AG
ARCHITECTS:
JSWD Architekten GmbH & Co. KG, Cologne
Atelier d'architecture Chaix & Morel et associés, Paris
TEAM LEADER LICHT KUNST LICHT:
Alexander Rotsch
Tanja Baum
COMPLETION:
2010
PROJECT SIZE:
170,000 m^2
OVERALL BUILDING BUDGET:
300 million euros
LIGHTING BUDGET:
9.8 million euros

总部大楼的整体呈现为被切割的立方体造型，这些被切割的体块仿佛悬浮在半空中，让人产生体量轻盈的视觉印象。其中假想的建筑切割饰面覆盖着蒂森克虏伯集团自己生产的香槟色钢板。

The headquarters building presents itself as a massive cube dissected into floating volumes, evoking an impression of weightlessness. The imaginary cut surfaces are clad in champagne-coloured steel panels from ThyssenKrupp's own production.

到了晚上，看似庞大的建筑群被室内照明的内透光影响，变成了通透明亮的半透明体块。楼梯井空间被设定为光线明亮的区域，香槟色的中庭饰面则通过擦墙光进行照明与表达。在室外，LED线形灯带勾勒出水轴的轮廓。

At night, the seemingly massive volumes are dissolved into sheer transparency by virtue of the interior illumination. The stairwells are displayed as luminous zones and the champagne-coloured atrium surfaces are displayed by means of graze lighting. Outside, the water edge of the pool is traced by LED lines.

总部大楼的中庭挑高有11层楼之高。连续布置的线形LED灯带隐藏在步行桥扶手中。这些扶手灯光从视觉上强化了步行桥的造型感，并将其标记为总部大楼各个区域之间连接的主要通道。

The atrium extends over eleven floors of the headquarters building. Concealed in the bridge handrails are continuous LED light lines. They visually enhance the bridges and mark them as connecting arteries.

低眩光的室外高杆灯沿着园区的主要动线提供照明。
图中右侧是集团的会议楼。

Low-glare exterior pole lights create light zones along the main campus routes. Shown on the right side is the Forum building.

会议楼主要为集团会议和项目提供工作场地和讨论空间。透过大楼首层通高的玻璃立面，可以看到员工餐厅和咖啡厅。位于较高楼层的穿孔金属立面也为大楼营造出半透明的视觉效果。

The Forum building offers space for meetings and project work. Its floor-to-ceiling glazing around the entire ground floor allows for views of the canteen and cafeteria. The perforated steel facade of the upper floors also creates a translucent effect.

强劲的擦墙光突显了食堂和咖啡厅的钛色饰面。在餐桌上方，设计师统一布置了底部带有磨砂玻璃罩的定制圆柱形吊灯。

Powerful graze light emphasizes the titanium-coloured wall paneling of the canteen and cafeteria. Custom cylindrical pendant luminaires with a frosted lower glass capping are centered above the individual tables.

访客餐厅的环形吊灯也是定制的，配有投向餐桌的直接重点照明和投向天花的间接柔光照明系统。此外，凹槽灯隐藏在天花夹层边缘，从而对整体空间的天花照明进行补充。

The circular pendant luminaires at the canteen restaurant are custom-made and have been fitted with a resonant direct light component for the dining tables, coupled with a gentle indirect lighting component. Cove lights have been added at the slab offset, thus complementing the ceiling Illumination along the entire length of the room.

“沉静之屋”旨在为员工提供一个冥想、小憩和静修的空间。在这个空旷的房间中部，一个巨型立方体悬浮在空中。这个立方体的内部包层打破了整体空间的一致性。该立方体的内径表面由30～40厘米长的钛板无缝拼接覆盖，只有人们站在立方体正下方时，才会对此有所察觉。

The 'room of silence' is intended as a space of meditation, introspection and retreat. It consists of a frugal space with a large cube floating at its centre. The consistently homogenous spatial expression is broken only by the cube volume's interior cladding. Its inward facing walls, only perceivable when standing directly underneath the cube, are fully shrouded with 30 to 40 centimeter long titanium shingles.

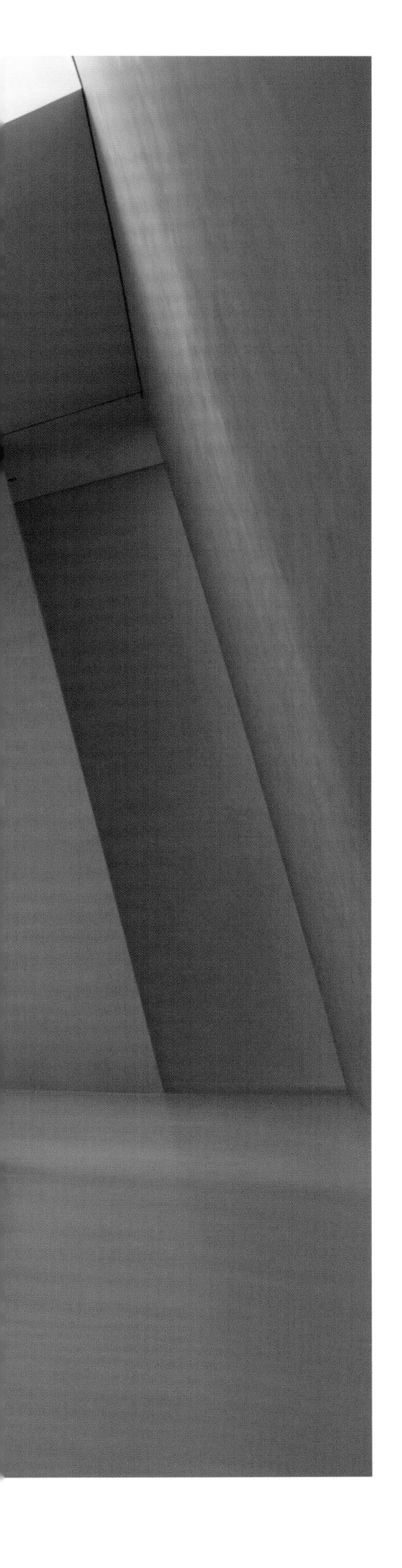

LED轻质发光天花凸显了这个立方体体块丰富的材料组成和造型感。可单独控制的RGB- LED光源构成了一个像素矩阵，可以呈现出或静态或动态的照明场景。

The material's eminence and the play of shapes are emphasised by a luminous LED ceiling. The individually addressable RGB-LED light diodes form a pixel matrix that can display an unlimited diversity of static and dynamic lighting scenes.

AWARD
WINNER

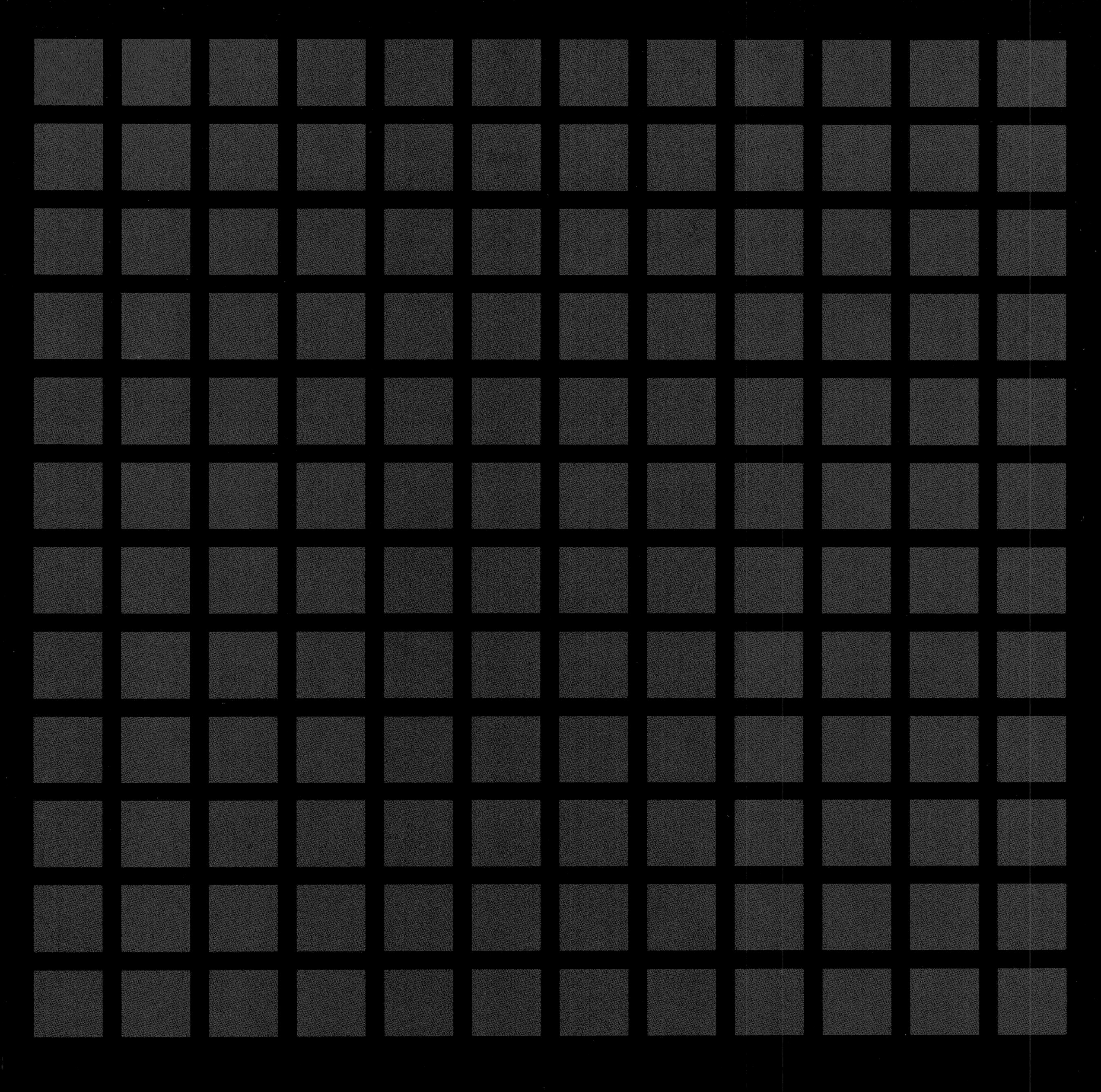

LICHT KUNST LICHT 4

LWL艺术与文化博物馆 明斯特

LWL-MUSEUM FÜR KUNST UND KULTUR MÜNSTER

LWL-Museum für Kunst und Kultur

空间和光线启发了新的体验视角——LWL艺术与文化博物馆

在经历了很长一段时间的翻修后，德国明斯特的LWL艺术与文化博物馆终于在2014年9月重新对外开放了。由施塔布建筑设计事务所设计的博物馆扩建部分与早在1908年建成的原始结构紧密相连。目前，博物馆内有51个展厅，展厅外围还设有图书馆、书店、影音室、休息区、餐厅以及连接不同功能区的通道空间。该项目的照明设计方案不仅为博物馆提供了灵活的展陈照明，同时也展现了建筑空间本身的魅力。

建筑与灯光的一体化设计

博物馆前院与明斯特市中心相连接，建筑空间以"院落建筑"的形式进行规划，参观者可以沿着一条城市公共小路抵达博物馆。这样的空间与建筑规划为博物馆作为公共文化场所融入城市文脉提供了最佳条件。建筑设计师通过6米高的落地窗，为博物馆内部空间与周围社区之间的相互联系提供了新视野。无论从博物馆内部还是外部观赏，这座建筑都彰显出开放且清晰易懂的设计风格。该项目的照明设计方案正是要突出建筑与空间的品质。因此，照明设计师决定将所有光源严格地集成到建筑结构中。照明设计的重点不是灯具本身，而是其在空间中所呈现的光线效果。

隐藏式灯具的安装位置和极具雕塑感的楼梯

博物馆大厅的屋顶由玻璃天窗和位于其下方的薄膜天花组件构成，挑高三层楼的空间向参观者展现出欢迎、好客的姿态。当阳光明媚时，大厅便会呈现出有趣的光影效果；即使是在阴天，这个空间也保留了与室外环境的视觉联系。为了不影响薄膜天花的表面视觉效果，博物馆大厅完全由集成在墙壁凹槽中的灯具来提供照明。聚光灯两两成对，在这个较高的空间中提供了强有力的直接照明。每组聚光灯可以单独开关和调光。这样一来，聚光灯就可以同时为博物馆的日常运营和在大厅举办的临时活动营造合适的光环境。

当参观者通过连接大厅和展厅的楼梯时，灯具光源仍然保持着隐蔽的状态。即便如此，光线依然有效地强调了楼梯的雕塑感特征。光线从一条看似是黑色的天花缝隙中投向楼梯。楼梯雕塑感的视觉印象是通过将光线聚焦在每级台阶上形成的。这种照明方式使楼梯本身深色的木质表面和每级台阶上明亮的光线形成带有律动感的视觉对比，使细长的楼梯显得格外庄重。

与墙壁轮廓相呼应的发光天花

照明设计师为该博物馆的展陈空间研发了一套特别的照明解决方案。沿墙壁轮廓向内延伸的发光天花框架不仅保证了天花造型的干净简洁，还提供了灵活的灯具安装条件，以便满足不同展陈形式的照明要求。这些量身定制的发光天花框架与常规的博物馆发光天花的不同之处在于，该照明方案不仅解决了常规照明方式下光线集中在墙壁边界的问题，同时也保证了垂直墙壁上均匀的照明效果。这种效果是通过在半透明薄膜背后精确分布的可调光式荧光灯管来实现的。照明设计师通过大量的模型测试，研发得出将灯管与灯管之间的距离朝着墙壁边缘方向逐渐增大的特殊排列方式。这种布灯方式，既避免了对墙壁顶部进行过度照明，又实现了在垂直表面上完全均匀的光线分布效果。

发光边框和重点照明元素共同谱写了丰富多样的光线"二重奏"

设计师在发光边框和天花中央区域之间的空隙处安装了三相式内嵌灯轨，便于提供灵活的轨道照明。窄光束角轨道灯为艺术品提供重点照明，宽光束角灯则可以为较大展品和区域提供照明。展厅可以单独或同时使用发光天花边框和轨道灯进行光环境布置。可选的光环境场景包括对展示墙的均匀洗墙照明、空间环境光与重点光的结合照明，以及仅仅采用重点照明形成戏剧性的灯光氛围。在灯具控制方面，发光边框以及每一盏LED灯具都可以单独调光，确保了在不同视觉任务和展品保护要求的情况下，照明系统均可以相对应地进行精确的调整。

自然光和人工光的巧妙结合

博物馆的顶层设有五个包含天窗的空间，居中设置的大型天窗让动态变化的自然光线进入室内空间。为了避免直射光所带来的不利影响，天窗的夹层玻璃中配备了棱镜分光层，仅允许柔和的自然光进入室内。此外，天窗还可以通过遮光卷帘来减少进入室内的光线，而另一层遮光帘则可以完全阻挡光线。面向室内空间的半透明薄膜天花作为额外的光线"过滤器"，使室内空间的光环境更为柔和。

为了使带天窗的空间拥有与仅采用人工光照明的展陈空间相统一的灯光氛围，照明设计师采用了上述发光边框与轨道式灯具相结合的灯光语汇进行照明。在这些空间中，人工照明的光通量会根据日光水平进行自动调节。用于该控制系统的日照实时数据由安装在博物馆屋顶上的日光传感器负责提供。发光边框内荧光灯管的光通量会根据传感器提供的实时户外照度水平，以及遮光卷帘的运行状况自动调节。

为餐饮空间设计的可调节照明方案

通过与建筑师和餐厅租户的沟通，照明设计师研发了一种可以为白天和夜晚的餐厅运营提供两种不同光环境场景的照明方案。沿着天井和餐厅立面的装饰性吊灯为白天的咖啡厅创造出友好、明亮的光线氛围。傍晚时分，吊灯会被调暗一些，以营造更加内敛、安静的就餐氛围。除了明显调暗的吊灯外，天花嵌入式可调节下照灯为桌面提供了重点照明。作为附加的第三种照明元素，设计师将LED灯条集成在工作台台面下方和吧台展示柜里，在墙壁上营造了柔和的垂直照明。

室外照明

室外照明方案特意避免了对新建筑外立面进行泛光照明。因为博物馆位于明斯特的"心脏"区域，照明设计师希望通过建筑立面大面积的玻璃窗户，给人留下由内向外散发光芒和活力的整体视觉印象。因此，设计师仅用灯光突出了大教堂前方老建筑一侧的立面。与门厅和天井一样，设计师在新建筑立面上设有安装洗墙灯的壁龛，以便柔和且均匀地照亮对面历史悠久的建筑立面。

新建筑的另一端是面向罗滕堡的广场，由多种照明元素强调。城市公共空间的环境照明由高杆灯提供，这里的高杆灯和周围其他街区所使用的灯具为同一类型。此外，博物馆外墙上由灯光艺术家奥托·皮恩创作的灯光雕塑作品——《银色频率》，以及与广场相邻的新餐厅的户外照明都被很好地融入了广场整体的室外照明气氛中。

与现存建筑拱廊形态相匹配的灯具形式

LKL的照明设计师还为现存建筑的拱廊空间研发了照明翻新策略。为了最大限度地减少对受保护的城市地标建筑的视觉干扰，设计师特意选择了造型优雅的灯具，为拱廊空间同时提供直接和间接照明。

业主：
威斯特伐利亚-利泊地区协会
使用人：
LWL艺术与文化博物馆
建筑设计：
施塔布建筑设计事务所，柏林
LKL项目经理：
玛蒂娜·维斯
展陈设计：
Space4有限责任公司，斯图加特
展陈照明：
LDE贝尔兹纳·福尔摩斯，斯图加特
竣工时间：
2014年
项目规模：
1800平方米
整体项目预算：
3870万欧元
照明预算：
180万欧元

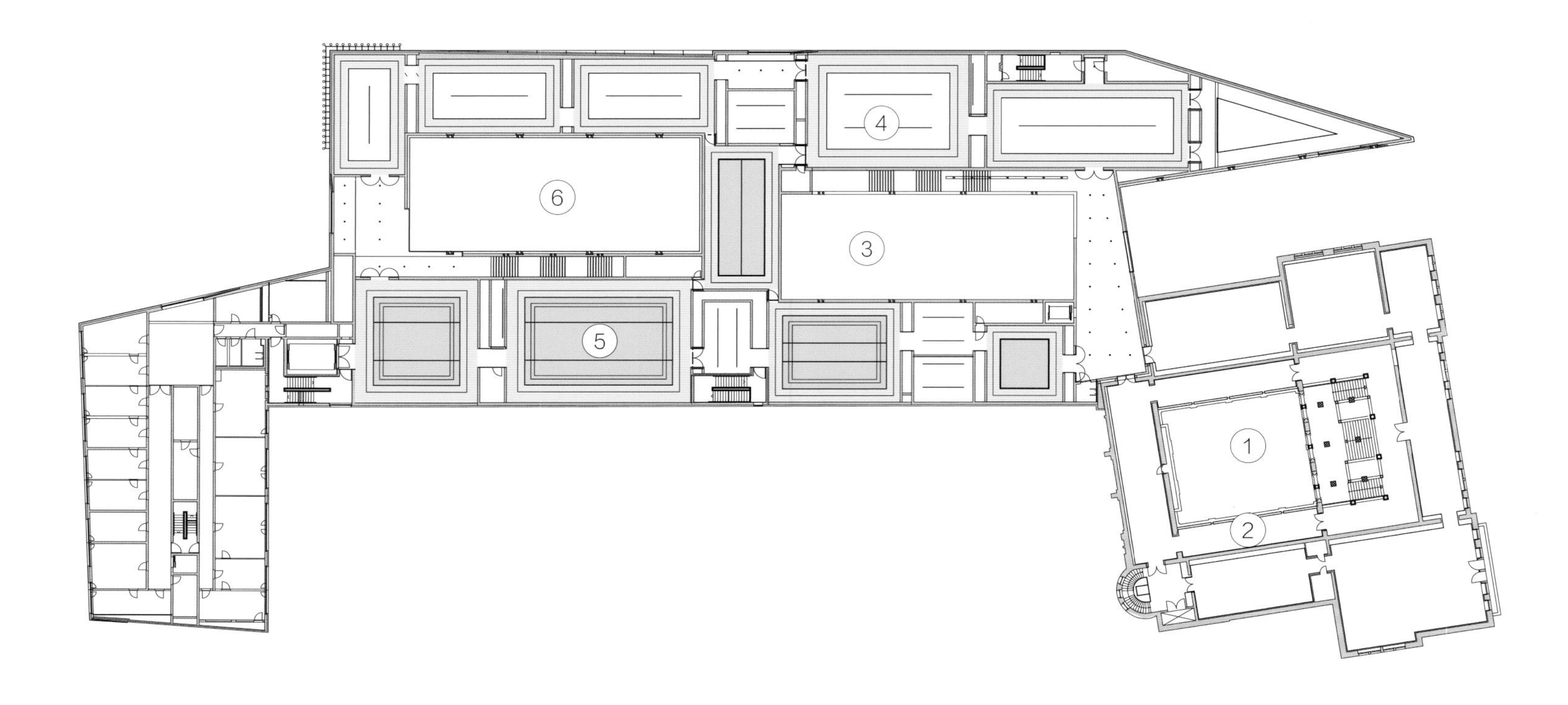

二层反向天花平面图

Reflected ceiling plan second floor

1.现存建筑
2.历史悠久的拱廊
3.大厅
4.展陈空间
5.带天窗的空间
6.天井

1. Existing building
2. Historic arcade
3. Foyer
4. Exhibition spaces
5. Skylight spaces
6. Patio

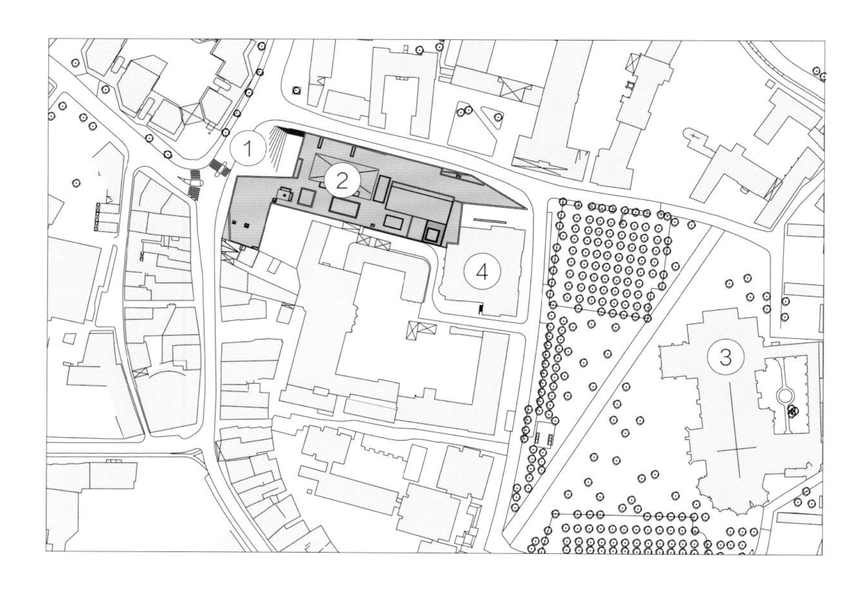

场地平面图

Site plan

1.罗滕堡前院
2.LWL艺术与文化博物馆扩建建筑
3.圣保罗大教堂
4.LWL艺术与文化博物馆现存建筑

1. Forecourt of Rothenburg
2. LWL-Museum extension building
3. St. Paul's Cathedral
4. LWL-Museum existing building

After a long-lasting construction period, the LWL-Museum für Kunst und Kultur in Münster was re-opened in September 2014. An extension designed by Staab Architekten has been connected to the existing structure built in 1908. The museum now displays its collection in 51 exhibition spaces which are flanked by a library, a bookshop, an auditorium, access and relaxation zones, as well as a restaurant. The lighting concept not only offers flexible exhibition illumination, but also showcases the architecture.

THE LIGHT IS INTEGRATED INTO THE ARCHITECTURE

Interlinked with Münster's city center through a sequence of forecourts, traversed by a publically accessible path, and organized in an 'architecture of courtyards', the museum offers optimal conditions for serving as a public cultural venue imbedded in an urban context. New visions of an interconnection between the museum and its neighbourhood arise from within by virtue of six meter high windows. Regardless of the perspective – whether from the interior or exterior – openness and clarity undeniably prevail in this building. The lighting concept was to underline these qualities. Consequently, the lighting designers opted for a stringent integration of all light sources into the architecture. It is not the luminaires that are at the design focus, but the lighting effect.

HIDDEN LIGHTING LOCATIONS AND SCULPTURAL STAIRS

The museum visitor is welcomed by a three-storey foyer which is spanned by a glass skylight and a membrane ceiling below. When there is direct sun light, an interesting play of light and shadow shapes the space, while even on overcast days, the visual link with the exterior exists. In order not to encumber the ceiling surface's visual effect, the foyer illumination was fully integrated into wall recesses. Grouped in pairs, the spotlights create powerful direct light in this tall space. Each group of luminaires can be switched and dimmed individually. This allows the spotlights to provide appropriate lighting for both the museum's operation and foyer events.

The light sources remain concealed from the observer when he passes the staircase connecting the foyer and the exhibition spaces. Nevertheless, the stair's sculptural character is emphasized effectively with light. It emanates from a linear ceiling recess that appears as a dark groove. This particular effect is achieved through light specifically focused on the steps. The result is a rhythmic contrast on the dark wood surface, which lends additional grandeur to the elongated stairs.

LUMINOUS CEILINGS FOLLOWING THE WALLS' OUTLINES

In the exhibition spaces, a unique lighting solution has been developed for this museum. A luminous ceiling frame following the perimeter walls contributes to keeping the ceiling uncluttered, yet provides a flexible light source to allow an unrestricted arrangement of the exhibits. The difference between these light frames and the luminous ceiling solution commonly used in museums, is not only the restriction of the luminous surface to the vicinity of the perimeter walls; the light frame also provides a particularly uniform illumination of the wall surfaces. This is achieved through a precise arrangement of dimmable fluorescent lamps behind the translucent membrane. Following extensive mock-up tests, the spacing between the light strips gradually increases towards the wall. Thus, excessive luminance levels at the wall top are avoided and an unusually uniform light distribution on the vertical surface is achieved.

LIGHT FRAMES AND ACCENT LIGHTS FORM A VERSATILE DUET

Following the gap between the light frame and the central ceiling area is a 3-phase-track, providing a flexible arrangement of track lights. The narrow beam fixtures accentuate works of art, while the wide beam fixtures illuminate larger objects lights areas. The museum can use the light frames and accent individually or together. The options include uniform wall washing of the display walls, a combination of ambient and accent lighting, and an introverted, dramatic light atmosphere only using accent lights. The ability to dim the light frames and each LED projector allows for an adaptation of the illumination for each visual task and conservation requirement.

DAYLIGHT AND ELECTRIC LIGHT CAREFULLY MATCHED

On the top floor, the museum houses five skylight spaces. Large, centrally located daylight ceilings allow the dynamics of natural light into the interior space. In order to avoid the detrimental effects of direct sunlight, a laminated glass sandwich has been fitted with a prismatic layer, only allowing diffuse skylight to pass through. The amount of light can be reduced through a roller blind, while another layer allows for a complete black-out. A translucent membrane ceiling facing the interior acts as an additional filter.

In order to achieve a lighting atmosphere in the skylight spaces identical to what is found in the exhibition spaces using only electric lighting, a combination of the aforementioned light frame and track mounted fixtures is employed. The luminous flux is automatically dimmed according to the level of daylight. The necessary data for this function is provided by a daylight sensor on the museum roof. Coordinated with the real-time exterior illuminance levels and the operation status of the roller blinds, the luminous flux of the light frames' fluorescent lamps is adjusted.

ADAPTABLE LIGHT FOR GASTRONOMY

In coordination with the architects and the tenants of the restaurant, a lighting concept has been developed that offers two distinct qualities of light for the day and night operation. Decorative pendant luminaires along the patio facade and the refectory generate a friendly and bright light atmosphere for the café use during the day. In the evening hours, the pendants are dimmed in order to create a more introverted atmosphere.

In addition to the strongly dimmed pendant luminaires, flush recessed adjustable downlights pinpoint direct light on the tables. As an additional, third light source, LED light strips have been integrated underneath the benches and bar furniture, producing a soft vertical illumination on the walls.

EXTERIOR ILLUMINATION

The exterior lighting concept deliberately avoids lighting the facade of the new building. The museum in the heart of Münster is to radiate from within, emanating light and life from the numerous window openings. Only the facade of the existing building, facing the square in front of the cathedral, has been accentuated with light. As in the foyer and patio, wall recesses in the new building's facade have been fitted with wall washers to softly and evenly illuminate the historical facade.

The square facing the Rothenburg on the opposite side of the new building is exhibited by various lighting elements. The urban space's general illumination is provided by pole luminaires, which can also be found in the surrounding neighborhood. Furthermore, the newly refurbished light sculpture 'Silberne Frequenz' (Silver Frequency) by Otto Piene on the museum's facade and the exterior effect of the new gastronomy flanking the square have been consciously integrated into the exterior lighting concept.

A LIGHTING PROFILE FOR THE EXISTING BUILDING'S ARCADES

Licht Kunst Licht has also developed the light refurbishment strategy for the existing building's arcade corridors. In order to minimize the visual disturbance to the listed landmark protected building, an elegant luminaire profile has been selected.

CLIENT:
Regional Association Westphalia-Lippe
OCCUPANT:
LWL-Museum für Kunst und Kultur
ARCHITECTS:
Staab Architekten GmbH, Berlin
TEAM LEADER LICHT KUNST LICHT:
Martina Weiss
EXHIBITION DESIGNER:
Space4 GmbH, Stuttgart
SCENOGRAPHIC EXHIBITION LIGHTING:
LDE Belzner Holmes, Stuttgart
COMPLETION:
2014
PROJECT SIZE:
1,800 m^2
OVERALL BUILDING BUDGET:
38.7 million euros
LIGHTING BUDGET:
1.8 million euros

施塔布建筑设计事务所设计新建的LWL艺术与文化博物馆是对这座可追溯到1908年的古老建筑的细腻且当代化的扩建。

The newly constructed LWL-Museum by Staab Architekten is a sensitive yet contemporary extension to the existing building dating back to 1908.

由灯光艺术家奥托·皮恩设计的著名灯光雕塑作品——《银色频率》，如今在博物馆立面上闪烁着光芒。

The notable, newly refurbished light sculpture 'Silberne Frequenz' (Silver Frequency) by Otto Piene, now gleams on the museum facade.

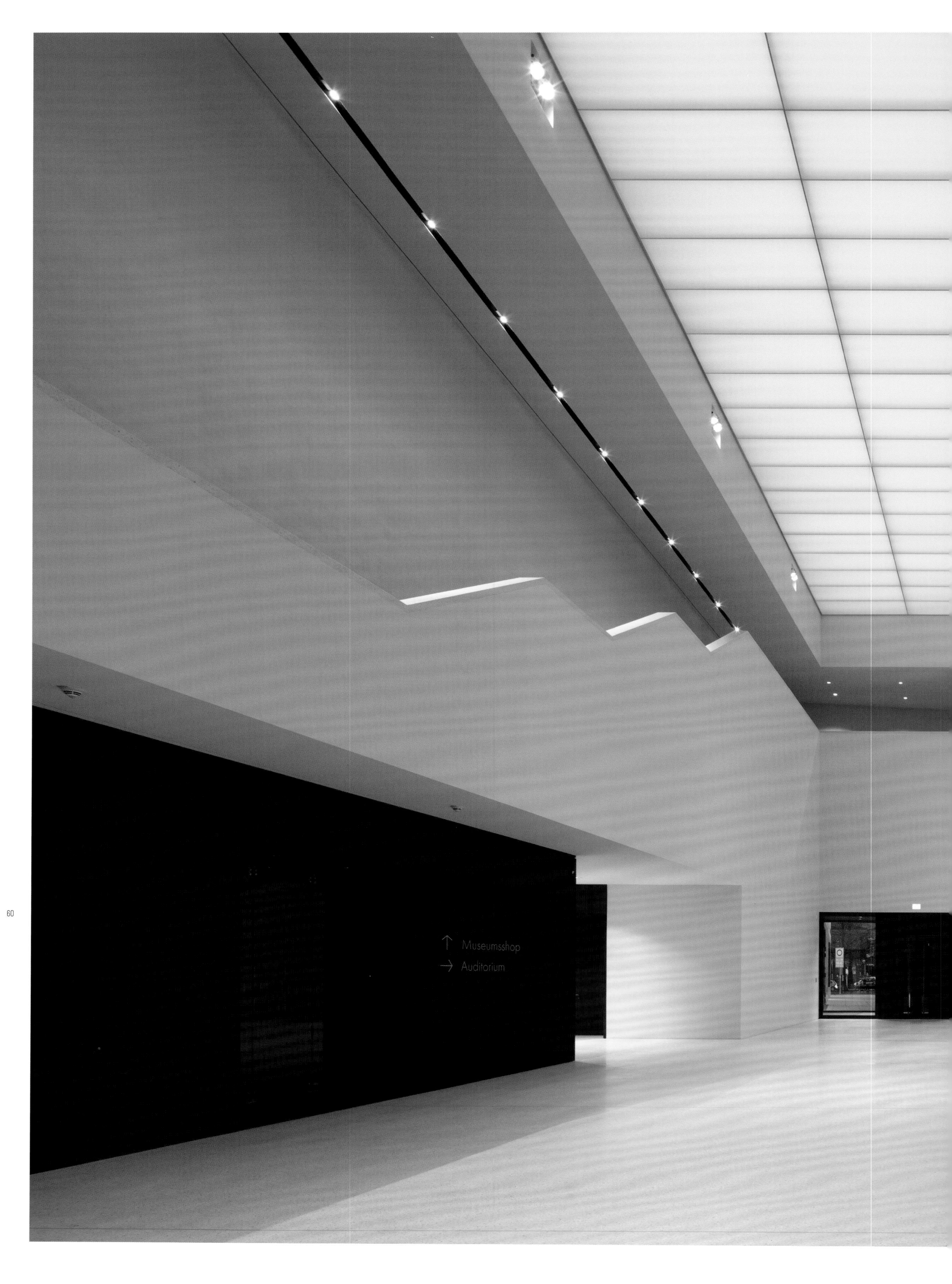
Museumsshop
Auditorium

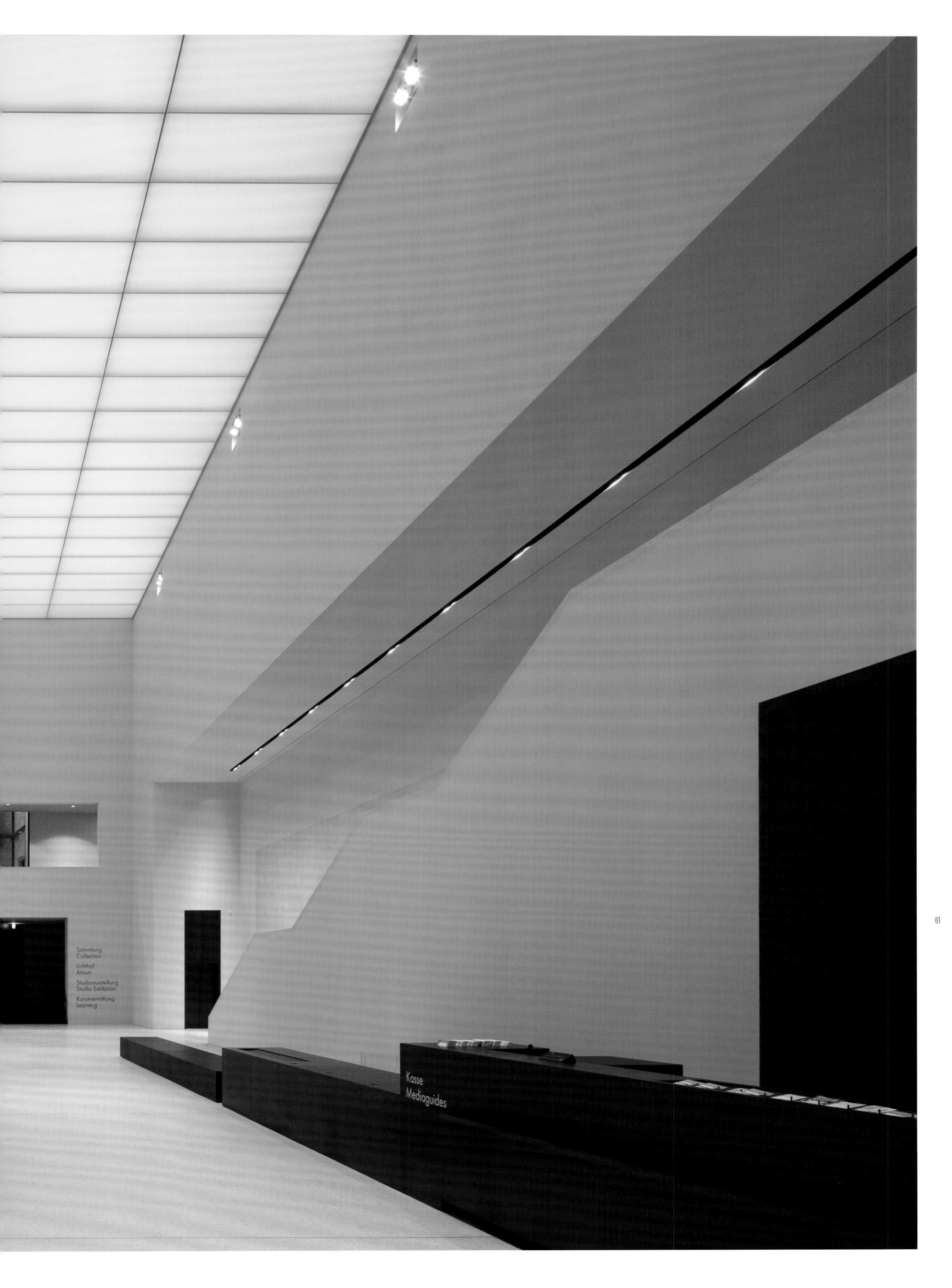
Sammlung
Collection
Lichthof
Atrium
Studioausstellung
Studio Exhibition
Kunstvermittlung
Learning
Kasse
Mediaguides

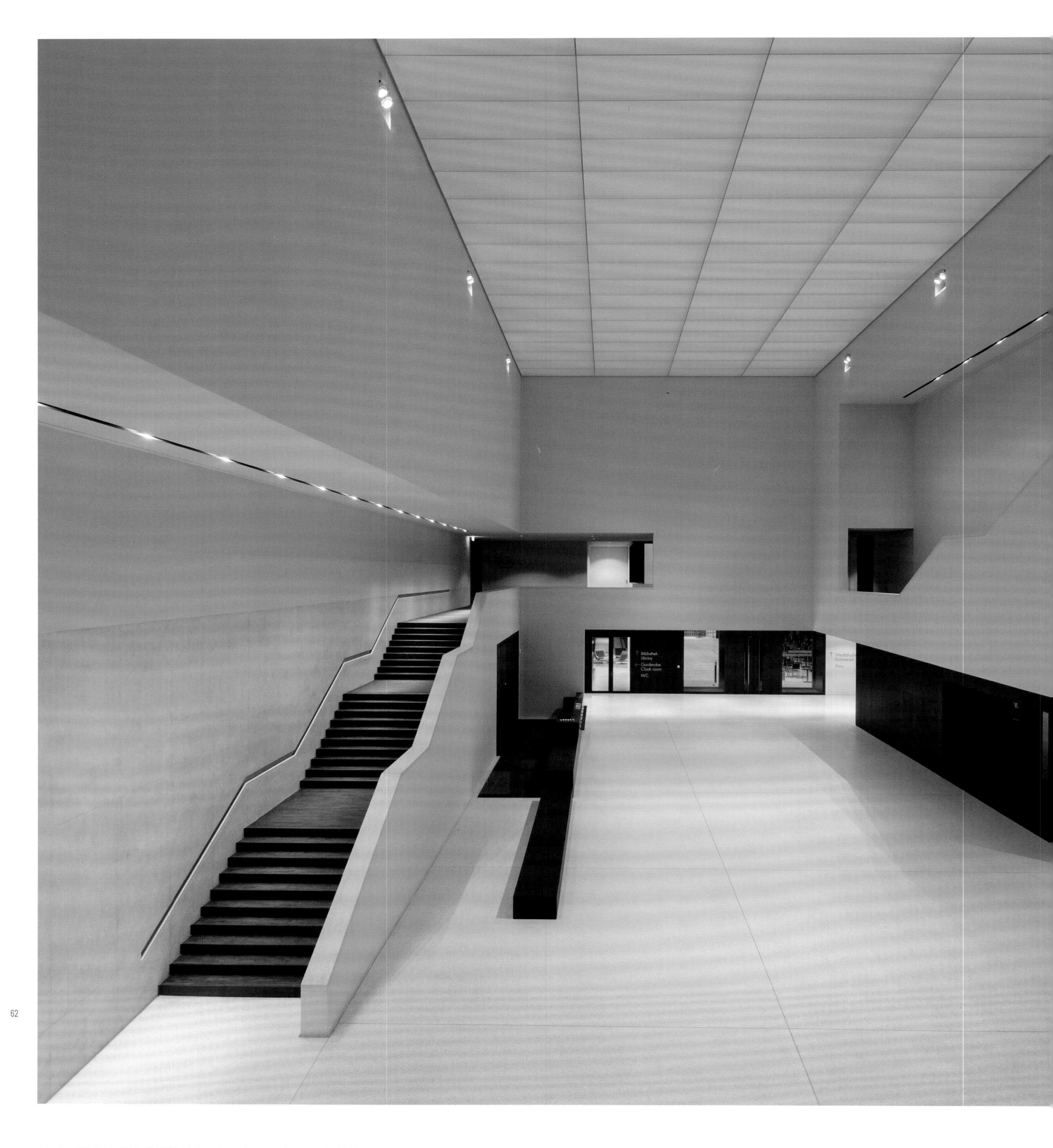

为了不影响天窗的视觉效果，大厅的照明灯具全部被集成在墙体的壁龛中。从天花条形凹槽中发出的光线着重强调了楼梯造型的雕塑感。

In order to not encumber the skylight's visual effect, the foyer illumination was fully integrated into wall recesses. The sculptural character of the stairwell is strikingly emphasized by light emanated from the linear ceiling recess.

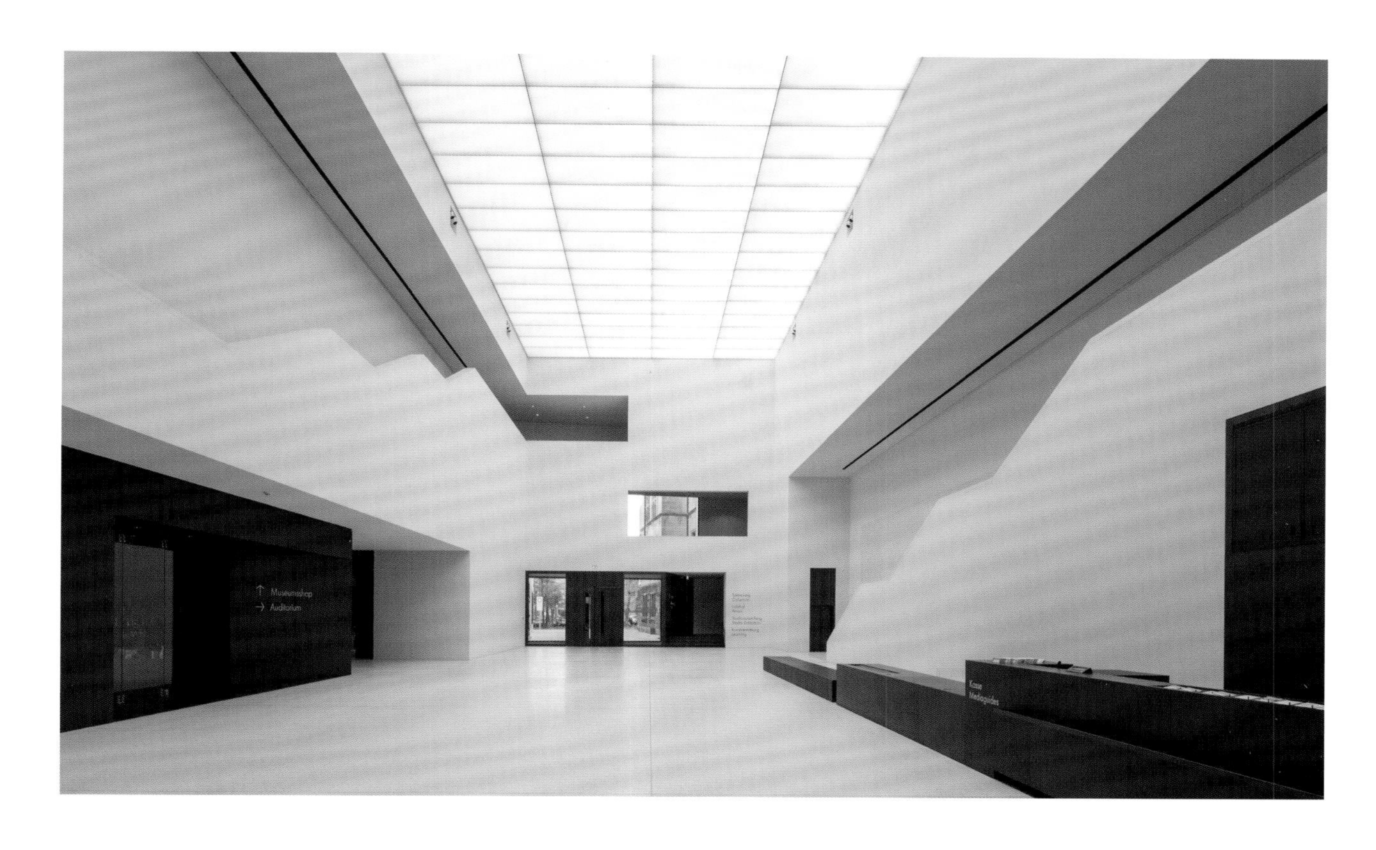

白天，大面积的天窗使整体空间都沉浸在生动的自然光线中。

Due to its large skylight, the space is immersed in vivid natural light during the day.

到了夜晚，天窗在视觉上后退为远景。人工光照明系统承担着环境照明的任务，并将空间的光环境从日间均匀柔和的气氛转变为夜间高对比度的视觉氛围。

In the nocturnal scenario, the skylight visually recedes. The artificial light assumes the task of the general illumination and transforms the space from being washed in light to an atmosphere rich in contrast.

在展陈空间中，照明设计师研发了一种特殊的照明解决方案。与天花相结合的轨道灯沿着墙壁顺势展开，可提供重点照明。

In the exhibition spaces, a unique lighting solution has been implemented. A ceiling-integrated artificial light frame has been specifically developed to run along the walls of the project.

周边的发光天花薄膜通过灯带进行背光照明。光源与光源之间的距离朝着墙壁边缘方向逐渐增大，确保了展陈墙面上获得完全均匀的照明效果。

The surrounding luminous membrane ceiling is backlit with light strips. Their spacing increases towards the perimeter, allowing for an unusually uniform illumination of the display walls.

如果情况需要，展陈空间也可以仅使用轨道聚光灯对陈列品进行重点照明。

If desired, the space can be solely illuminated with accent lighting via track-mounted spotlights.

Medienkunst und Po
Media Art and Po

Nam June Paik gehört zu den
kunst. Sein Werk »Mongolian
Präsentation im deutschen Pavi
Venedig, die mit einem Golde
wird. Es symbolisiert das kulture
tum zeitgenössischer Künstler.
mit seiner Kunst dagegen die S
rialien zeigen, dass dieser Zus
gegensätzlicher Kräfte entsteh
festen Rahmen gespannt.

Nam June Paik is one of the p
1993 installation »Mongolian
sentation at the German pavili
which was awarded the Gold
restless cultural nomadism of c
Ruthenbeck, on the other hand
in his work. This condition is ac
opposite forces represented by
instance, soft textile stretching c

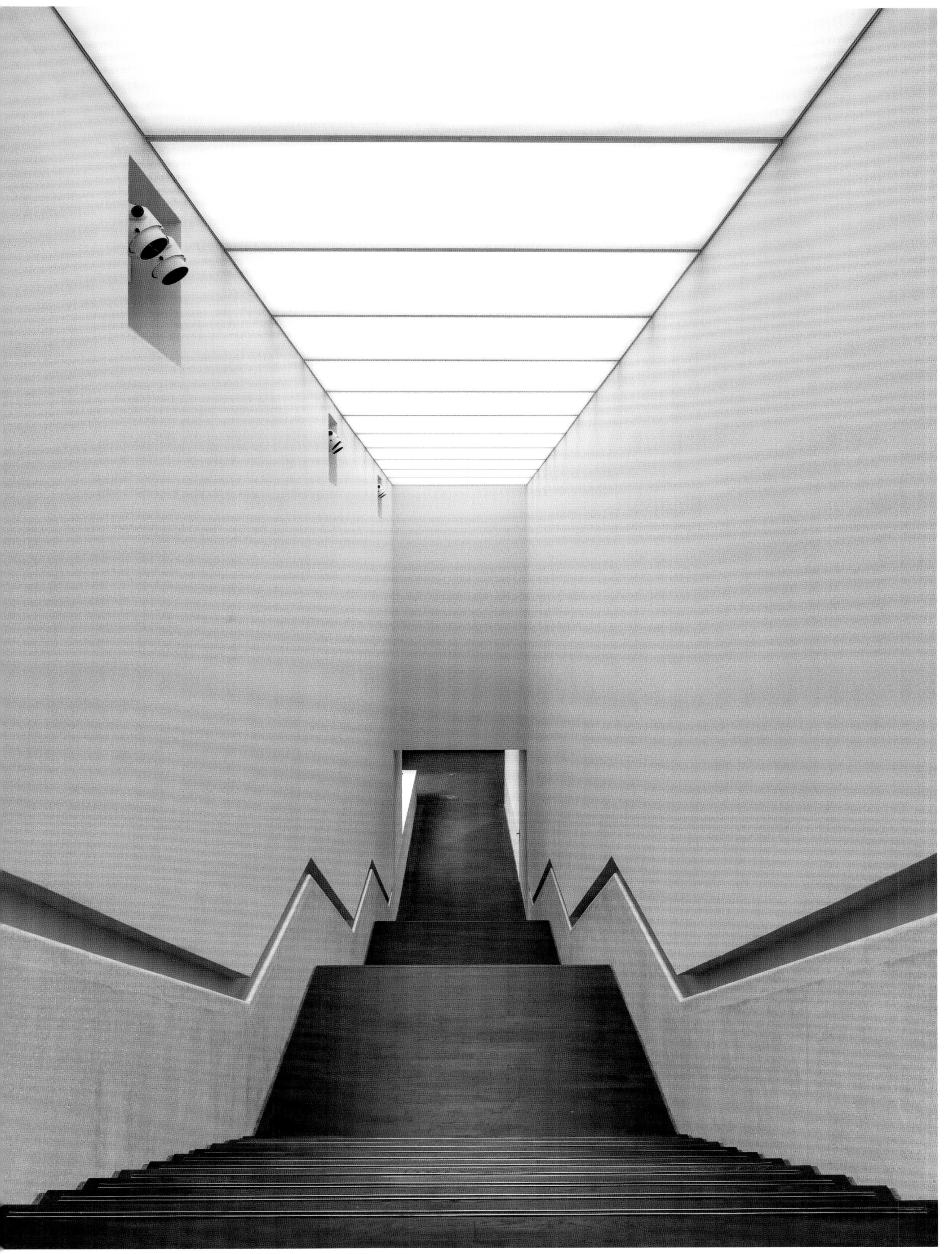

在顶层有自然光照射的空间中，人工光天花边框的照明形式依然被保留了下来。这样可以使下层的展陈空间与上层的天窗空间拥有相同的照明氛围，特别是在夜间或是遮光卷帘完全封闭的时候，整体空间的光环境仍可以高度协调。

In the sky lit spaces of the top floor, electric light frames are implemented as well. This allows the same light atmosphere in the exhibition spaces below to be re-created above, specifically at night or when the black-out is closed.

白天，位于空间中央位置的天窗可以将动态变化的自然光引入室内空间。天窗夹层玻璃中的分光棱镜可以保护室内展品免受阳光的直射。

During the day, the centered skylights allow the dynamics of natural light into the interior space. A prismatic layer in the glass sandwich offers protection from the detrimental effects of direct sunlight.

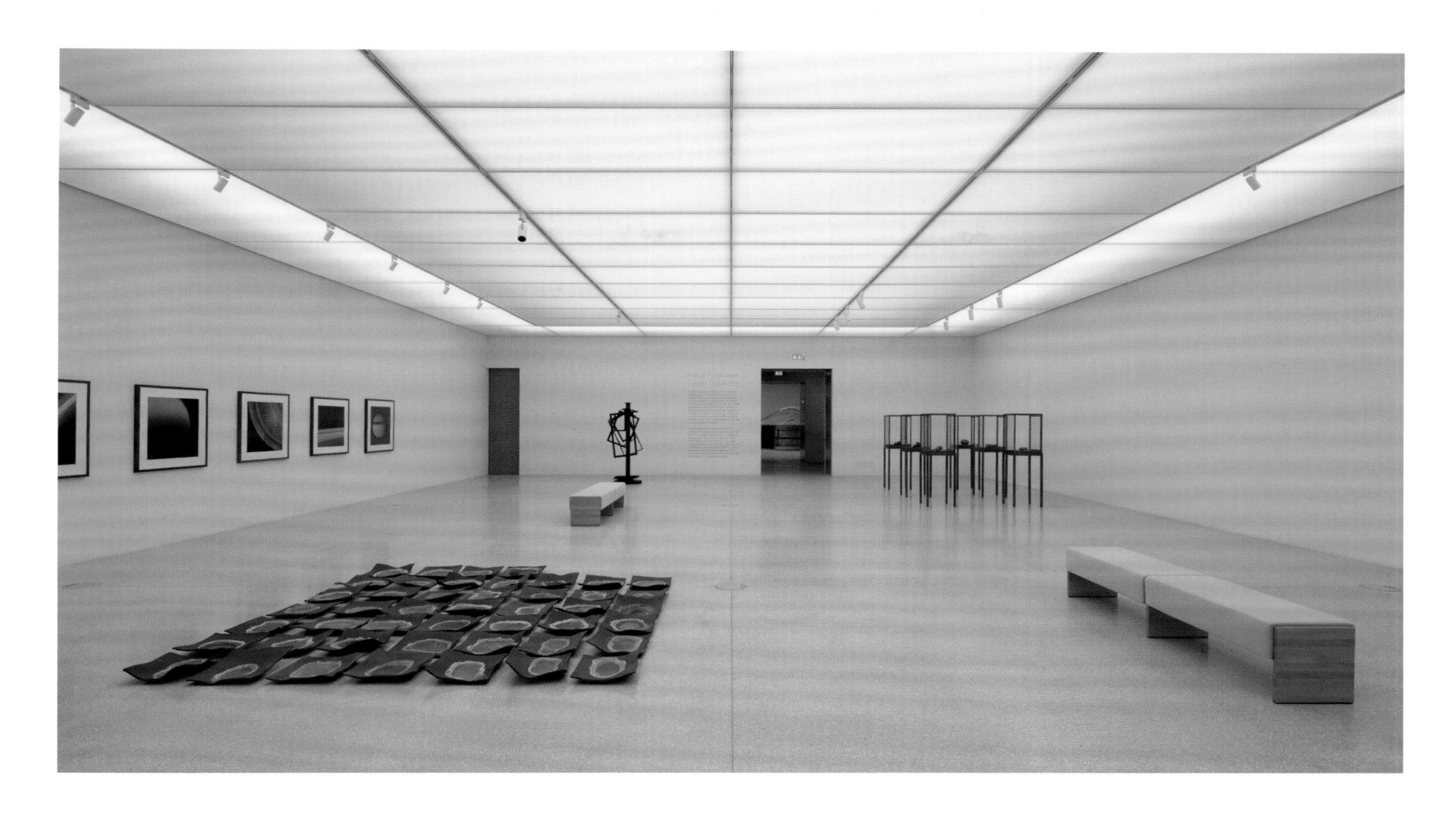

自然光和展陈墙面上人工光的结合为展陈空间提供了细腻的光线，强调了空间的视觉秩序。

The combination of natural daylight and the illumination of the display walls with electric light provides spatial arrangements with carefully tuned luminances.

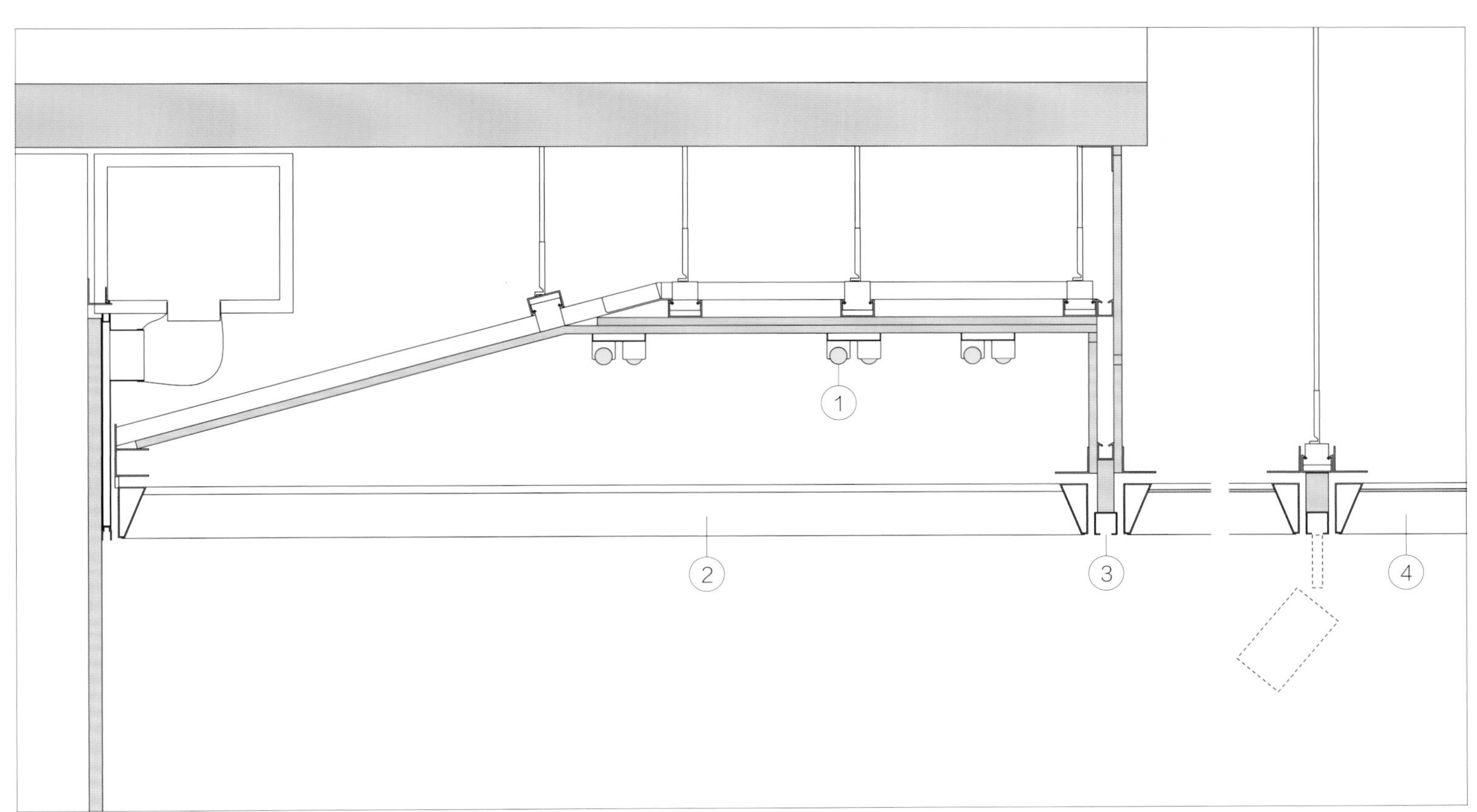

天花结构的细节剖面大样图
1.灯带
2.背光天花
3.轨道灯轨
4.天窗

Detail section of ceiling structure
1. Light strips
2. Backlit ceiling
3. Lighting track
4. Skylight ceiling

博物馆餐厅的空间中心是一组排列有趣的吊灯。卡座背面的彩色灯光柔和地渲染着空间的色彩。

A cluster of playfully arranged pendant luminaires centers the spatial arrangement of the museum cafeteria. Coloured light behind the benches gently immerses the space colour.

与大厅和新建筑立面的照明方式相似，照明设计师在天井墙体的壁龛内设置了可调节聚光灯，以便对雕塑作品和其他物体进行照明。

Similar to the foyer and the new building's facade, wall recesses in the patio have been fitted with adjustable spotlights, in order to illuminate sculptures and objects.

在餐厅四周，天花吊灯柔和地照亮桌面，而吧台区域则通过定向照明和彩色灯光营造出极具戏剧感的空间氛围。

At the cafeteria's perimeter, tables are gently illuminated by pendant luminaires, while the bar is dramatically emphasized by use of directional and coloured light.

地标性拱廊始建于20世纪初，这个受文化遗产保护的拱廊在两层楼高的空间中勾勒出一个宽敞的中庭。如今，中庭被玻璃拱顶覆盖。

Built in the early 20th century, the landmark protected arcades outline a generous courtyard on two levels. Today, the atrium is covered by a barrel vault glass roof.

作为连接空间的照明元素，照明设计师在这个拱廊中布置了一条连续且精致的条形灯具。灯具悬挂在拱廊中央，与立柱的柱头高度齐平。这个条形灯具在为拱顶表面提供间接照明的同时，也包含了数个用于直接重点照明的小型可调节聚光灯。

As a connecting light element, a continuous, delicate profile luminaire has been implemented in the sequence of ribbed vaults. It is suspended in the center of the arcades, at the same height as the columns' capitals, and provides indirect illumination for the vaults while containing small adjustable luminaire heads for direct accent lighting.

AWARD
WINNER

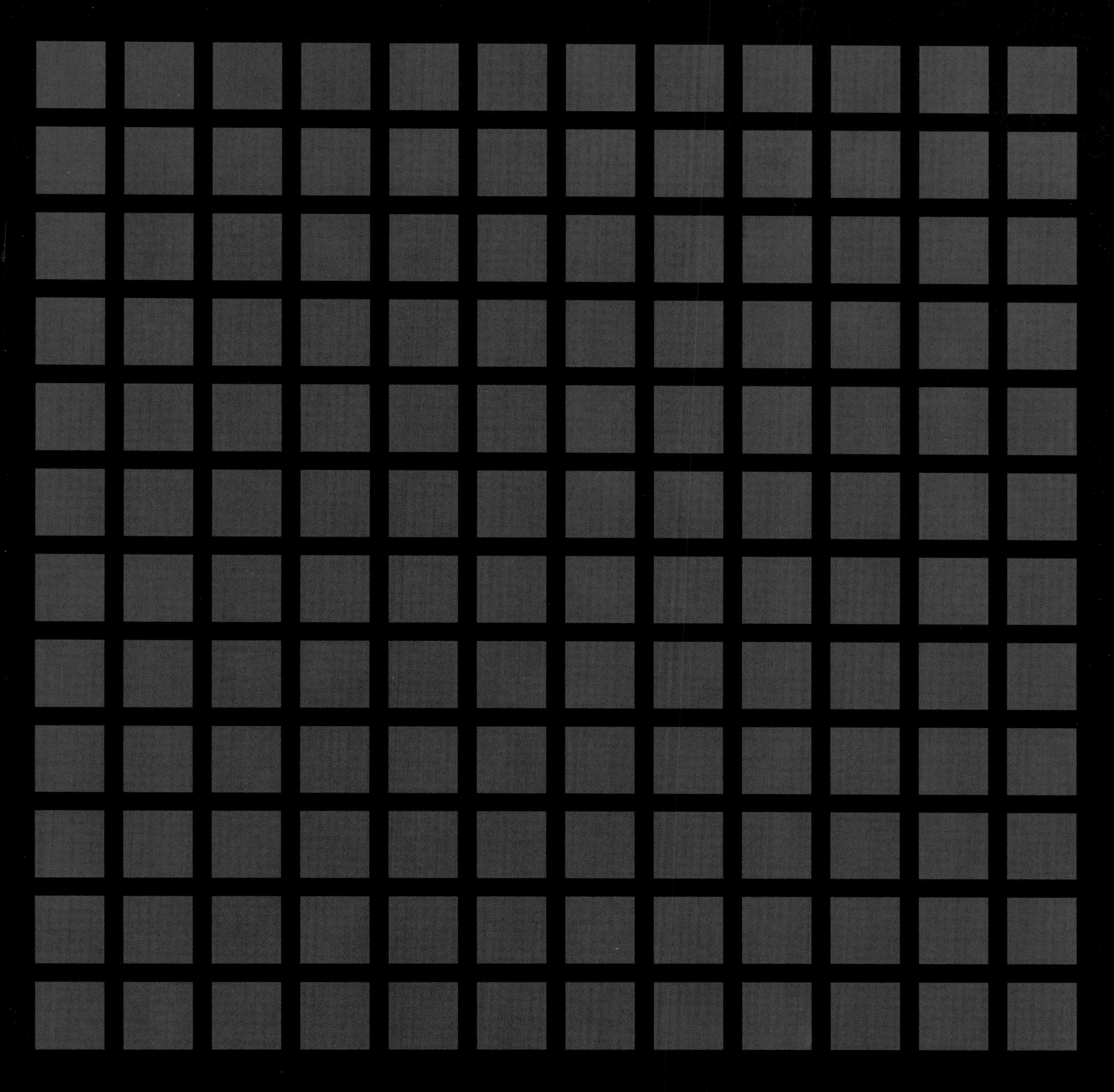

LICHT KUNST LICHT 4

赛恩钢铁铸件馆 本多夫

SAYN IRON WORKS FOUNDRY BENDORF

ERMES 1968

旧钢铁厂的新照明——赛恩钢铁铸件馆的照明设计

位于德国本多夫小镇的赛恩园区坐落在莱茵河畔宁静的小山谷中，与科布伦茨市相邻。山谷中分布着赛恩堡垒、宫殿和公园，沿着这个细长的山谷，两侧山坡上树木茂密，从如此景象中很难让人联想到它的历史：这里曾经是19世纪德国铸铁产量最大的工业地区之一，当地的经济发展和命运沉浮都曾由铸铁产业的兴衰决定。赛恩钢铁厂位于小镇的北端，委身在城堡山坡附近一个非常不起眼的地方。在这片面积不大的区域中，除了简易的房屋、大片的砾石地和朴素的办公楼外，还隐藏着一座设计精美的新哥特式铸造大厅，它散发着一种令人意想不到的脱俗的气息。

在普鲁士统治期间，赛恩钢铁厂专门负责生产轻型的大跨度建筑支撑结构。因此，于1830年竣工的铸造大厅首次以工厂车间批量生产的预制铸件作为铸造大厅的建筑核心结构，对其进行了新用途的创新探索。最终，这座铸造大厅未被视为一个单纯的功能性建筑，而成了一座由铸铁构建的拥有三个通道长廊结构的工业展厅。它仿佛一座钢铁大教堂，成为欧洲工业化大型工程建筑结构的典型样本。

在这座建筑中，宗教式的建筑风格与世俗的工业生产特征之间的对比塑造了铸造大厅的整体视觉印象。其中，交叉穹棱的桁架造型主导了西侧新哥特式玻璃立面的视觉特征，而在大厅另一端，则放置了铸铁熔炉的大型设施。

在建筑两端之间，挑高的中殿区域与两侧较低的侧廊空间由多立克式的立柱分隔，其中排布着工厂车间。在带有精美装饰结构的屋顶下，一条铸铁运输通道横贯整个大厅。这条通道由鱼腹式大梁支撑，经过设计师数次调整后，还配备了悬臂式起重机。与立柱相连的是带回转支承的可转动式起重机，起重机的作用是将铸铁元件传送到侧廊区域。

自从赛恩钢铁厂于1926年关闭后，这片区域就沦为了废墟，当地政府部门甚至曾计划在1973年将其彻底拆除。好在本多夫的村民们同心协力将铸造大厅保留了下来。在随后的几年，赛恩铸造大厅因其建筑设计声名远扬，走出本多夫小镇，并受到了建筑界的广泛认可，历时多年的建筑翻新计划也因此得以顺利展开。2010年，德国联邦工程师学会授予赛恩钢铁厂“德国艺术工程的历史性地标建筑”的荣誉称号。

为地标建筑和文化中心量身打造的多功能照明方案

这座铸造大厅经过精心翻修后，于2015年6月向游客开放。这里曾经是普鲁士时代辛克尔愿景下的铸铁工厂，熔化的铁矿在车间传送带中流动。如今，此处变成了一个多功能活动中心，优雅地融合在这个文化性地标建筑中。

根据铸造大厅空间的新用途，照明方案需要在兼顾整体建筑环境的同时，表现出建筑元素的特色和结构。此外，照明设计还需满足该空间作为不同文化活动场所的视觉需要。为此，LKL研发的照明方案中全部采用了LED灯具，使之能够对建筑元素逐一进行强调，并允许操作人员独立调整和控制所有照明设备。

通过触控面板，操作人员可以轻松访问照明“DALI”控制系统。使用者还可以根据不同的空间用途，调取设计师精心预置的多种照明场景，从而为夜间的建筑空间营造细致的光环境。

由外向内望向建筑深处

夜间，从室外看向赛恩铸造大厅时，建筑的西立面首先吸引了人们的注意力，并将大家的视线引导至建筑深处。在彩光渲染的屋顶之下，建筑中轴线顺着当年的熔铁传送通道一直延伸至后墙。这个曾经放置铸铁熔炉的空间如今成了凸显建筑金属结构的背景板。室内的翻新设计强调了带有教堂仪式感的建筑空间的通透性。在夜间，铸造大厅从室内向室外散发出温暖的光芒，建筑整体的夜间光环境与周围曾经开采矿石的山峰远景形成了鲜明对比。

在铸造大厅内部，空间的纵向延伸和建筑的几何形态给人留下了深刻的印象。悬臂式起重机的轨道在视觉上进一步增强了较低空间中坚实的垂直立柱与屋顶空腔中水平延展的花丝纹理之间不同形式的对比。如今，照明设计师在这些轨道上安装了灯具，灯光将空间进行了横向分层，更加增强了空间的视觉对比度。

强调空间对比和反差的和谐照明

环境照明中宽光束角的聚光灯为中殿和侧廊区域披上了一层温暖的光线，并营造出统一且连续的空间光环境背景。在整体的光环境中，间接反射的光线轻轻拂过垂直墙面和屋顶天花表面，温柔地表达其特征。此外，安装在立柱顶端、左右各一盏的窄光束角聚光灯则通过较低色温的光线精准地对立柱进行重点照明。

设计师采用了与悬臂式起重机相同的窄光束角聚光灯对鱼腹式大梁进行强调。这样一来，空间中的光线色温会产生微妙的对比，形成抑扬顿挫的照明韵律感，使空间整体和其中的细节都变得清晰明了。

位于熔炉后方的高砖墙是这个空间中唯一采用直接照明手法的垂直立面。地埋式上照洗墙灯将高砖墙上不规则的表面纹理表现得淋漓尽致。由此一来，大厅中明亮的后墙为空间创造了一个东向的空间边界，并与前方深色的铁艺建筑结构形成了视觉对比。

抬头向上望去，可以看到中殿区域令人叹为观止的屋顶。设计师采用了隐蔽的线形RGB灯条，并对铸造空间中包罗万象的建筑元素进行勾勒。灯具配套了经严格计算的定制遮光板，可以很好地隐藏在凹槽之中，不被下方的参观者发现，同时也能避免相邻的空间表面出现彩色杂散光。设计师在屋顶部分用红光进行照明，红光的色调是特意根据空间中砖墙和木质横梁的色彩倾向专门调配的，营造出过去熔铁生产车间的光色印象。由此不难推断，根据在该空间所举办活动的不同需求，空间的光色环境可以通过照明控制系统进行调整，以营造合适的色彩氛围。

空间中所有暴露在外的灯具配件都选自同一灯具系列，保证了所有与灯光相关的视觉元素都拥有统一的设计语汇。为了不影响作为文化遗产受保护的建筑结构，设计师精心安排了每盏灯具的安装位置和固定方式。通过与当地文化产业保护部门的协调，设计师将所有灯具及其配件的完成面统一喷涂了与建筑结构颜色相同的涂层。此外，所有直接照明的聚光灯都配有蜂窝状遮光格栅和遮光罩，以确保精准且低眩光的光线分布。

通过照明设计展现空间的形式和功能

德国现存最古老的铸铁大厅在设计师细致且精准的照明方案的烘托下，呈现出空间的独特性和对比性。可调节的预置光环境场景不仅可以适应不同的空间用途和活动类型，而且还能够始终为其打造细致且醒目的光线氛围。夜间，整个建筑从内部发光，散发着温暖而坚实的光芒，引发人们对曾经铸铁黄金时代的畅想。

业主：
本多夫市政府
使用人：
赛纳钢铁厂基金会
LKL项目经理：
约翰尼斯 · 罗洛夫
斯蒂芬妮 · 乔赫姆
竣工时间：
2015年
项目规模：
1150平方米
照明预算：
10万欧元

横向剖面图

Cross section

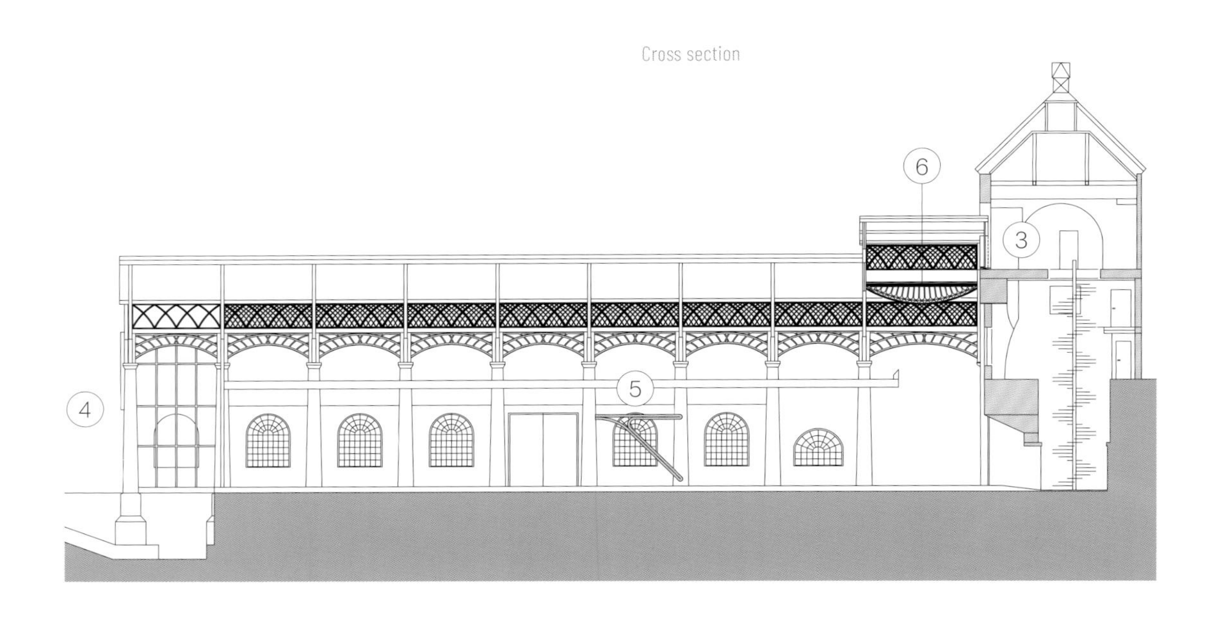

纵向剖面图

Longitudinal section

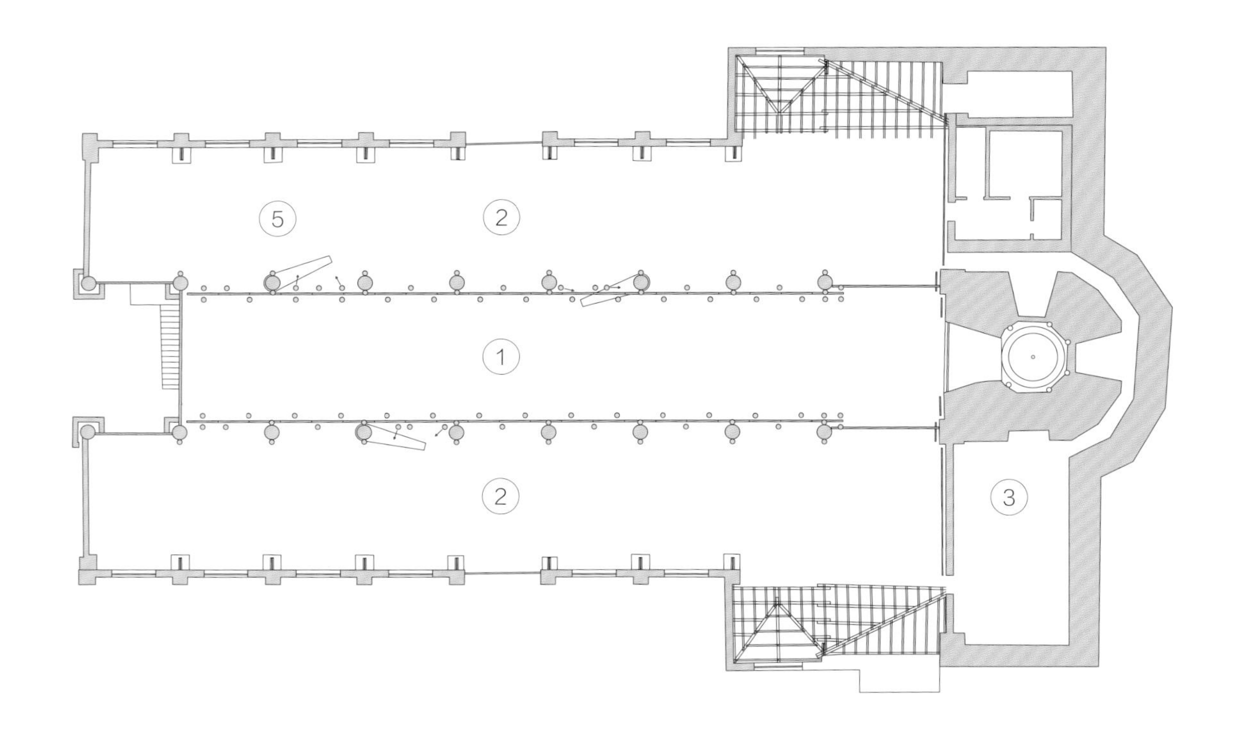

平面图

Floor plan

1.中殿
2.侧廊
3.熔炉房
4.西立面
5.悬臂式起重机
6.鱼腹式大梁

1. Central nave
2. Side aisle
3. Furnace building
4. Western facade
5. Jib crane
6. Fish-bellied girder

Located in a quiet secondary valley of the river Rhine near Coblenz is the Bendorf locality Sayn. Wooded slopes flank the longitudinal village with its fortress, palace and park, and nothing alludes to the fact that this used to be one of the most prolific sites of industrial cast iron production in the 19th century and that its economy and fate were once determined by iron. Hidden away at the northern end of the village, near the castle hill, the Sayn iron works facility can be found. Here, on the brink of an area with simple residential homes, graveled expanses and frugal halls, is a delicate, neo-gothic hall of unexpected airiness and elegance: the foundry.

Under Prussian reign, the iron works of Sayn were specialized in the production of light-weight far-spanning structures. Thus, the foundry building, completed in 1830, featured the novel use of pre-fabricated cast elements from serial production. The result was not a sober functional building, but an industrial hall taking the form a three-nave cast-iron basilica, an iron cathedral, a prototype for the great engineered buildings of European industrialization.

This contrast between the sacral and the secular characterizes the foundry's visual impact. While the groined girders rule the glazed neo-gothic western façade, a massive construction, housing the furnace, is located at the opposite end of the hall.

In between spans a factory workshop with its elevated central nave and the lower side aisles, separated by cylindrical Doric columns. Underneath the roof with its filigree structure, a transport lane spans the entire length of the hall, held by fish-bellied girders and after several alterations – fitted with travelling cranes. Attached to the columns are rotatable cranes on ball joints that once lifted the cast-iron elements into the side aisles.

After the iron works of Sayn were closed in 1926, the area fell into ruin and was even planned to be demolished in 1973. Only the interception of the Bendorf citizens saved the iron works. During the following years the Sayn foundry achieved architectural recognition outside of Bendorf and the long refurbishment of the facility was successfully initiated. In 2010, the Federal Chamber of Engineers awarded the title "Historisches Wahrzeichen der Ingenieurbaukunst in Deutschland" (historical landmark of the art engineering in Germany) to the cultural monument.

MULTIFUNCTIONAL ILLUMINATION FOR A LANDMARK AND CULTURAL CENTER

Since June 2015 the foundry with its elaborately refurbished architecture has been open to visitors. Where once molten ore flowed in channels and Schinkel's iron visions for Prussia were cast, an event center for flexible uses has been elegantly integrated into the cultural landmark.

This new use called for an illumination that features both the entire ensemble and individual elements or structural details of the architecture. Furthermore, the illumination was to allow for frequent conversions and various events. For this purpose, Licht Kunst Licht devised an illumination with LED luminaires only, emphasizing architectural elements one by one and permitting an independent adjustment and control of all lighting devices.

A DALI control system was integrated that can be comfortably accessed via touch panel. The occupant can trigger sensitively pre-programmed lighting scenes for various uses, thus reproducing nuanced light scenographies for the nocturnal edifice.

A VIEW INTO THE ARCHITECTURE

When looking at the Sayn foundry from outside at night, it is the western façade that attracts the view and guides it into the building's depths. The center line aligns with the roof in its coloured illumination, following the path of the iron melt from the backwall that once housed the furnace and now forms the backdrop for the metal structure. The airiness and transparency of the basilica-like layout are emphasized, the foundry hall emanates warmth from within and the nocturnal elevation contrasts with the mountain scenery from where the ore was once extracted.

Inside the foundry hall, it is the longitudinal expanse and spatial geometry that manage to impress. The formal contrast between the powerful vertical pillars of the lower building volume and the filigree, horizontally sprawling roof cavity is further enhanced by the crane tracks for the travelling cranes. These now house the luminaires and by virtue of the illumination offer a horizontal divide between the spatial layers, thus underlining the inherent contrasts.

OPPOSITES AND CONTRASTS, HARMONIOUSLY ILLUMINATED

The direct wide-beam spotlights of the general illumination add a warm light layer to the floor areas of the central nave and side aisles and create a consistent luminous background, that gently features the wall and roof surfaces with reflected light. Additionally, columns are concisely accentuated by narrow-beam spotlights in a cooler light colour, placed on either side at the top of the pillars.

The swivelling cranes and fish-bellied girders are emphasised by the same narrow-beam spotlights. Thus, a subtle contrast of colour temperatures emerges and creates a differentiated light orchestration that makes the space palpable, both as a whole and in detail.

The tall brick wall of the former furnace is the only directly illuminated wall surface and its irregular texture is deliberately featured by floor recessed wall washers. As a result, the hall's bright rear wall creates an eastward spatial boundary and a contrasting backdrop for the dark iron structure.

Sprawling above is the impressive roof volume of the central nave, that is featured as an all-encompassing element by means of concealed linear RGB-strips. The luminaires are shielded from views from below by means of bespoke and precisely calibrated shields, in order to avoid coloured halos on adjacent room surfaces. The red hue, used for the roof illumination is adapted to the colour of the bricks and wooden beams and creates a formal bracket, evocative of the path, the molten iron once took. Evidently, the control system can adjust the colour to different hues, if the respective event requires it.

The visible light fittings originate from the same luminaire family, thus creating a homogenous design language in all elements related to lighting. Each luminaire was placed and installed with great diligence, in order not to interfere with the listed building structure. In coordination with the heritage department, all light fixtures and accessories have been coated in the same colour as the structure. Additionally, all direct spotlights have been fitted with honeycomb louvers and snoots, to ensure a precise and low-glare light distribution.

FORM AND FUNCTION GAIN PRESENCE THROUGH LIGHT

The oldest preserved fabrication hall in iron is presented in its uniqueness and dichotomy by a sensitive and sophisticated lighting concept. The adjustable, pre-programmed atmospheres cater to different uses and event types while always offering a nuanced and striking illumination. The building appears to glow from within and is thus reminiscent of its former use by emanating a warm, yet rugged ambience.

CLIENT:
Municipality of Bendorf
OCCUPANT:
Foundation Sayn Iron Works
TEAM LEADER LICHT KUNST LICHT:
Johannes Roloff
Stephanie Jochem
COMPLETION:
2015
PROJECT SIZE:
1,150 m^2
LIGHTING BUDGET:
0.1 million euros

较长轴向的视图展现了高炉建筑与悬挑铸造大厅的整体样貌。宽敞的玻璃幕窗彰显了建筑结构，同时散发出温暖好客的光芒。

The view of the long side shows the ensemble of the blast furnace building with the self-supporting foundry. The generous glazing displays the structure of the building while emanating a welcoming warmth.

夜色下，铸造大厅在后方深色山脉远景的映衬下，通过带有精致花丝竖框的西立面，呈现出开阔、明亮的建筑特征。

The west facade's delicate mullions present the cast house amid the iron works as open and luminous architecture in front of the backdrop of the dark mountain.

完成布展方案后的室内景象阐释了照明方案不仅表达了充满戏剧感的建筑结构，同时最为重要的是，灯光设计也充分满足了活动策展方和参观者的视觉需求。

The view of the completed furnished exhibition illustrates that the lighting concept not only presents the dramatic architecture, but also and above all provides adequate lighting for the user as well as the visitor.

悬臂式起重机的回转支承安装在空心铸铁立柱上，每根立柱顶端都设有窄光束角聚光灯，对其进行重点突出。较低色温的聚光灯确保了重点照明与环境照明之间形成微妙的对比。

Jib cranes on ball joints are mounted on the hollow cast-iron columns, both of which are featured by narrow-beam lights. The cooler accenting light colour ensures a subtle contrast to the general lighting.

AWARD
WINNER

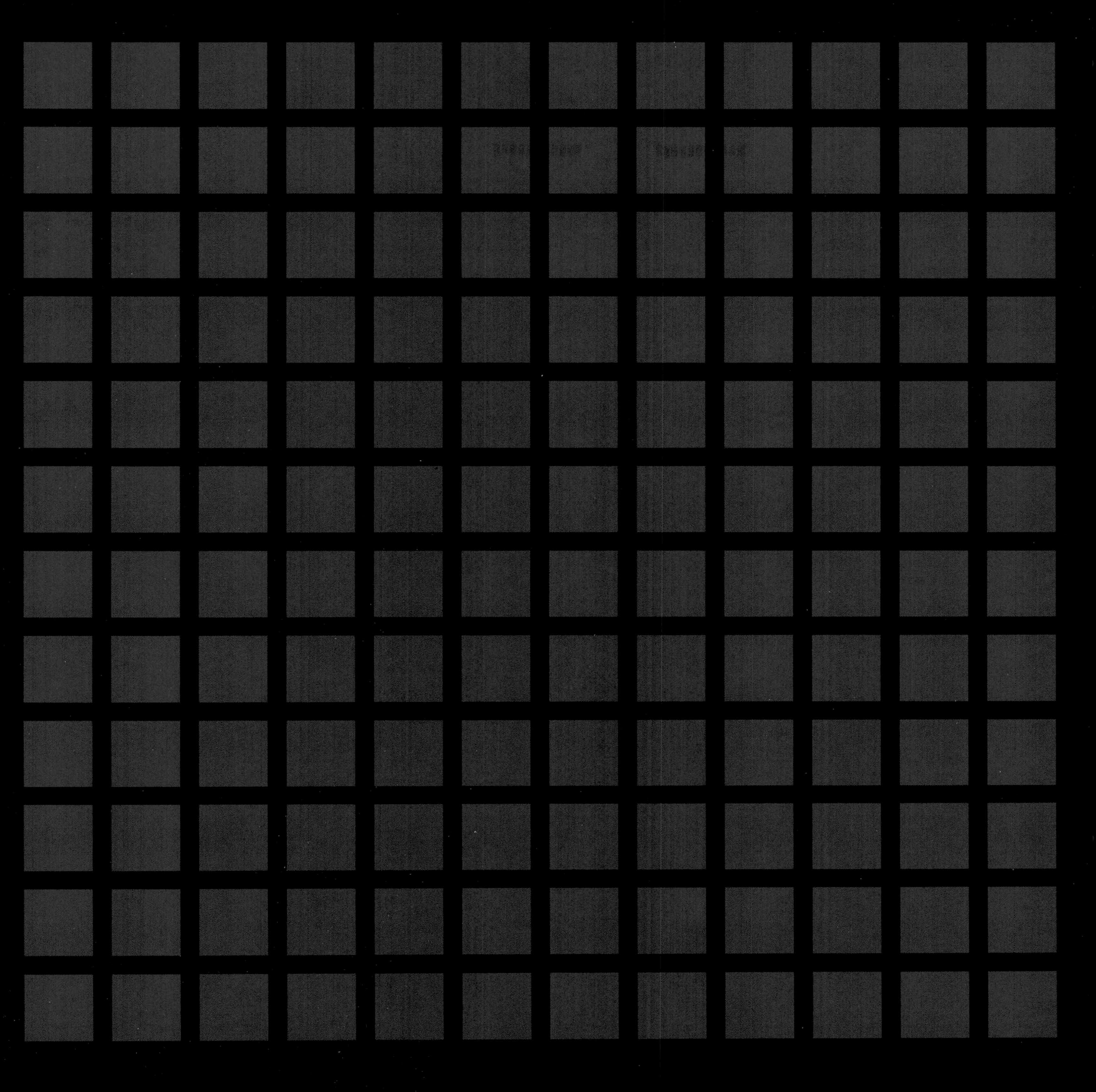

LICHT KUNST LICHT 4

阿伦斯霍普艺术博物馆 阿伦斯霍普

AHRENSHOOP MUSEUM OF ART AHRENSHOOP

融入当地的风土人情——为阿伦斯霍普艺术家聚居地建造的现代博物馆

从当代的观点来看，阿伦斯霍普一直以“艺术家的聚居地”而闻名，而不是传统概念中“波罗的海的海滨度假胜地”。这个区域标签已经有120多年的历史了。也许是因为早在19世纪末期，在第一批做水疗的游客还未前往阿伦斯霍普带动当地旅游业之时，画家们就已经在狭长的菲施兰- 达斯- 辛思特半岛上发现了这个小镇，并以这个小镇为灵感来源进行艺术创作。阿伦斯霍普坐落在波罗的海和萨勒博登湖之间，紧邻一片海滩和壮丽的峭壁悬崖。狂野而浪漫的自然景观配以当地明艳且变化丰富的自然光线，令人印象深刻，并常常给人带来创作启发。因此，艺术家们从一开始就深受当地建筑风格的影响，这种与当地风土人情之间的联系对他们的艺术作品有着潜移默化的影响。此后，几乎所有与德国现代艺术浪潮相关的知名艺术家们都相继参观了阿伦斯霍普及其周边地区，并为其协会的艺术收藏事业做出了贡献。如今，协会已经收藏了包含绘画、摄影和雕塑艺术等约作品500件。

120多年后，终于迎来了属于自己的博物馆

这座小镇虽然有众多画廊和艺术工作室，但令人惊讶的是，直到2013年夏天才拥有属于自己的艺术博物馆。等待往往是值得的。来自柏林的施塔布建筑设计事务所与协会里一群乐于奉献的艺术爱好者展开了卓有成效的合作，由此打造了一座现代博物馆。博物馆的结构和轮廓让人不禁联想起当地的传统房屋，但室内布局又不像传统房屋那样随意。与此同时，建筑设计师对整个建筑体块的划分也呼应了传统房屋的室内氛围，这种氛围与建筑的地理位置特征十分契合，这很大程度上得益于精巧的日光设计方案。

独立又似曾相识——遵循历史且令人信服的现代化设计

这座博物馆建筑成功地将当地建筑的过去和现在联系了起来。设计师大胆地通过对传统建筑材料和形式的重新诠释，将其转变为一个全新但符合当地传统的建筑群。多个建筑单体组成的联合体以过去带有倾斜屋顶的渔民木屋为原型。设计师特意将它们以不同的间距聚集在一起，形成一个完整且具有雕塑感的建筑群，并将中心区域扁平的多边体门厅包裹起来。这个多功能建筑群可以提供近900平方米的展览空间，另设有400平方米作为办公室、储藏室和设备间。在这5栋建筑中，其中4栋用于分别展示来自不同历史时期的艺术家聚居地的作品，并举办一些永久性或临时性展览。剩下的那栋建筑虽然从室外看起来并不突出，但也有两层楼高。首层被称作“Kabinett”，是一个无自然光的研讨室，二层则设置为行政办公室。

建筑的“皮肤”

除了标志性的整体造型外，建筑的黄铜立面同样吸引了人们的注意力。除了首层的少部分立面采用相对低调的透明玻璃外，黄铜立面仿佛是建筑外层精细的波纹状“皮肤”，将整个博物馆包裹起来。铜本身的光泽感极易消逝，因此建筑设计师预想出该金属表面氧化之后的样貌，氧化后的颜色与传统茅草屋顶的颜色相似。外墙立面上不规则的、生动的定制波纹包边进一步强化了这种视觉效果。建筑表皮的色彩则使人联想起干枯的芦苇。随着时间的推移，建筑仿佛芦苇般优雅地老去，并与周边环境和景观融为一体。

对日光的利用在阿伦斯霍普地区有着悠久的传统

每栋小屋和大厅的扁平屋顶均由钢筋混凝土建造而成。除了1栋设有研讨室和办公室的建筑外，其余小屋均为大开间。室内采光充足，均设有1个大型的屋顶天窗。在项目的早期规划阶段，照明设计师就与建筑设计师紧密合作，共同研发了结合展厅天窗的日光设计策略。所有展陈空间均沿其隔断的屋脊线在屋顶区域配备了细长的水平天窗。为了在室内实现几乎没有阴影的光照分布，照明设计师在天窗玻璃上集成了光学棱镜系统。这个棱镜系统既可以阻挡强烈的直射阳光，又可以让柔和的漫射光线自由通过。在需要时，玻璃下方的可操作电动遮光板可以用来降低入射日光的强度。

在日照条件不足的情况下，人工光辅助照明可以通过天窗开口模拟自然光线，为空间提供基础照明。为此，设计师将灯具安装支架集成在天窗开口的裙板内，将光线巧妙地遍布整个室内空间。对于环境照明而言，空间的照度水平可以根据不同的要求，通过独立的开关进行灵活调控。展品的重点照明由天窗下方的轨道式LED聚光灯提供。

重点照明、环境照明和聚焦照明

在研讨室中，采用LED背光照明的方形发光面板营造了均匀的展陈环境照明。灯具调光系统可以根据不同的展览需求，在50～300勒克斯的范围内调节照度。照明设计师还在方形发光面板中嵌入了可调节LED聚光灯，对展品进行重点照明。这两个照明组件都可以单独开关和调光。

前厅主要由下照灯和洗墙灯提供照明。通过对竖直墙面的均匀照明，照明设计师将空间的形态变得生机勃勃，营造出舒适宜人的室内氛围。此外，墙面的均匀照明还为艺术展览创造了额外的展陈立面。通过对灯具进行调光，立面上的展品可以获得合适的照度水平。窗户前方的细长空间位于两栋建筑之间，由三盏聚光灯照亮。在这里，带有方向性的聚焦灯可以将多个雕塑作品各自的特征分别进行突出与强调。

在接待台上方，与天窗结构相结合的可调光灯条营造出了宽敞明亮的光环境氛围。礼品店的展架和寄存处的衣橱均由家具内部集成的隐蔽式灯带进行照明。这样的设计既确保了光线能够突出商品，又保证衣物的合理照明。

让光线成为“主角”

在入口区域，3盏相邻的LED下照灯组成一组，达到了较高的空间照度水平。它们确保了建筑的室外拥有良好的视觉效果，同时为前厅提供照明。3盏下照灯与室内透出的灯光相呼应，引导参观者进入这个由内而外散发着光芒的展厅。休息室的照明也采用了同样的设计语汇，5盏下照灯营造出温暖的光环境，将入口处热情好客的空间氛围逐渐过渡到安静放松的状态。集成在展陈区域家具上的隐蔽式灯带温柔地照亮了展架，强化了空间的光线场景感。

建筑设计师和照明设计师成功将这座新建博物馆融入周边环境。建筑与环境的联系不仅体现在外观设计上，设计师也将其打造成连接当地历史和现代文明的重要纽带。日光作为不同时代艺术家共同的创作灵感来源，在这个博物馆中也成功通过天窗系统的设计与应用得到了很好的体现。照明方案通过宽敞且特别的天窗形态，为展厅空间增添了微妙且宁静的气息。

业主：
阿伦斯霍普艺术博物馆
使用人：
阿伦斯霍普艺术博物馆
建筑设计：
施塔布建筑设计事务所，柏林
LKL项目经理：
迈克 · 沙尼亚克
竣工时间：
2013年
项目规模：
900平方米
整体项目预算：
640万欧元
照明预算：
10万欧元

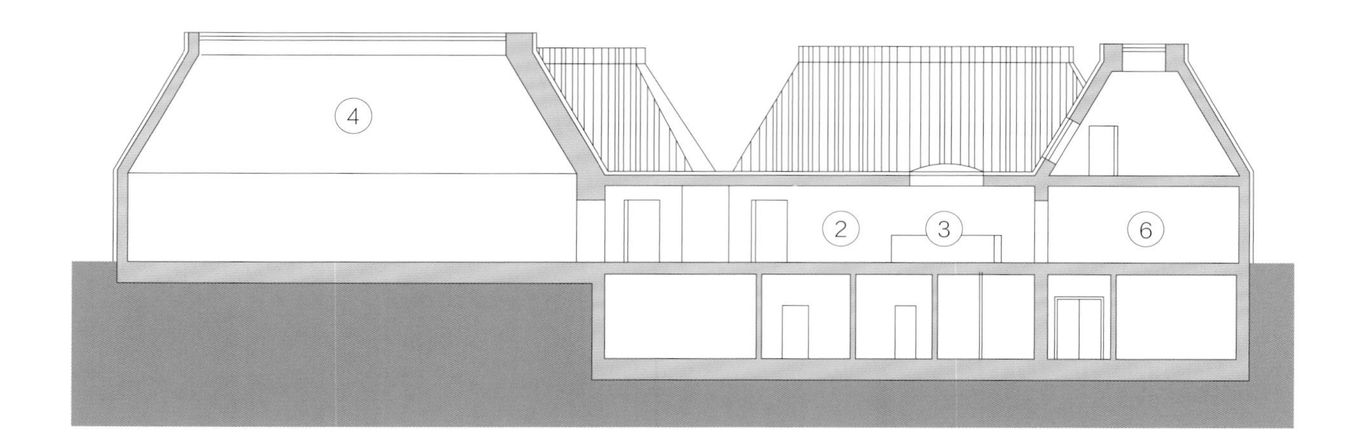

剖面图

Section

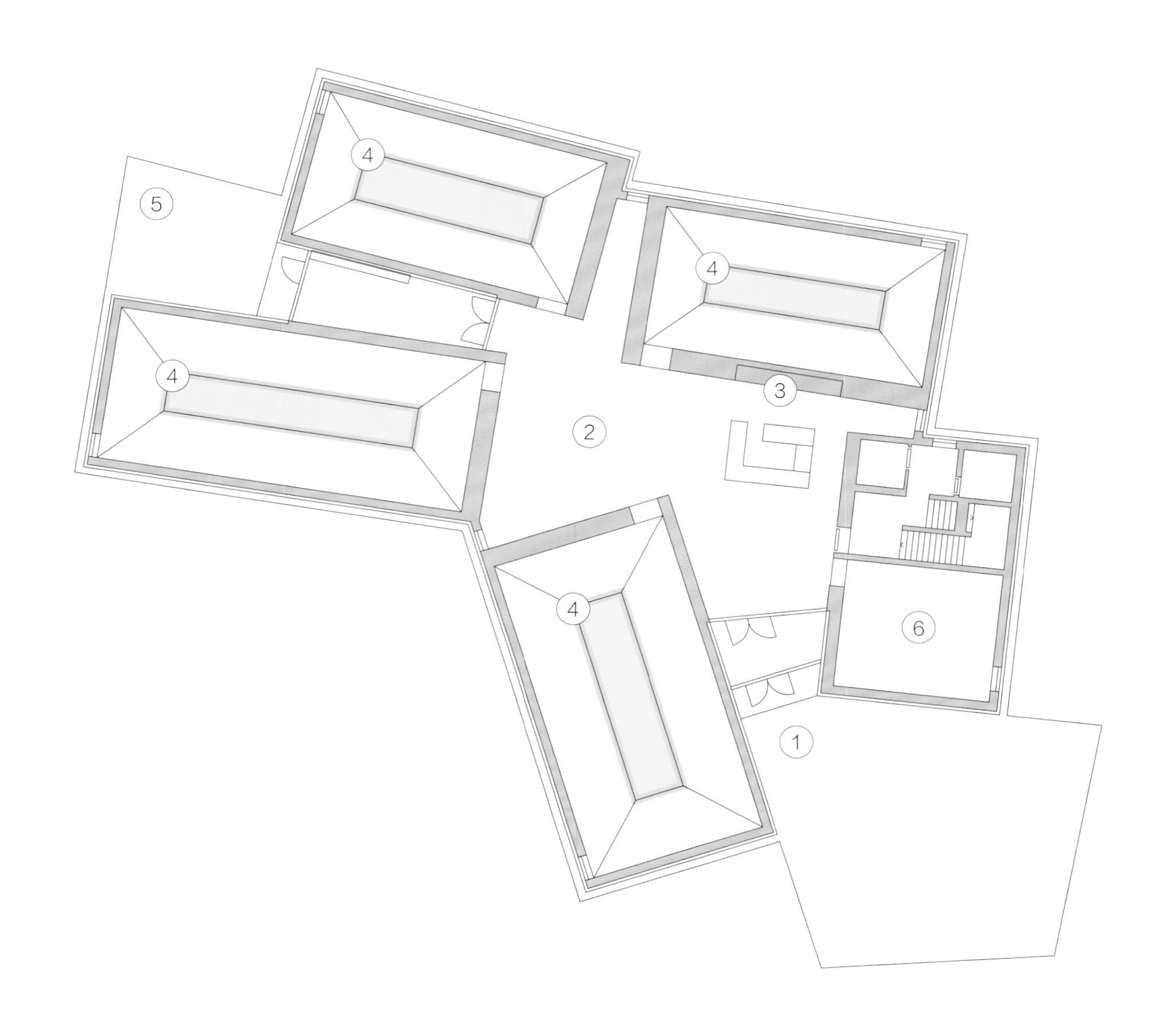

平面图

Floor plan

1.主入口
2.大厅
3.票务、礼品店
4.2号展馆~5号展馆（展陈空间）
5.露台
6.1号展馆的研讨室

1. Main entrance
2. Foyer
3. Tickets / Shop
4. Buildings 2-5: exhibition spaces
5. Terrace
6. Cabinet at building 1

In its contemporaries' view, Ahrenshoop has always prevailed as an 'artists' colony' rather than a 'Baltic seaside resort'. It has been that way for over 120 years, perhaps because, even before the arrival of the first spa visitors, painters discovered the town for themselves on the elongated peninsula Fischland-Darss-Zingst towards the end of the 19 th century. Located between the Baltic Sea and the Saaler Bodden, in close proximity to a beach and a cliff line, the picturesque landscape is inspiring, especially with its intense and varying light. From the beginning, there was a deep connection between the artists and the local building traditions – a connection that was subsequently reflected in their art. Countless renowned artists of virtually all relevant German modern art movements have since visited Ahrenshoop and its surroundings, and have contributed to the growth of the friends' association's collectibles, now numbering 500 paintings, graphics and sculptures.

FINALLY A MUSEUM OF THEIR OWN-AFTER MORE THAN 120 YEARS

Thus, it seems rather surprising that this town, otherwise rich in galleries and ateliers, has only had its own art museum since the summer of 2013. However, the wait has been worthwhile. In a fruitful collaboration with the dedicated art lovers of the friends' association, the Staab Architekten office in Berlin has created a contemporary exhibition building with a form and silhouette reminiscent of, but not mimicking the traditional houses and their apparently random layout. At the same time, the design of the grouped volumes alludes to the traditional interior atmosphere – being keenly appropriate to the location – it is largely derived from a clever daylight concept.

INDEPENDENT, YET FAMILIAR. HISTORICALLY-CONSCIOUS AND CONVINCINGLY MODERN

The museum building succeeds at interconnecting formal ties with past and present local architectural history and transforming them into something new, yet authentic by courageously reinterpreting materials and shapes. Modeled after old fishermen's cottages with their hipped roofs, an ensemble of individual buildings was developed. They are purposefully set closely together at irregular distances and form a complete, sculptural building volume, engulfing a flat polygonal foyer in their centre. The complex offers approximately 800 m^2 of exhibition space and another 400 m^2 for offices, storage, and utility rooms. Four of the altogether five buildings are assigned to individual eras of the artists' colony and house permanent and temporary exhibitions. The remaining building is, though not obvious from the exterior, two-story tall. Its ground floor accommodates the so-called Kabinet, a daylight-free seminar room, while its upper floor contains administrative offices.

A SKIN FOR THE BUILDING

In addition to its iconic overall form, the building's brass façade also attracts attention to itself. Like a finely corrugated skin, it covers the museum's entire volume, with the sole exception being the less emphasized glass surfaces on the ground floor. The initial splendor of bronze is obviously short-lived; the designers envisioned the final, weathered colouring, which resembles the traditional thatched roofing. This effect is further reinforced by the irregular, lively, and custom designed folding of the façade cladding. With a colour reminiscent of dried reed, the building envelope will mature gracefully and blend into its built environment and landscape.

APPRECIATION OF DAYLIGHT HAS A LONG-STANDING TRADITION IN AHRENSHOOP

The individual cottages and the foyer's flat roof have been built in massive reinforced concrete. With the exception of the building accommodating seminar and office spaces, the one-room houses have been designed with daylit interiors, each with one large roof opening. This skylight integrated daylight strategy has been developed in close collaboration with the architects from the early stages of the project. All exhibition spaces are equipped with elongated horizontal daylight openings, following their severed ridge line. In order to generate a distribution of light that is perceived as virtually shadow-free, prisms have been integrated into the skylight glazing. They block the direct sunlight, while letting the diffuse light pass unhindered. If required, the daylight intensity can be reduced through an electrically operable sun screen layer below the glass

The supporting electric illumination for overcast days and the darker part of the year, was to emulate the daylight situation by entering the space through the same skylight opening. For this purpose, the lighting profiles are integrated in the perimeter skirting of the skylight, distributing their light openly in the room. Via individual switching channels, the illuminance levels can be flexibly controlled according to requirements. For accent lighting, a 3-phase-track for LED spotlights has been implemented below the skylight profile.

ACCENTS, AMBIENT ILLUMINATION AND FOCAL POINTS

At the Cabinet, a backlit LED light square creates uniform illumination. Dimming the luminaires allows for the light levels to be adjusted as required, ranging from 50 to 300 lux. Located inside the square are recessed adjustable LED spotlights that accentuate the exhibits. Both components can be individually switched and dimmed.

The foyer is largely illuminated by downlights and wall-washers. By emphasizing the walls, the spatial boundaries are brought to life and create a pleasant interior atmosphere. Beyond that, the uniform wall illumination offers an additional display surface for art exhibits, which can be shown at appropriate light levels by virtue of the luminaires' ability to be dimmed. The slender space in front of the window between two individual buildings is illuminated by three spotlights. Here, sculptures and objects can be accordingly featured with directional, focused light.

Integrated in the skylight above the counter, dimmable light profiles generate ample brightness in the area. The shelves in the shop and the coatroom are illuminated by light strips integrated and concealed in the millwork. This ensures an appropriate illumination of the merchandize.

LIGHT TAKES THE 'LEADING ROLE'

At the entrance, three closely grouped LED downlights create high illuminance levels. They support the exterior appearance of the building and also supply the forecourt illumination. In conjunction with the interior lighting, these three lights guide the visitor to the museum entrance, which radiates from within. Similarly in the lounge, the warm atmospheric light from five downlights extends an invitation to linger and relax. The scenario is supported by concealed light profiles, gently illuminating the shelves.

The architects and lighting designers have succeeded at making the new museum building an integral part of its immediate surroundings, not only in its exterior appearance, but also turning it into a vital link between history and the contemporary. The successful integration of daylight, a tool greatly appreciated by artists of all eras, shapes the generous cross-sections of the exhibition rooms into interiors of subtle public intimacy.

CLIENT:
Kunstmuseum Ahrenshoop e. V.
OCCUPANT:
Kunstmuseum Ahrenshoop e. V.
ARCHITECTS:
Staab Architekten GmbH, Berlin
TEAM LEADER LICHT KUNST LICHT:
Maik Czarniak
COMPLETION:
2013
PROJECT SIZE:
900 m^2
OVERALL BUILDING BUDGET:
6.4 million euros
LIGHTING BUDGET:
0.1 million euros

借助波纹状的黄铜立面，建筑群的造型容易让人联想到过去渔民的茅草小屋。

By virtue of its folded bronze façades, the ensemble of individual houses is reminiscent of thatched fishermen's cottages.

在夜间，博物馆从内向外散发着光芒，与其在日间的视觉印象形成鲜明的对比。参观者迎着光明的方向，在灯光的“欢迎”和引导下进入展厅。

At night, the museum radiates from within and poetically inverts the day-time appearance. The visitor is invited and guided by light.

礼品店的展架被隐蔽式灯条照亮。天花上低眩光的嵌入式下照灯以十分有趣的形式分布着，为空间营造了明亮的光线氛围。

The shelves at the shop are illuminated by concealed light profiles. A playful layout of low-glare regressed downlights creates a brilliant light atmosphere.

礼品店收银台由上方的天光系统照亮。在夜间，集成在天花凹槽内的可调光式灯带将天花模块变为迷人的发光面。

The cashier's desk at the shop is appropriately emphasized by a skylight. At night, integrated dimmable light strips turn the ceiling module into a compelling light surface.

每个展馆都配备了一个大型的细长天窗。天窗下方集成了棱镜系统，能完全阻挡会对展品造成损害的入射日光，仅让柔和的光线进入室内空间。

Each building is equipped with a large, elongated skylight. A prism inlet in the skylight glazing transmits only diffuse light into the space, while damaging sunlight is fully deterred.

人工光照明同日光一样，也通过这个天窗开口进入室内。灯具被集成在天窗系统周围的裙板之中，根据不同的情况，对日光进行补充或者替换。

The electric light enters the space through the same opening as the daylight. It is integrated into the perimeter skirting of the skylight and supplements or replaces the daylight as required.

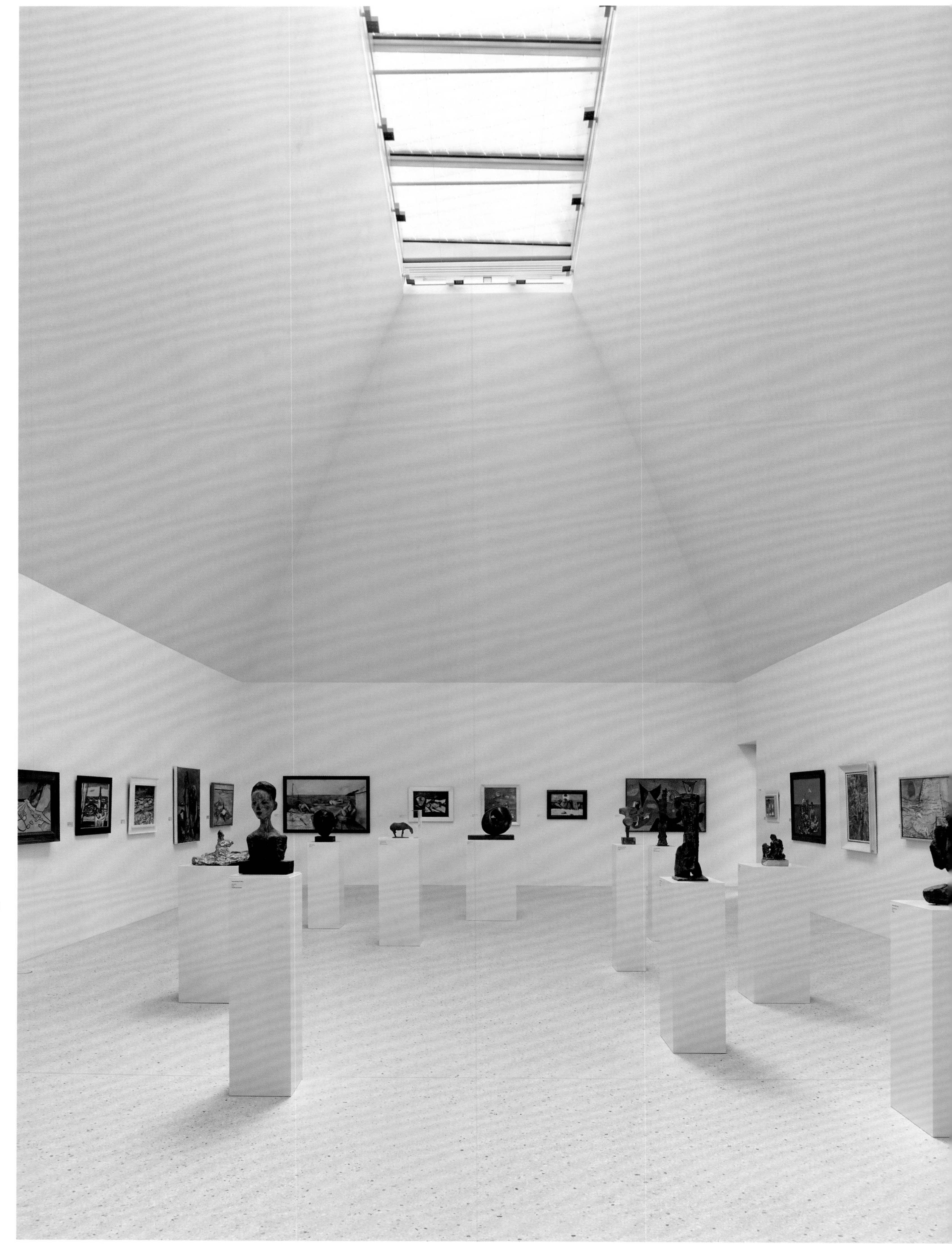

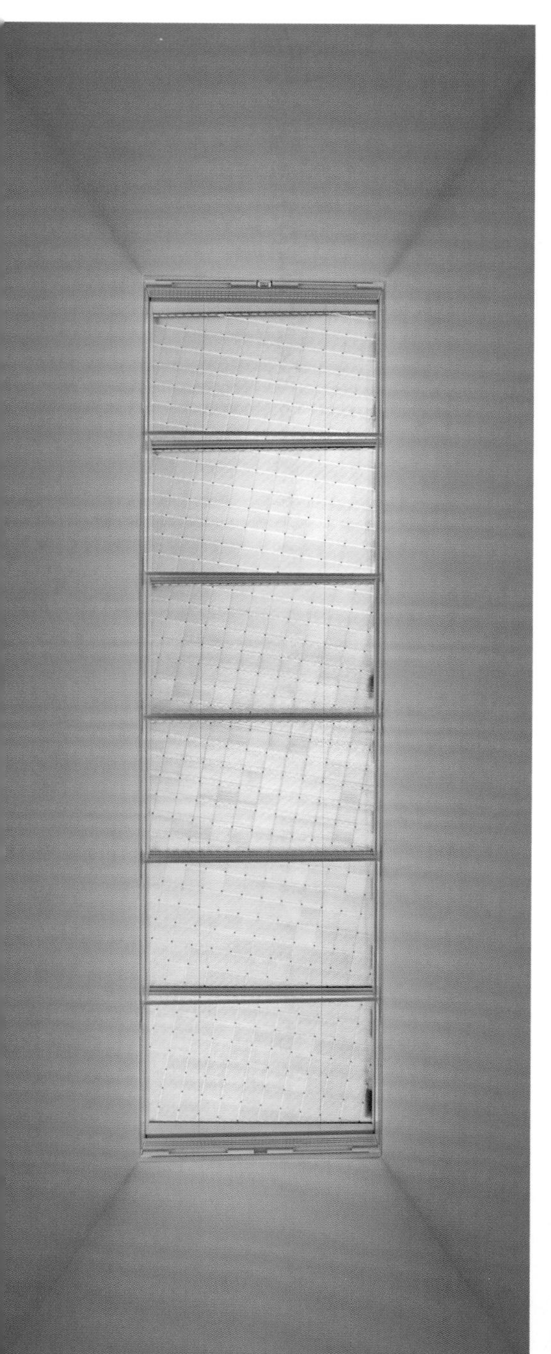

多云天气下的天窗系统，无人工光辅助
Skylight on a cloudy day, no electric light

晴朗天气下的天窗系统，有人工光辅助
Skylight on a sunny day, with electric light

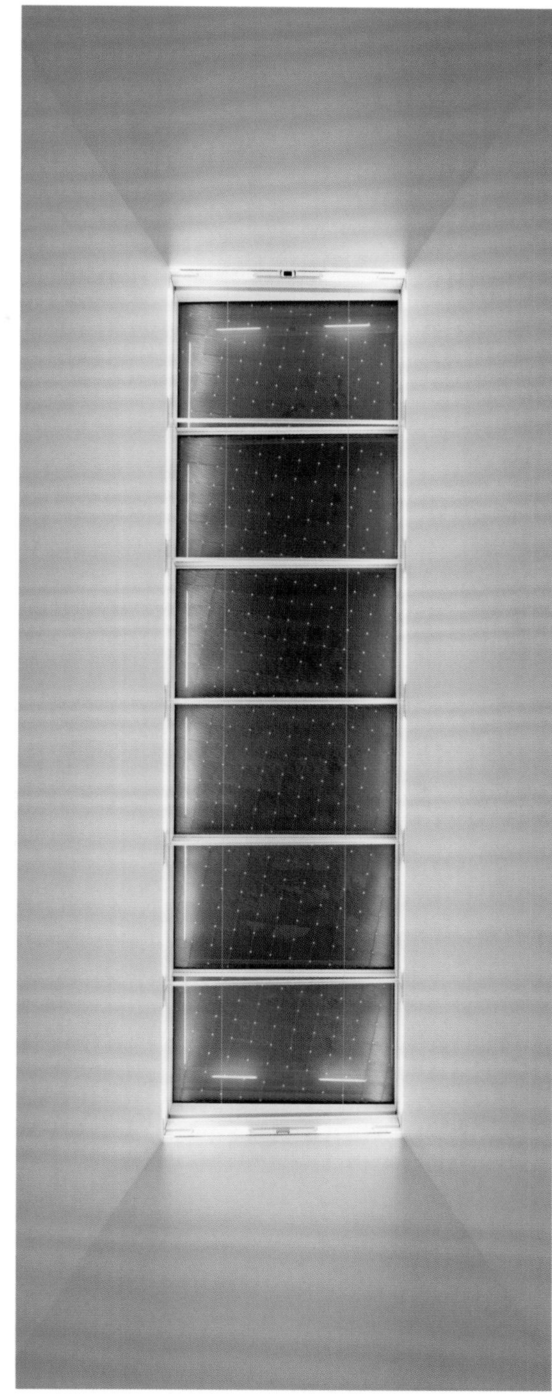

夜间的天窗系统，有人工光辅助
Skylight during dark hours, with electric light

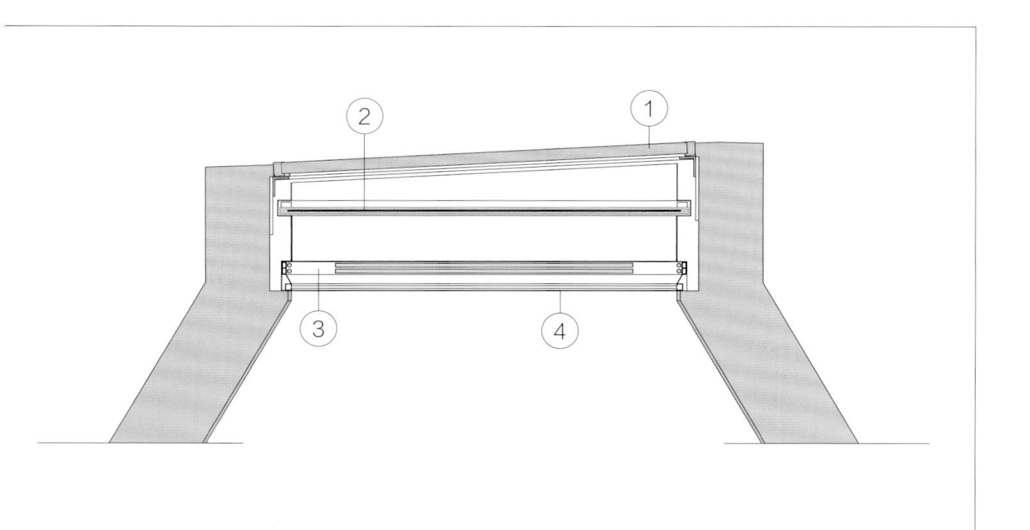

天窗系统细节剖面图
1.多层玻璃结构
2.可遥控遮光板
3.单独控制的T16荧光灯
4.嵌入式轨道

Detail section of skylight
1. Multi-layer glass structure
2. Operable sunscreens
3. Individually switchable T16 fluorescent lamps
4. Recessed track

Der besondere Gast

AWARD
WINNER

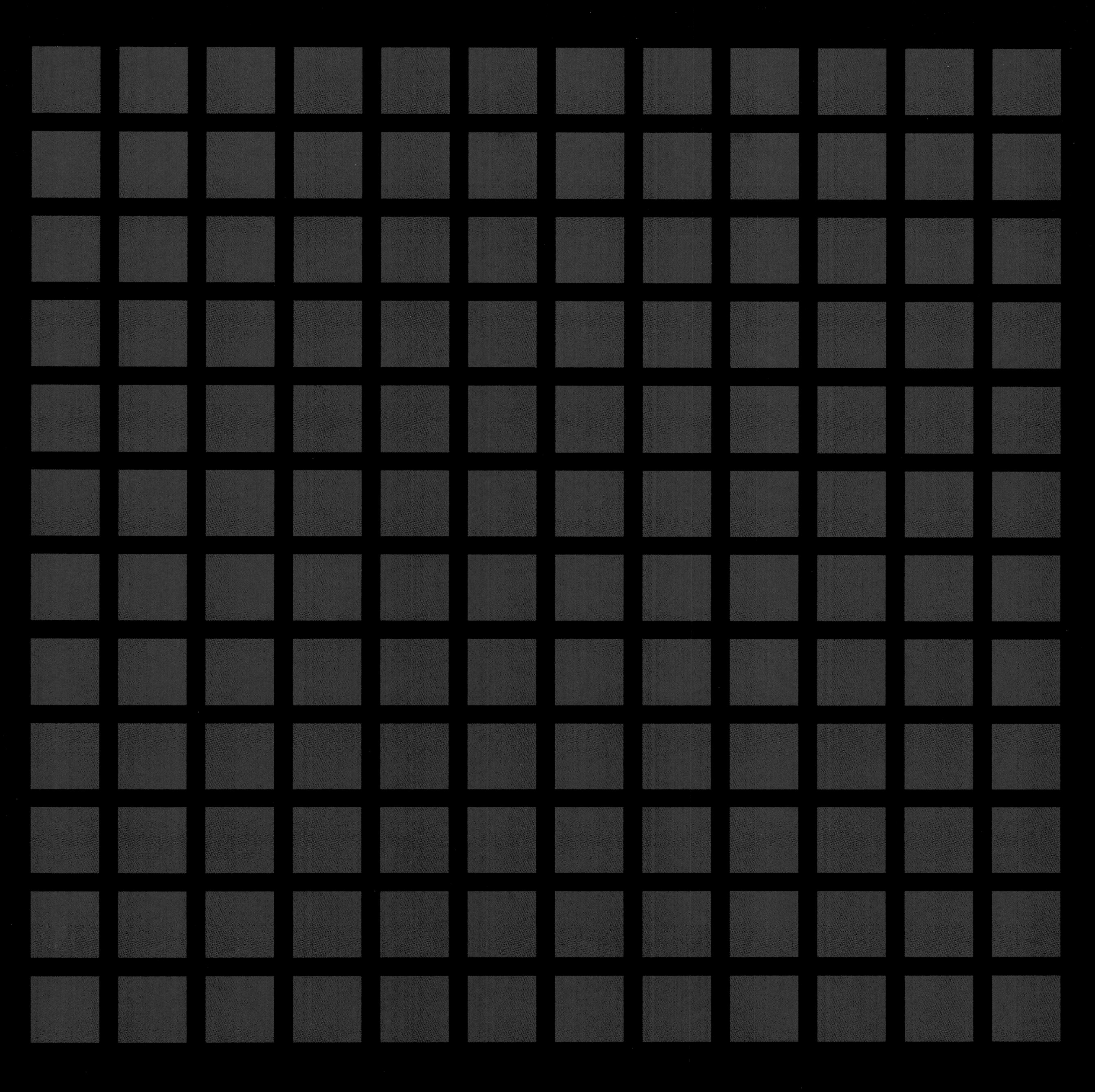

LICHT KUNST LICHT 4

理查德-瓦格纳广场　莱比锡

RICHARD-WAGNER-PLATZ LEIPZIG

城市大门处的都市活力——莱比锡理查德-瓦格纳广场翻修项目

这个重量级的购物中心开发项目形成了德国布吕尔区一个名为“锡罐”（Blechbüchse）的东部扩展建设区，引起了莱比锡市民的热烈讨论，以及对其建设过程的关注。

“锡罐”是当地的标志性建筑，牢牢扎根于莱比锡的城市形象建构之中。该建筑由艺术家、金属设计师哈里·穆勒于1966年设计。自1968年起，这里就设有当时德意志民主共和国最大的购物场所——Konsument am Brühl。

理查德-瓦格纳广场位于这座建筑的前方，作为城市的“门面”，它不仅具有城市大门般的通行功能，最为重要的是，该广场的设计需要与其后方建筑“锡罐”所受到的市民欢迎度和关注度相匹配。因此，在莱比锡市政当局的组织下，许多景观设计事务所参与了这个顶级广场翻修的设计竞赛。在这个过程中，当地政府提出的设计任务是将翻修后的广场打造为城市的重点区域，使之成为市民经常聚会的场所。最终，来自德国汉诺威的艾琳·洛豪斯·彼得·卡尔景观建筑设计事务所的设计作品脱颖而出，赢得了比赛。其广场的设计方案包括一个滑板场和一个迷人的传统喷泉。

一个由图案投影构成的灯光图层

对照明设计师而言，针对完全不同的建筑风格来研发一套连贯性的照明解决方案是充满挑战的。在这个项目中，建筑外立面的一部分是现存建筑综合体中极具艺术感的建筑侧翼表皮，另一部分是格伦图赫·恩斯特建筑设计事务所从莱比锡城市建筑发展历史中提取形象元素，并以此为基础设计的当代购物中心立面。在不影响广场正常使用的前提下，每个建筑立面都需要受到照明设计师的特别关注。因此，照明设计师使用了一种新型的照明系统，通过投影技术将特定的图案精准地投向引人注目且可以较好承接光线的建筑正立面，并延伸至立面边缘。该照明设计旨在实现一种温柔的、几乎令人察觉不到的灯光图层，让人们感觉不到建筑立面是从外部被照亮的，给人留下光线是从立面内部向外散溢出来的视觉印象。由此一来，建筑前方的广场就有了合适的城市背景环境。

用于广场多样化灯光方案的模块化照明系统

对于广场本身的照明，设计师选用了可以单独响应多种空间布局和用途的照明系统。模块化的高杆灯可以配置具有不同光束角的灯具，并满足广场对重点照明和环境照明的要求。因此，滑板场的照度相对较高，以满足使用者较高的视觉需求；其他区域则采用了较低的照度等级，更偏向于气氛照明。

新设计获得了当地市民和业界专家的正面反馈

新开发的城市广场和建筑街区受到了莱比锡市民的极大欢迎。从远处看，该建筑于20世纪60年代后期设计的充满艺术感的外立面在灯光的精心诠释下，让人眼前一亮。走近广场时，人们的注意力会自然而然地转移到广场本身。广场上的景观小品和设施也让人无比好奇，并让人产生探索的欲望。此外，该项目也在国际上引起了广泛关注，被由业界专家组成的评审团授予2014年度“城市·居民·灯光”奖。

业主：
莱比锡城市交通和土木工程办公室
使用人：
公众
景观建筑设计：
艾琳·洛豪斯·彼得·卡尔景观建筑设计事务所，汉诺威
LKL项目经理：
LKL团队
竣工时间：
2013年
项目规模：
7200 平方米
整体项目预算：
230万欧元
照明预算：
20万欧元

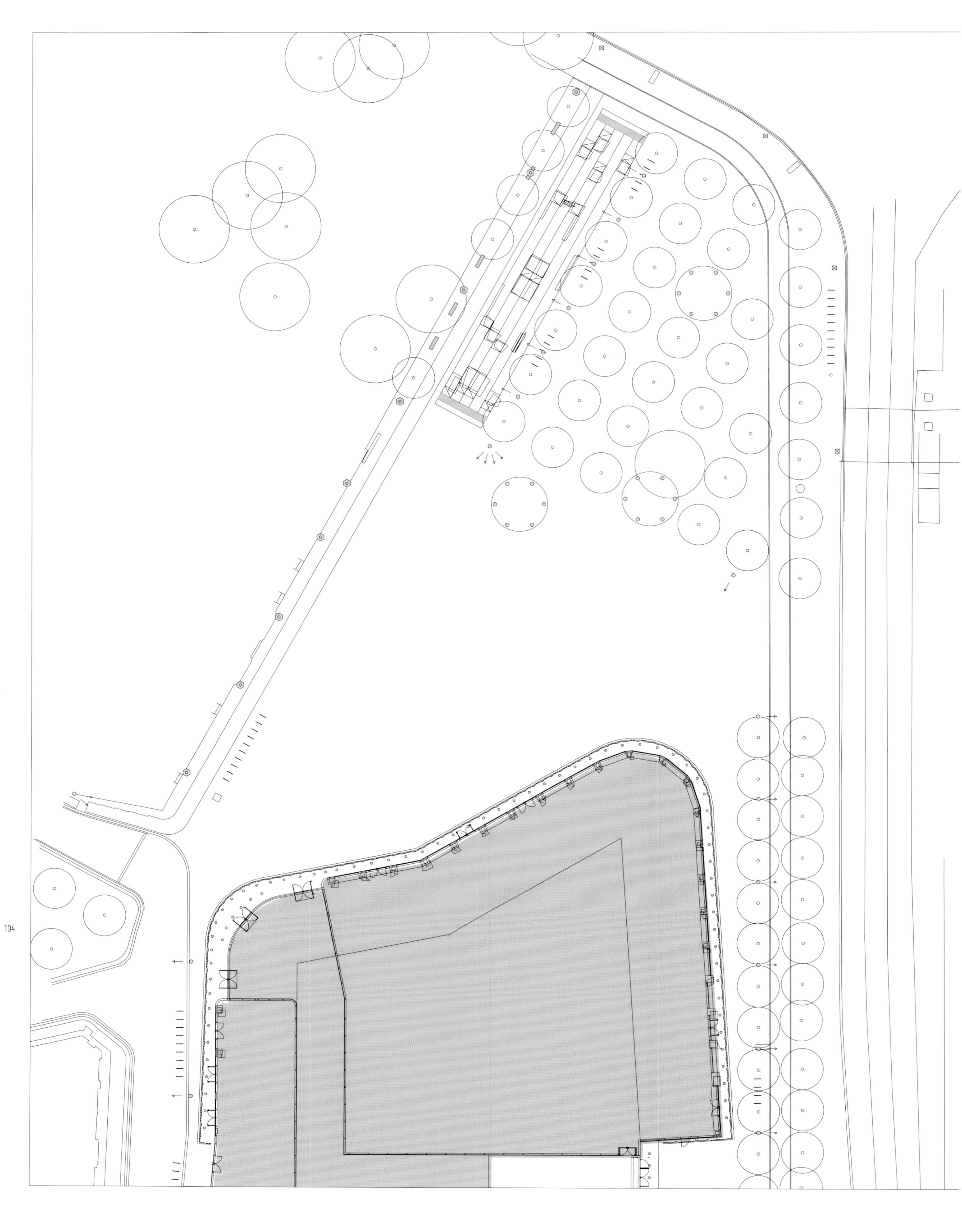

理查德- 瓦格纳广场场地平面图

Site plan of Richard-Wagner-Platz

The very ambitious shopping center development project, forming an eastern extension of the Brühl district's so-called 'Blechbüchse', has caused spirited discussions among the citizens of Leipzig, as well as an attentive intervention in the building process.

The 'Blechbüchse' (tin can) is an iconic building that is firmly rooted in Leipzig's city self-image. It was designed by the artist and metal designer Harry Müller in 1966. Since 1968, it accommodated the 'Konsument am Brühl', the GDR's largest shopping venue of its time.

Situated in front of this building, Richard-Wagner-Platz not only has a gateway-like function as an entry to the city, but must also and above all, respond to the popularity and level of attention this building receives. Therefore, a top-class competition for the redevelopment of the square was organized among landscape designers and with the involvement of Leipzig's municipal administration. In the process, requirements were established to generate a prominent location and to make the site available for lively use by the Leipzig population. Hence, the awarded design by Irene Lohaus Peter Carl Landschaftsarchitektur from Hannover, includes a skate park and a charming historic fountain.

A LIGHT LAYER FROM GOBO PROJECTION

For the lighting designers, the development of a consistent illumination concept was challenging. On one hand was the prominent artistic envelope of the existing building complex, and on the other, the contemporary shopping center façade design by Grüntuch Ernst Architects, quoting elements from Leipzig's architectural history. Each were to be given special attention without detracting from the square. Correspondingly, the striking and light-responsive facade of the front building has been illuminated with a novel system that uses gobos in a projection technique that precisely lights the building to the edges of its elevation. The goal was to apply a gentle, almost imperceptible light layer; not for the building to appear as if it is flood-lit, but rather to give the impression of it radiating from within. The square in front of the building thus has an appropriate setting in its urban context.

MODULAR LIGHT FOR A DIVERSIFIED PLAZA ILLUMINATION

For the square itself, a lighting system was selected that is able to respond individually to the various spatial layouts and uses. A modular pole system allows the utilization of luminaires that have the ability to service to the square with different light distributions, ranging from accentuated to uniform. As a result, the skate park could be furnished with relatively high illuminance levels, thus meeting its use's elevated requirements. Different parts of the square, on the other hand, were supplied with a lower and more atmospheric light level.

POSITIVELY ACCEPTED – BY CITIZENS AND JUDGES

The newly developed urban ensemble of square and building blocks enjoys great acceptance among the citizens of Leipzig. From afar, the artfully designed facade from the late 60's presents itself as a carefully choreographed eye-catcher; upon approaching the plaza, however, the observer's perception shifts towards the horizontal plaza and makes one curious to explore the attractions and installations on site. Internationally, the realized project has aroused a great deal of attention and has been recognized with the 'city.people.light Award' 2014 by an experienced jury.

CLIENT:
City of Leipzig, Traffic and Civil Engineering Office
OCCUPANT:
Public
LANDSCAPE DESIGNERS:
Irene Lohaus Peter Carl Landschaftsarchitektur, Hannover
PROJECT LEAD LICHT KUNST LICHT:
Team Licht Kunst Licht
COMPLETION:
2013
PROJECT SIZE:
7,200 m^2
OVERALL BUILDING BUDGET:
2.3 million euros
LIGHTING BUDGET:
0.2 million euros

模块化的高杆灯为广场提供了不同等级的照度。“锡罐”建筑的重要外立面被灯光赋予了视觉优先权。广场上其他的活动区域，例如滑板场，也通过灯光得到了强调。

Modular pole luminaires provide differentiated lighting levels to the square. As a result, the prominent facade of the 'Blechbüchse' (tin can) is given visual precedence and more animated areas of the plaza, such as the skating rink, are also accordingly emphasized.

喷泉的动感水舞由水下灯具提供重点照明。相比之下，周围人行道区域的照明设计则比较克制。

The dynamics of the fountains are accented with submersible luminaires. The surrounding pedestrian areas, by contrast, are illuminated with restraint.

MLP
DIS

RISTORANTE
VALENTINO
PIZZERIA
CAFE & EIS
PAULANER

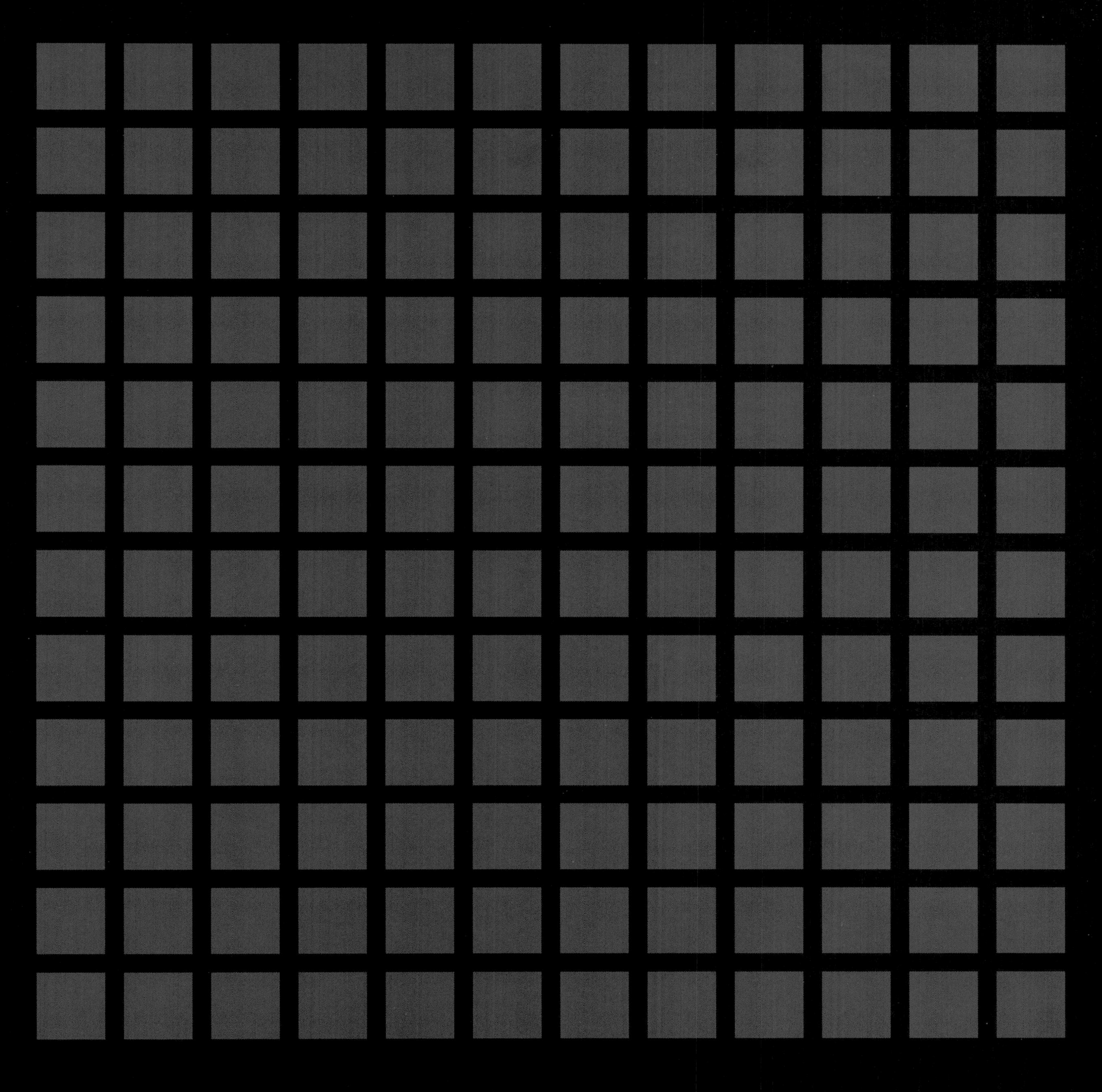

柏林大道购物中心　柏林

BOULEVARD BERLIN BERLIN

丰富的灯光指引——柏林大道购物中心

位于斯蒂格里茨地区施洛斯大街上的柏林大道购物中心是目前德国最大的购物中心之一，其租赁面积约为76 000平方米，包括用于零售、服务和餐饮相关的近160个出租单元。LKL对该项目的设计跳出了购物中心的常规照明模式。最为重要的是，该项目的照明方案是服务于建筑造型及其空间环境，并为顾客带来印象深刻的空间体验的。

购物中心的天花造型由凹槽灯重点勾勒，既创造了宽敞且易于理解的空间环境，又避免了像寻常购物中心那样过度使用照明灯具。与常规购物中心的照明相比，一个包含局部反转性光线语言形式的照明方案强化了这种大方整体的空间印象。照明设计师特意采用更加内敛的隐藏式下照灯对从购物区通往其他区域的通道进行照明。通过这种方式，购物区雕塑般的天花造型在灯光的衬托下能吸引顾客返回于这片区域。

室内中庭空间的分层照明

在室内中庭区域，三级天花暗槽结构呈阶梯状向内交错排列，天花凹槽的灯光从内侧向外对环形购物区域进行照明。由于照明设计师将该照明策略贯穿应用在所有楼层，人们习以为常的凹槽灯光语汇出乎意料地展现出巨大的视觉冲击力，特别是当顾客抬头看时，这种视觉冲击感尤为强烈。为了使照明效率最大化，设计师在间接照明的光源上增加了非对称配光透镜。根据不同的天花槽口长度定制的T16荧光灯管可以满足凹槽中灯具的多种重叠长度组合。在所有凹槽转角区域，特殊的转角构件可以确保天花转角没有暗区。从楼上向下可以看到灯具的区域，设计师采用轻微磨砂的亚克力玻璃覆盖其上，确保建筑干净清爽的外观。在横跨整个中庭的大型天窗下方，设计师采用HIT聚光灯提供了强有力的定向照明。

在开放式平台层尽情享受欢乐

天花凹槽构成了一个视觉连接元素，通过统一的设计语汇，将不同的建筑区域联系在一起。因此，在开放式平台层也能看见这些视觉连接元素的身影。天花吊顶暗槽和间接灯光的设计从建筑本身造型出发，通过选择合适的照明设备安装位置，创造出富有表现力的设计语汇。设计师引入多条环绕空间立柱的灯带，营造出欢乐的氛围，让人感觉既不俗气也不张扬。环形照明区域之外的空间则由松散排布的下照灯进行照明。

长廊空间中的视觉反差和优雅氛围

从首层看，新建中庭与现存韦特海姆大楼相连接的两个长廊展现出与购物中心大部分区域完全不同的视觉印象。虽然明亮轻快的天花完成面主导了购物中心整体的视觉色调，但这两个长廊的天花和立柱表面却是由黑色石材包裹的，由此戏剧性地创造了一个极具优雅感的大型空间。这种视觉效果在精准的定向下照灯和从天棚向下的重点照明中得到增强。与天花和立柱的深色表面形成鲜明对比的是，设计师在长廊的中央区域，包括天花板下方和从首层到地下层天花开口的垂直护墙表面，统一设置了明亮的背光照明。高功率的LED线条灯被安装在半透明薄膜后方，在白色背墙上营造了掠射光的效果。四周的长廊通过商铺橱窗周围反复出现的凹槽灯光连为一体。与中庭照明不同，这里的凹槽灯并不是交错排布的。

韦特海姆大厅：耀眼的自动扶梯成为空间的视觉焦点

在作为历史文化遗产受保护的韦特海姆建筑立面的后方，室内空间已被完全拆毁并进行了重建。来自柏林的奥特纳和奥特纳建筑艺术设计事务所在此设置了一个开放式混凝土天花龙骨，将这个空间与其他建筑区域区分开来。在开放式天花龙骨中，每个花格镶板的中心都设有一个柔和的光源，同时配备额外的定向照明组件。冲孔黄铜板包边的自动扶梯是韦特海姆大厅中的视觉焦点。冲孔黄铜板和半透明天花中的背光照明进一步增强了这种视觉效果。

作为连接空间的漫步长廊

这条供行人漫步的长廊曾经是两条街道之间的通道。该市规定即使在夜间，这条人行通道也应保持开放。因此，设计师在购物中心人行通道的顶部覆盖了玻璃屋顶。此外，与这条通道相邻的建筑立面也被设计为旨在唤起人们对这里过去作为户外空间的气氛联想。因为这个空间如今是购物中心的主入口，所以照明设计在增强户外空间感的同时，营造了温馨、好客的灯光氛围。在室内种植橄榄树的构想给灯光设计带来了新的挑战。为了满足此项特殊要求，需要1500勒克斯的额外照明来补足。

基于此，该空间的照明设计方案中包含了一组可调式聚光灯，光源采用能发出适合植物生长的光谱波段的150瓦HIT光源，可直接投向人行通道和指定树木。这些灯具安装在定制的模块化边框内，其中也包含了应急照明组件。灵活性较高的投光灯可以调整投光角度，在空间中发出定向光，在一定程度上遮掩了建筑顶部边缘较为明显的设备安装细节。

业主：
柏林多威斯特股份有限公司
使用人：
克莱佩尔
建筑设计：
奥特纳和奥特纳建筑艺术设计事务所，柏林
LKL项目经理：
埃德温 · 斯米达
竣工时间：
2012年
项目规模：
76 000 平方米
项目总预算：
3.9亿欧元
照明预算：
120万欧元

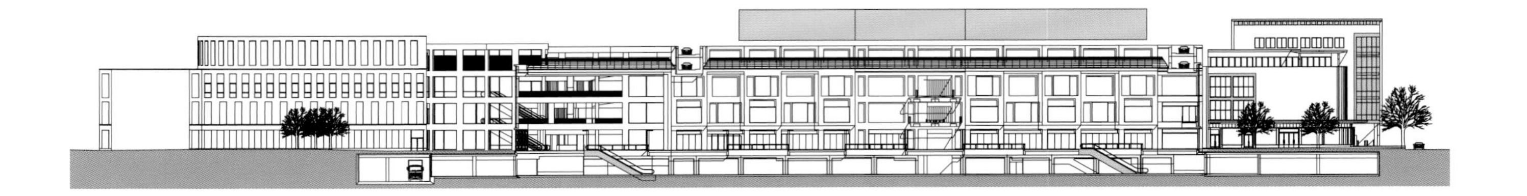

长廊纵断面剖面图 Longitudinal section of the promenade

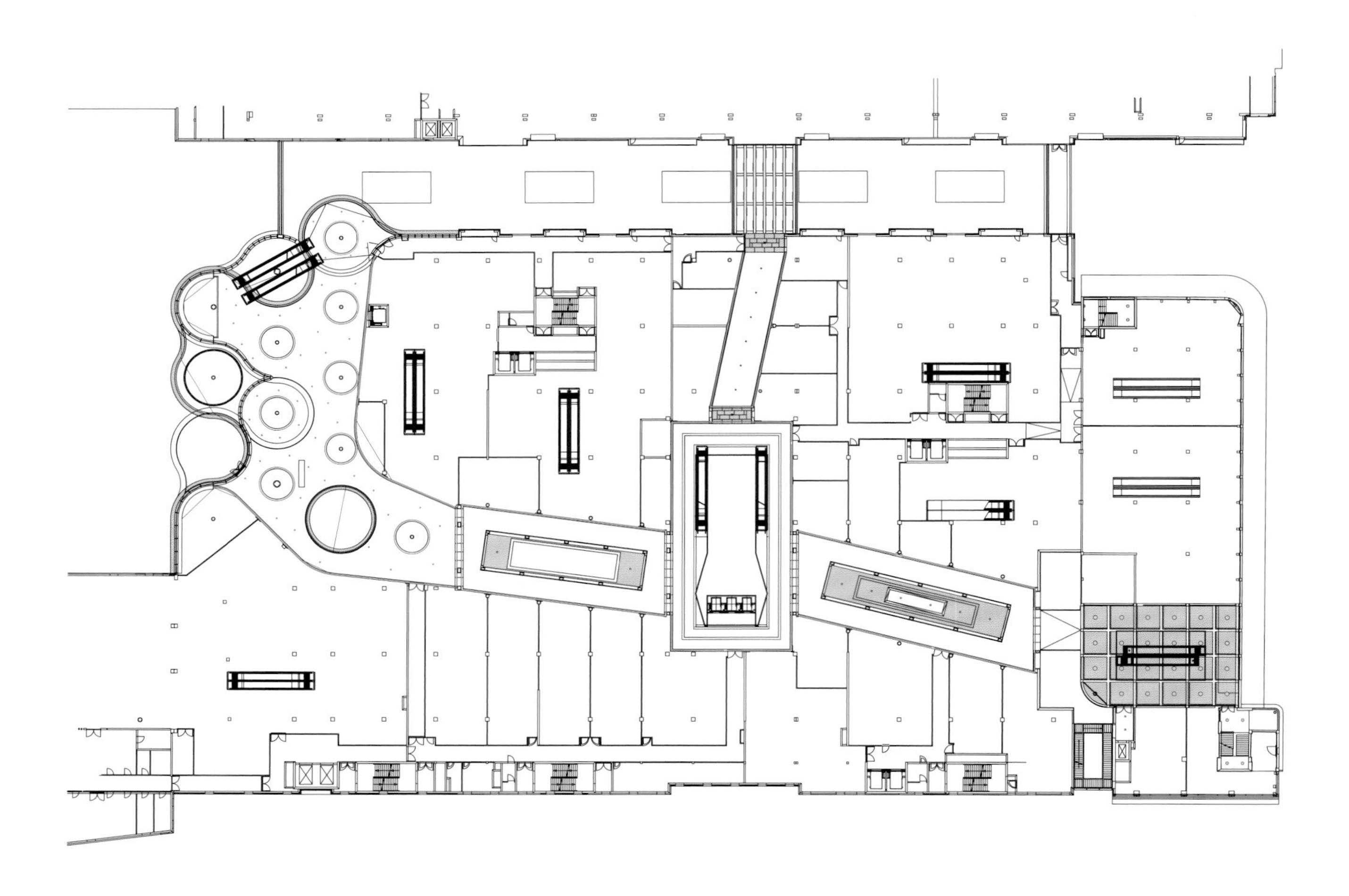

二层反向天花平面图 Reflected ceiling plan of first floor

With 76,000 m^2 of rental space, consisting of nearly 160 rental units for retail, service stores, and gastronomic concepts, Boulevard Berlin on Steglitz's Schlossstrasse is now one of Germany's largest shopping malls. The lighting concept by Licht Kunst Licht goes beyond the usual lighting of shopping malls. Above all, it serves the architectural space and its ambience and manages to gain a memorable spatial experience for the costumers.

The ceiling slabs are featured by cove lighting and create a comprehensible and spacious atmosphere without using the usual over-designed luminaires. This generous spatial feeling includes a partial reversal of the system – passageways to other areas are deliberately lit in a more introverted fashion with concealed downlights. This way, the costumer is drawn back to the shopping areas where the illuminated ceilings unfold an almost sculptural effect.

LAYERED LIGHT AT THE INNER COURT YARD

At the inner court yard, three staggered coves are stepped inward and illuminate the orbiting mall areas from within its structure. By virtue of its consequential use across all floors, the familiar concept of cove lighting unfolds an enormous visual impact, especially when viewed from the bottom to the top. In order to achieve maximum efficiency, the indirect light sources have an asymmetric light optic. A custom made cove lighting fixture using 16 fluorescent lamps allows various overlap lengths, depending on the length of the cove recess dimensions. Special corner elements ensure continuous illumination in all corner situations. In areas where the lamp can be seen from the upper levels, slightly rosted acrylic glass covers ensure a clean appearance. Under he large skylight spanning the courtyard, HIT spotlights provide powerful directional light.

LAYFUL SERENITY AT THE TERRACE HOUSE

he coves form a connecting element that ties together the istinct architectural zones with a common design language. ccordingly, they are also found in the terrace house. The ceiling ff-sets and indirect lighting locations are selected to specifically reate an expressive design language emerging from the architecture itself. Introducing variety with light lines that circle the olumns achieves a playful serenity which feels neither tacky nor verdone. The remaining areas outside these circle segments are t by loosely interspersed downlights.

ONTRAST AND ELEGANCE IN THE HALLS

iewed from the ground floor, where the court yard connects to the xisting Wertheim building, yet another spatial impression emerges. Vhile bright and crisp ceiling surfaces dominate the mall, the eilings and pillars in the halls are covered with black stone. The esulting effect generates a dramatically large space with extreme legance, which is enhanced by sharp directional downlights and ccent lights from above. Contrasting with the dark surfaces, acklit friezes are located at the centre of the halls, integrated in he balustrades flanking the ground floor voids, or directly under he ceiling. Powerful LED profiles mounted behind a translucent ilm create a grazing light effect on the white back walls. The urrounding galleries are linked by the familiar cove lighting near the shopping window, but here, unlike in the courtyard, the coves occur without staggering.

WERTHEIM FOYER: DAZZLING ESCALATORS AS THE FOCAL POINT

The interior space behind the historically protected Wertheim building façade was completely gutted and rebuilt. The architectural office Ortner & Ortner Baukunst from Berlin chose an open concrete ceiling grid, which significantly distinguishes this section from the other building areas. Each centre of these coffers is fitted with a diffuse light object as well as an additional directional light component. The escalators, clad in punched sheet brass, are the most prominent focal point in the Wertheim foyer. This effect is further enhanced by backlighting both the brass panels and the translucent ceiling.

THE PROMENADE AS A CONNECTION SPACE

The promenade once was a road link between two streets. Statutory requirements of the city dictated that this pedestrian connection be kept open, even during night hours. Therefore, this part of the mall was covered with a glass roof, and the adjacent facades were designed to evoke an association with the former outdoor space. The lighting design was to support the impression of an outdoor space while creating a warm and inviting lighting atmosphere, as this space functions as the mall's main entrance. An additional challenge arose with the intention to plant olive trees, which called for supplementary light levels of 1,500 lux.

The solution consisted of a layout of adjustable spotlights with 150 W HIT lamps of suitable light spectrum, where the light sources can be directed at the circulation areas or the trees. These luminaires are attached to a custom-made gimbal frame, which also houses an emergency lighting component. The high flexibility and directional light emanating from the gimbal projectors compensate for the design compromise of a relatively ostentatious surface mounting detail at the upper building edge.

CLIENT:
Multi Veste Berlin GmbH
OCCUPANT:
Klépierre
ARCHITECTS:
Ortner & Ortner Baukunst, Berlin
TEAM LEADER LICHT KUNST LICHT:
Edwin Smida
COMPLETION:
2012
PROJECT SIZE:
76,000 m^2
OVERALL BUILDING BUDGET:
390 million euros
LIGHTING BUDGET:
1.2 million euros

SATURN
ecco
SALE SALE SALE

韦特海姆大厅的中央区域设有冲孔黄铜板包边的自动扶梯，呈现出背光照明的效果。

The Wertheim Foyer's centerpiece is the backlit, perforated sheet brass cladded escalator.

设计师在所有楼层统一使用了简洁的凹槽灯，打造令人印象深刻的视觉效果。

The simple idea of using a cove light consistently across all floors creates a remarkable visual impact.

虽然明亮轻快的天花完成面主导了购物中心整体的视觉色调，但长廊的天花和立柱表面却是由黑色石材包裹的。

While bright and crisp ceiling surfaces dominate nearly all areas of the mall, the ceilings and pillars in the halls are clad in black stone.

穿过数个大型天花的开口，购物中心的一部分地下空间会与首层的漫步长廊相连接。

Through large ceiling apertures, a part of the mall's basement connects to the promenade at the ground floor.

MNÄCHST ERÖFFNET HIER
H&M

ZARA
COLLECTION

AWARD
WINNER

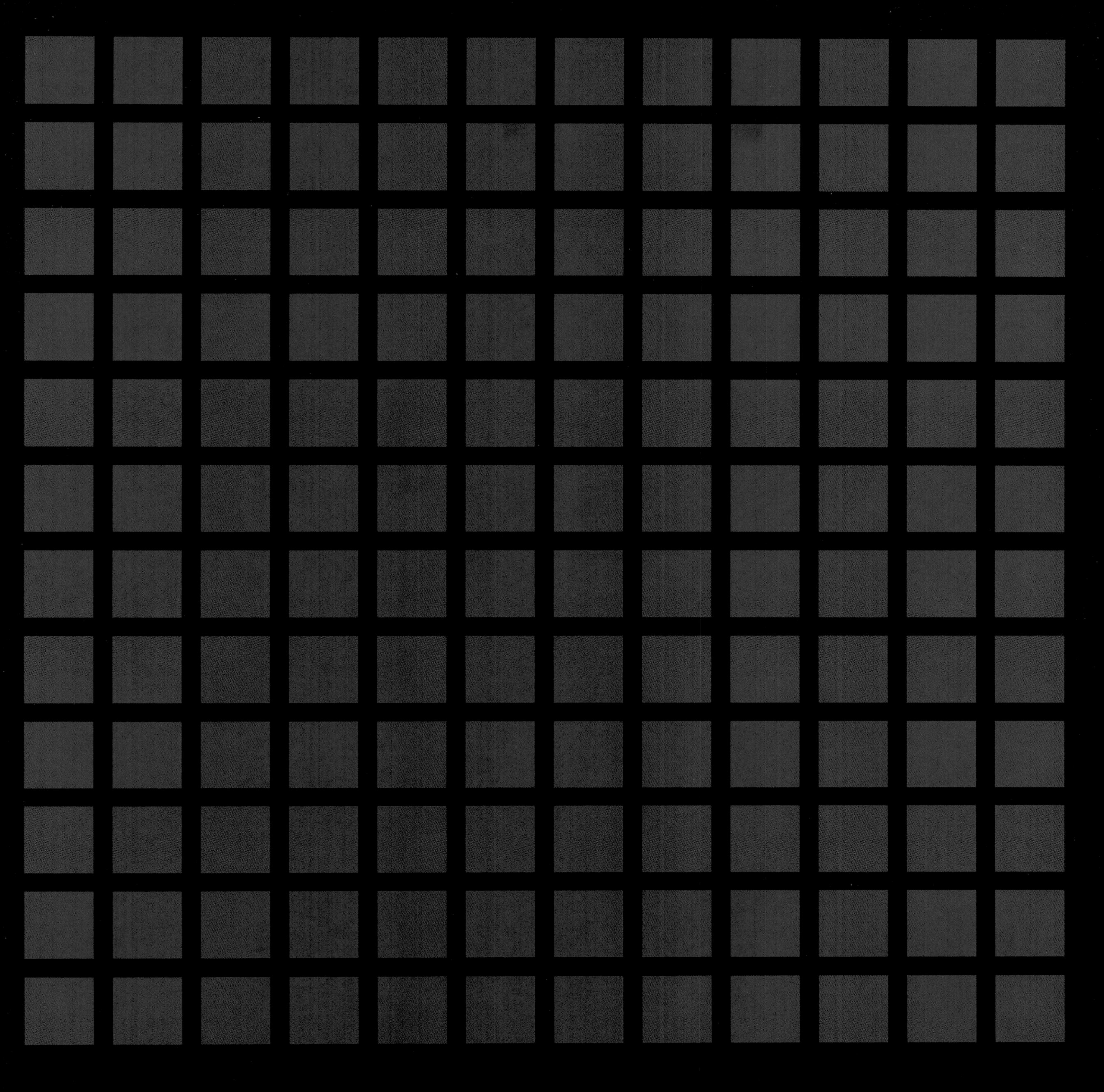

LICHT KUNST LICHT 4

龙岩山高原餐厅 克尼格斯温特尔

DRACHENFELSPLATEAU KÖNIGSWINTER

可以俯瞰城市全景的地标建筑——龙岩山高原餐厅的照明方案

为一座旨在吸引游客欣赏其外部风景的建筑进行照明设计，似乎不是一项值得重点关注的工作。但如果这座建筑的外立面是一个玻璃立方体，那么其立面上的反光将会成为影响观景效果的因素，因此，这个项目需要专业的照明设计团队。除此之外，业主对照明设计的要求还包括利用灯光来打造一个地标建筑，注重灯光对周围自然环境和野生生物的影响，以及营造适用于多种场景的灯光氛围效果。如此多的设计要求给照明设计方案的实现带来了严峻的挑战。最终，通过新建龙岩山高原餐厅的灯光设计，LKL的照明设计师们证明了他们基于这些看似相互矛盾的设计要求所提出的照明解决方案，不仅在技术上是可行的，而且还创造了令人印象深刻的视觉效果，营造出餐厅热情好客的灯光氛围。

莱茵河畔的浪漫主义与野兽派建筑风格

龙岩山位于克尼格斯温特尔和巴特洪内夫之间的莱茵河河畔，是德国最热门的登山地点之一。这里流传着一个传说，尽管与尼伯龙根传说关联不大，但据传有一条神龙常年盘踞于此。这片带有古堡遗址的龙岩山脉形成了壮丽的景观，这里曾是凸显德国浪漫主义文明的重要场所。1816年，乔治·戈登·拜伦的一首诗作使莱茵河畔的陡峭山坡、古堡遗迹和龙岩宫殿声名鹊起。从拜伦的英国同胞开始，因其诗作前往此地观光游览的人们络绎不绝。20世纪0年代，龙岩山年均客流量高达90万人次。

然而，近年来龙岩山的客流量却未能达到过去的辉煌。客流量回落的数据强有力地推动了克尼格斯温特尔市政府对其进行旅游开发的规划，因此当地政府将克尼格斯温特尔龙岩山观光项目归为“2010年度地区规划”中的A类项目。由此，政府投入了100万欧元来提高龙岩山的文化和旅游吸引力。在众多建筑规划之中，山顶的酒店设施建设是项目的重点。这套设施由一个始建于20世纪30年代的宾馆和拥有约40年历史的餐厅组成。

这座野兽派风格的建筑始建于1976年，从一开始就遭到了业界的批判。餐厅的水平混凝土横梁影响了游客欣赏古堡遗址和龙岩城堡，其向高原山脉南端延伸的建筑结构也阻挡了龙岩山最为宝贵的财富——面向莱茵河的观景点。为了改建这座建筑，当地政府开展了景观提升方案的竞赛招标，最终普兰德雷景观建筑设计事务所和佩普与佩普建筑设计事务所中标。

玻璃立方体建筑和观景台阶

通过这次改建，前往龙岩山的游客可以从场地景观、当地历史和现代化酒店及餐厅的体验中感受到连贯整体的设计理念。在这个开放空间中，能够俯瞰莱茵河谷、古堡遗迹和七峰山的阶梯式观景平台是景观设计的核心。招待宴请场所嵌入阶梯式平台，其中包括一个新餐厅和前往老宾馆的连接走廊。餐厅的体量与山顶自然景观及现存历史建筑相协调，并且完美地融合在一起。为了尽量减少该建筑物外露在山顶平地之上的可见部分，建筑设计师在山顶平地下方的斜坡上设置了两个完整的楼层，作为与后勤相关的空间。用餐空间仿佛从山顶拔地而起，由一个透明玻璃立方体和浅色外露混凝土结构构成。

新建筑和历史建筑之间通过一个高度灵活的空间布局连接。整体化的空间结构可满足旅行团的餐饮需求——在一天之内服务数千名客人。此外，餐厅也能为举办会议和晚间活动提供场地。

室外全景视角和室内氛围照明

玻璃立方体建筑内的餐厅照明设计方案必须满足空间最大灵活性的要求。因此，研发一套魅力十足的照明解决方案至关重要。它既要保证餐厅的日常营业，又要满足夜间活动的场景氛围需求。从浪漫的烛光晚餐到大型宴会活动，室内灯光需要适用于桌椅的多种摆设。在这个项目中，照明设计师们面临的挑战是，既要创造出舒适的室内照明氛围，又要将客人的目光引向室外美景。

面对这个挑战，照明设计师采用了直接照明的下照灯，借助餐厅定制金属天花板上的圆形开孔提供照明。这些下照灯与空间中以暗色调为主的家具相结合，打造出一个个热情好客的“发光岛屿”。下照灯配有万向调节支架，可根据不同的用餐活动适当调节灯具的投光方向。餐厅工作人员通过专业工具，站在地面就可以准确地调整灯光的投向。照明装置的直射光线分布搭配整个空间中低反射率的表面，可以防止在餐厅落地玻璃窗上产生令人厌烦的反光。由此一来，室内光线就不会干扰客人欣赏窗外山谷的全景。

从谷底仰望可见的、带有可控制彩光氛围的地标建筑

龙岩山餐厅的直接照明元素满足了餐厅经营的灵活性要求，并在不干扰客人欣赏室外风景的同时，营造出迷人的空间氛围。为了适当增强远距离欣赏该观光胜地的视觉效果，照明设计师对向下投射的直接照明进行了补充和丰富。在夜间，为了便于游客从莱茵河谷抬头观赏这一地标建筑，照明设计师用灯光元素增强了从低处向高处仰视视角下的天花视觉效果。由于龙岩山餐厅位于自然保护区内，因此照明设计方案必须谨慎地深化与实施。人工光照明不能对保护区内的动植物造成干扰。

为此，照明设计师采用了可变色的灯光设计方案。即使在光通量输出较低的情况下，带有色彩倾向的光线也能达到想要的视觉效果。设计师在餐厅方格天花的金属穿孔板背面安装了RGB- LED灯带。这些LED灯带发出的漫射光为上方的天花方格提供了柔和的照明。餐厅工作人员可通过集成了所有室内外照明设备的控制系统，调节天花照明的光色。除了塑造室外的视觉印象外，这一照明系统还有助于增强室内的用餐氛围。无论是节日晚会所需的暖色调光线，还是商业活动中定制的代表企业形象的光色，龙岩山餐厅都能从容应对，并根据不同的活动需求营造理想的光线氛围。

业主：

克尼格斯温特尔市经济发展和房屋建筑公司

使用人：

赫尔曼·J.诺登餐饮和活动场所

建筑设计：

佩普与佩普建筑设计事务所，卡塞尔

LKL项目经理：

卡尔·玛丽亚·雷格

竣工时间：

2013年

项目规模：

1500 平方米

整体项目预算：

930万欧元

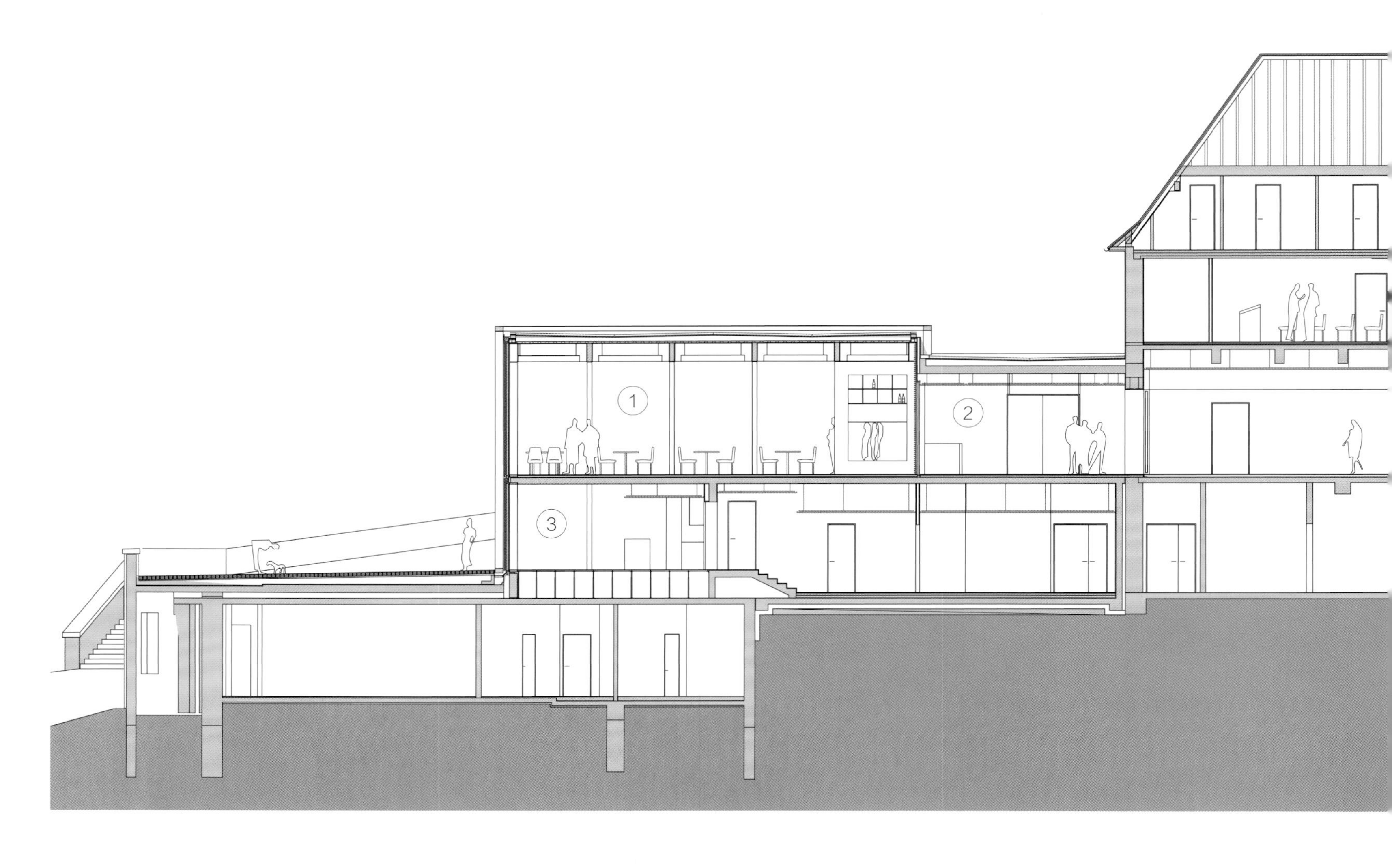

新建和现有建筑的纵剖面图

Longitudinal section of new and existing building

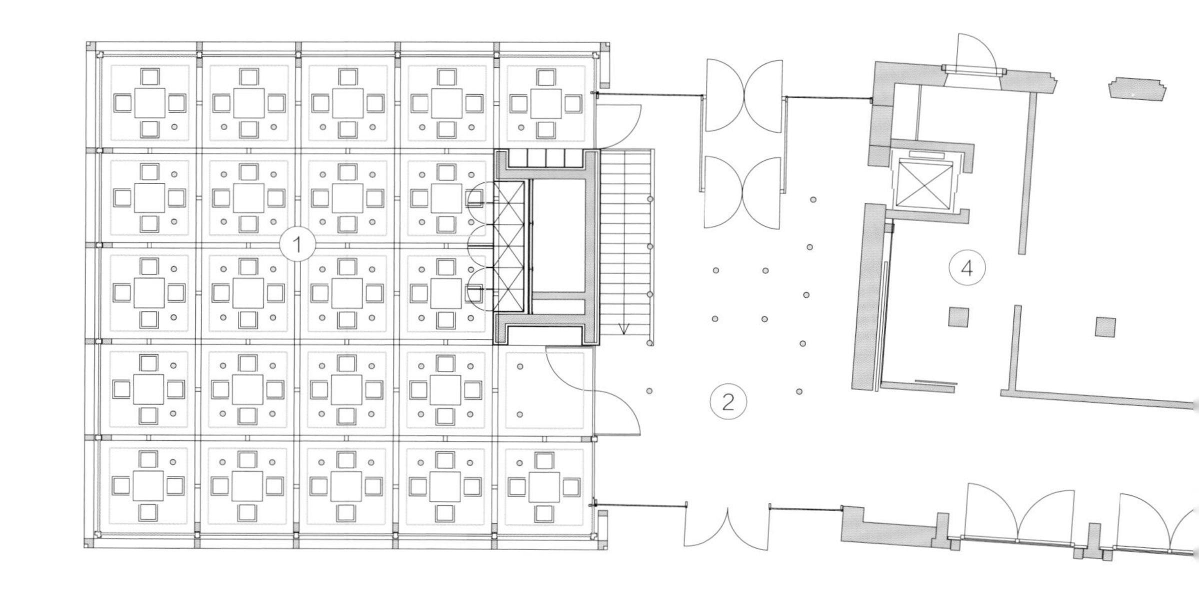

平面图

Floor plan

1.餐厅
2.门厅
3.商店
4.厨房

1. Restaurant
2. Foyer
3. Shop
4. Kitchen

Designing light for a building that attracts visitors for its view outside does not appear to be a rewarding task. If the building is a glass cube, thus making reflections on the panes a view impairing factor, an elevated level of expertise is required. Other client design requirements, including using light to generate a landmark, being sensitive to the surrounding nature and wildlife, and creating an atmospheric lighting effect that is also flexible for a gamut of uses, raise the question of whether all this is possible. With its lighting solution for the new restaurant on the Drachenfelsplateau, the lighting designers of Licht Kunst Licht prove that the resolution of hese apparent contradictions is not only technically feasible, but s impressive and inviting at the same time.

RHINE ROMANTICISM AND BRUTALISM

he Drachenfels, located on the Rhine River between Königswinter and Bad Honnef, is considered to be the most frequently climbed mountain in Germany. Its legend of a resident dragon is – although loosely – connected to the Nibelungen Legend. The sublime landscape with its castle ruin was a pivotal backdrop to German Romanticism. In 1816 the steep mountain slope on the banks of the Rhine, the castle ruin and the Drachenburg Palace all ascended to fame through a poem by George Gordon Byron. Beginning with his fellow British countrymen, his poem triggered an onrush of tourists hat has yet to subside. In the 1960s the Drachenfels had up to 00,000 visitors per year.

More recently, however, such records have failed to materialize. eclining visitor attendance figures were a motivation for the city f Koenigswinter to apply for a classification as Status-A-Project f the 'Regionale 2010' with the project 'Holistic Perspective oenigswinter – Drachenfels'. As a result, 21 million euros were provided to improve Drachenfels' cultural and touristic appeal. Among other criteria, the focus was on the gastronomy infrastructure on he mountain top. It consisted of a hotel from the 1930s and a large, nearly 40-year-old restaurant.

Built in 1976 in the Brutalism style of architecture, the destination restaurant was always met with criticism. Its horizontal concrete bbons not only interfered with the perception of the castle ruin and the palace, its sprawling architecture extending to the southern end of the plateau also limited Drachenfels' greatest asset: the ew of the Rhine. In order to redefine the context, a competition was held, in which the designs from landscape architects plandrei andschaftsarchitekten and architects pape + pape architekten ere implemented.

GLASS CUBE AND SEATING STEPS WITH A VIEW

Since its redevelopment, visitors to Drachenfels experience a consistent design throughout the site's landscape, local history, and contemporary gastronomy. A fundamental element of the landscape design is tiered seating overlooking the Rhine valley, the castle ruin, nd the Seven Hills. The gastronomic venues are embedded in the epped gradient, including the interconnected volumes of the old otel and a newly built restaurant. The restaurant's dimensions end well with the landscape and existing historical building. In order to minimize the visibility of the above-grade portion of the building, two entire back-of-house related floors were located below the plateau slope. The restaurant emerges above the slope as a transparent cube of glass and light-coloured exposed concrete.

The connection between the new building and the historical volume has created a highly flexible spatial layout. The ensemble fulfils the requirements of tour group gastronomy with the ability to service several thousand guests during the course of one day. Furthermore, hosting conferences and evening events is also possible.

PANORAMIC VIEWS TO THE EXTERIOR AND ATMOSPHERIC ILLUMINATION WITHIN

The illumination concept for the restaurant inside the glass cube had to be designed to cater to the requirement of maximum flexibility. An attractive solution for everyday operation, and glamour for nocturnal events was imperative. Both configurations were to function with diverse table arrangements, from romantic dinners to large banquets. The lighting designers at Licht Kunst Licht were challenged with creating a pleasant light atmosphere in a space where guests enter with the intention of gazing outside.

They responded to the challenge by using direct downlights emanating their light from circular apertures in the restaurant's custom-made metal ceiling panels. In combination with the predominantly dark surfaces in the space, they create inviting luminous islands. A gimbal frame allows for an appropriate luminaire adjustment for each table configuration. The restaurant staff can accurately aim the projectors from the ground with the assistance of a special tool. The luminaire's direct light distribution combined with low-reflectance surfaces throughout the space, prevent unwanted reflections in the floor-to-ceiling restaurant glazing. Thus, the panoramic views of the valley remain unimpaired.

A LANDMARK VISIBLE FROM THE VALLEY WITH CONTROLLABLE COLOURED LIGHT

The direct illumination of the Drachenfels restaurant fulfils the gastronomy business's flexibility demands and establishes an attractive atmosphere with an undisturbed view for the guests. To support the long-distance external visual effect appropriate for this prominent and popular location, the downward directed light has been complemented with an additional component. In order to experience the landmark status building from the Rhine valley at night, the lighting designers decided to enhance the slab view from below with light. This had to be approached with care, as the Drachenfels restaurant is located inside a nature reserve. Flora and fauna were not to be disturbed by the illumination.

The use of coloured light was the solution for this challenge. Even with a low luminous output, strong visual effects can be achieved. Therefore, RGB-LED light strips were installed behind the perforated metal panels of the restaurant's coffered ceiling. Their diffuse LED light softly illuminates the square ceiling coffers above. By virtue of a control system directing all indoor and outdoor lighting systems, the restaurant staff can adjust the ceiling illumination's colour. Apart from its exterior visual impact, it additionally contributes to enhancing the interior atmosphere. Whether an especially warm light for a festive evening party, or a corporate identity colour for an office event, the Drachenfels restaurant stands ready to create the intended lighting atmosphere.

CLIENT:
WWG Wirtschaftsförderungs- und Wohnungsbaugesellschaft mbH der Stadt Königswinter
OCCUPANT:
Restaurant & Eventlocation Hermann J. Nolden
ARCHITECTS:
pape + pape architekten bda, Kassel
TEAM LEADER LICHT KUNST LICHT:
Karl Maria Reger
COMPLETION:
2013
PROJECT SIZE:
1,500 m^2
OVERALL BUILDING BUDGET:
9.3 million euros

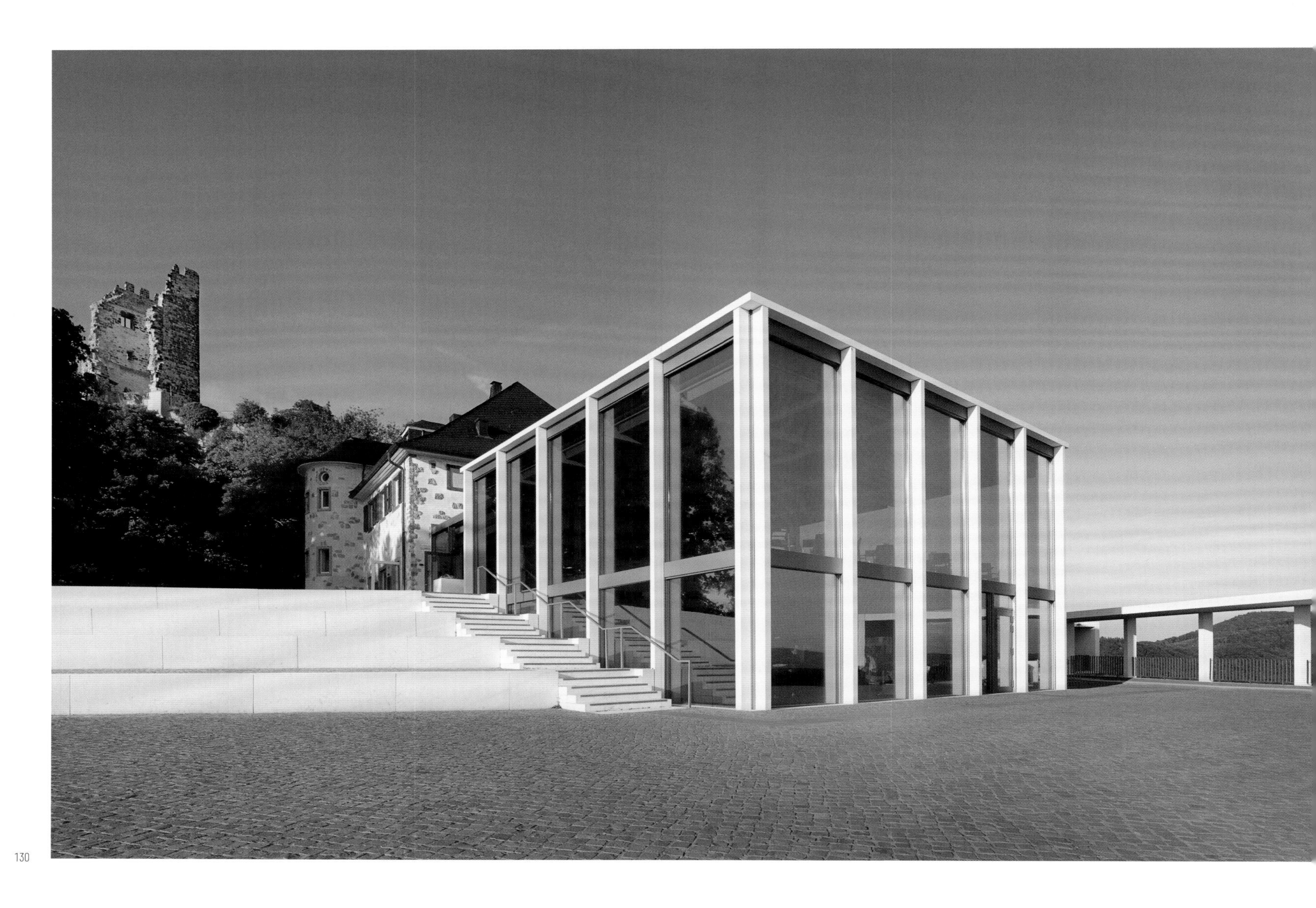

通过这次改建，前往龙岩山的游客可以从场地景观、当地历史和现代化酒店及餐厅的体验中感受到整体连贯的设计理念。

Since its redevelopment, visitors to Drachenfels experience a consistent design throughout the site's landscape, local history, and contemporary gastronomy.

为了使这个地标建筑在夜间的莱茵河谷中清晰可见，设计师通过灯光元素增强了玻璃立方体天花板的视觉效果。

In order to make the landmark tangible from the Rhine valley at night, the glass cube's ceiling slab has been enhanced with light.

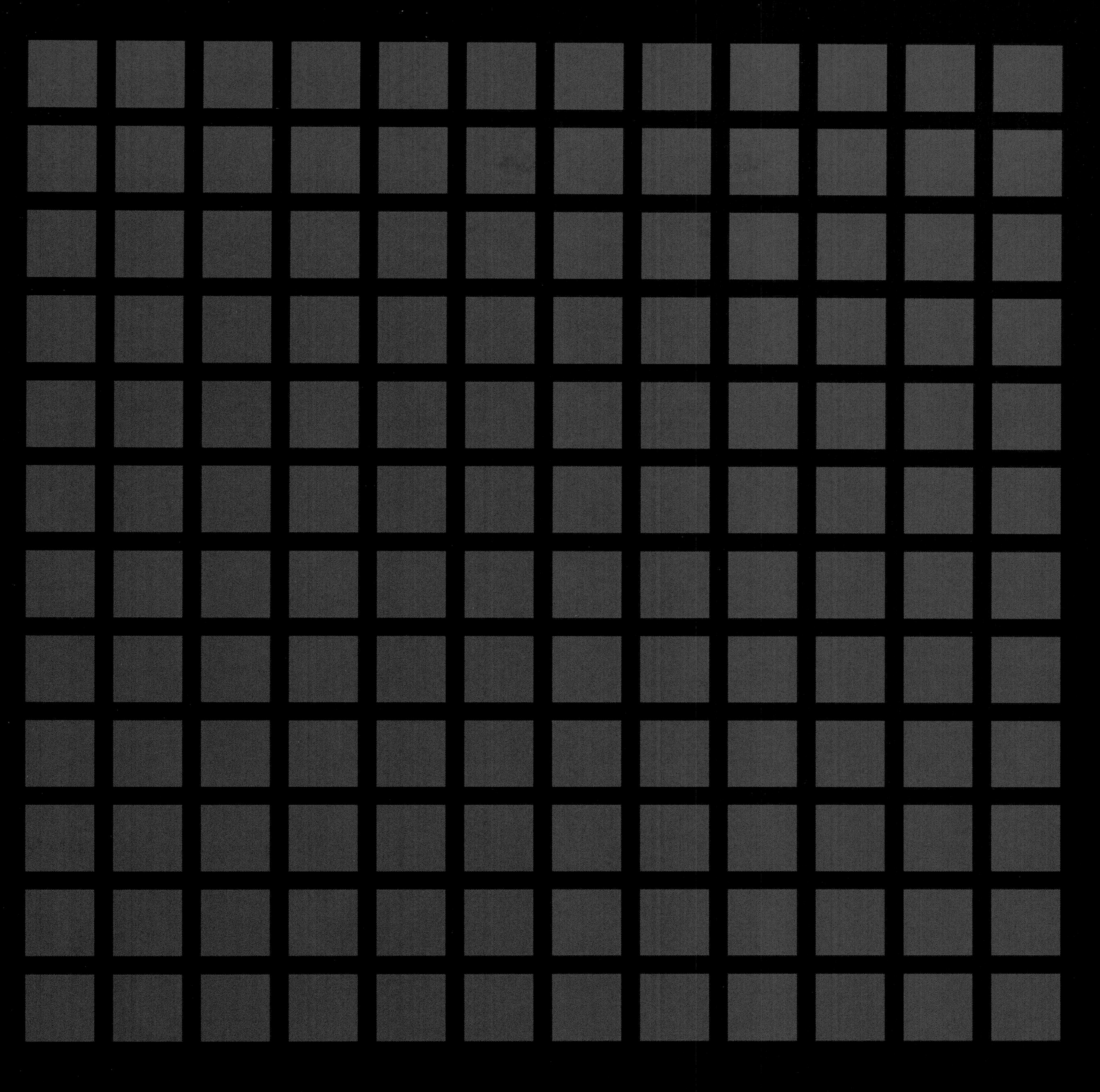

LICHT KUNST LICHT 4

科隆大教堂文化交流中心 科隆

DOMFORUM COLOGNE

面向公众的通透空间——新设灯光下的科隆大教堂文化交流中心

大教堂大楼是科隆天主教会的信息交流与文化活动中心，地处科隆市的核心地段。它位于科隆大教堂对面，经过马提尼建筑设计事务所的大规模改建和翻新后，该文化交流中心呈现出别样的光线效果和空间质感。LKL照明设计团队成功为其扩建空间创造了一个兼具功能性和氛围感的照明设计方案。

800多年前，中世纪最早的建筑理论家之一——法国圣德尼修道院院长苏格曾说：“光线传达信仰！”这条名言在今天看来仍然是正确的，但在现代化社会中，光线条件已经变得更加复杂和多元。在这个充满视觉干扰的社会环境下，不太容易建立起给人提供反省和宗教冥想空间的休息场所。尽管如此，大教堂文化交流中心仍然成为一个令人信服的案例。作为科隆大教堂的游客中心，它成为德国游客最多的教堂门厅。这有力地证明了即使在当今社会，通过室内设计和照明设计手法，也可以为宗教场所增添现代化的表达。

大教堂的接待室——邂逅和默观的界面

大教堂文化交流中心既欢迎朝圣者前来朝拜，也允许个人和团体游客参观，并为他们提供导览服务和精神关怀。此外，大教堂文化交流中心还是一个多主题活动场所，会举办宗教性和社会性相关的主题活动，涉及多种活动形式和内容（例如，精神激励的邂逅、小组讨论、宗教教育和文化活动等）。

透明和开放

现在作为大教堂文化交流中心的建筑物是一个受保护的历史遗迹，由弗里茨·沙勒（1904—2002）出资建于1951年至1954年间，该建筑起初是为公共经济银行而建。在1995年到1996年间，沙勒的儿子克里斯汀·沙勒将其改建为天主教会的文化交流中心。由于空间的日益紧缺，加上这栋建筑所面临的结构与服务相关的新要求，尤其是人们对空间多功能使用需求的日益增长，科隆大主教为大教堂文化交流中心的改建工程举办了一场建筑设计竞赛，最终马提尼建筑设计事务所脱颖而出。

该建筑方案旨在创造一个公共友好的空间，传达开放和透明的姿态。建筑设计的另一个目标是在对现存历史建筑充分尊重的基础上，兼顾大教堂文化交流中心更多的空间诉求。来自波恩的马提尼建筑团队将主要设计任务定义为建筑空间需要向外界尽量开放，同时保留其宗教色彩。与此同时，室内的翻修设计理念也是为了再次突出幕墙结构的现有质感。由此一来，翻新的大教堂文化交流中心被设定为一个多功能的聚集场所，其宗教背景只有在人们深入观察时才会被发现。

三面全玻璃的建筑外立面设计使游客可以多角度地从室内、户外或者房间过道中获得良好的视野。这样的设计也将一个具备自然采光特色的大型室内空间与其前方大教堂广场的户外环境连接起来。这样一来，游客们可以一目了然地对整个空间产生初步印象。尽管这座建筑很高，但其内部空间视野清晰，在视觉上给人以横向较宽而非纵向较高的空间印象。每一个独立的空间设施都被设计得十分简约，白色是主要的材质颜色。灰色钢筋混凝土大梁构成了空间的结构性元素，营造了一个明亮、开放空间所必要的中性背景环境，使其能够同时兼具活动舞台、礼堂和休息室的多重功能。在接待台后方，风格突出的十字架以其自然的木质纹理，与墙壁其他部分的包层图案形成对比，提醒游客这个空间蕴含的宗教内涵。

多功能的空间组织和座椅布置

多功能空间的室内设计包括三个主要元素：接待台、合适尺度的多功能壁柜和咨询岛。作为一件固定家具，新的接待台是空间中的独立物体。其曲线造型使它成为大教堂文化交流中心游客的主要会合点。接待台被细分为信息咨询区和协调团体参观事宜的区域，其由可丽耐实体面材无缝拼接而成，成为一件能够定义空间的雕塑作品。

室内设计的另一个特色元素是柜台后侧的中央背景墙，该墙面囊括了多种空间功能。壁柜包括一个带升降功能的座椅储存空间、一间厨房、一个配备复印机的电脑工作站和一个寄存处。它还为宣传册和文献资料提供了额外的存储空间。这种墙面让门厅看上去不显杂乱，并重新划分了柜台后方的工作区域。原先通往楼梯井的走道被保留了下来，并与壁柜设计融为一体。咨询岛位于活动中心的西侧、接待台的后方。它在门厅空间中，为前来进行精神关怀的人群创造了独立的区域，在视觉和听觉方面具有一定的隐私性。此外，空间还可以根据不同活动中观众席的布置需求，将座椅统一朝向正面或者进行分散式布置。活动中心日间陈设的弧形座位岛和用于夜间活动的移动座椅均采用胡桃木制成。

功能性照明——氛围照明

空间的照明设计方案是基于空间多功能使用的要求进行研发的。除了临时展览外，门厅还需要举办小组讨论会、音乐和戏剧演出以及多种讲座活动。同时，灯光还需要强调清晰简洁的建筑语汇。设计师的任务是引入一种既能达到一般照明效果又能满足重点照明需求的照明元素，还可以对现有的空间结构做出回应。因此，灯光布局重复了地面对角线的连接图形，这种连接图形也延续到了室外的大教堂广场上。

该空间的天花定制灯具包含一个周围带有阴影缝隙环的柔光磨砂亚克力灯罩，它巧妙地与天花结构融为一体。灯罩有两个圆柱形槽口，其中内嵌了提供直接照明的可调式下照灯组件。因此，定向照明的LED聚光灯在游客的大多数视角下都能保持隐蔽。为了满足不同场景的照明需求，定制灯具中的两个下照灯组件设计为可单独控制和调光的。

表现力丰富的可丽耐长桌是定义空间的主要设计元素，其上方的定向聚光照明突出了这一设计元素。在这些可调式聚光灯的烘托下，展陈家具和自由摆放的座位岛都可以根据不同的需求灵活排布。柔和的光线确保了展演区拥有均匀的照明。灯光设计师和多媒体设计师相互配合，在天花系统上集成了可远程控制的可调节LED投光灯，使该区域的活动能够伴随着动态和彩色的灯光氛围。

白天，这个采光均匀的友好型空间吸引了众多游客前来驻足。黄昏时分，门厅的视觉外观与日间恰好相反：门厅由内向外散发光线，吸引人们的注意力。照明系统的控制是通过一个嵌在接待台的触控屏来实现的。根据不同的需求，工作人员可以对设计师预设的灯光场景单独进行调整。

业主：
科隆大主教
使用人：
科隆大主教
建筑设计：
马提尼建筑设计事务所，波恩
LKL项目经理：
劳拉·苏德布罗克
伊莎贝尔·埃姆
竣工时间：
2010年
项目规模：
326平方米
整体项目预算：
60万欧元
照明预算：
10万欧元

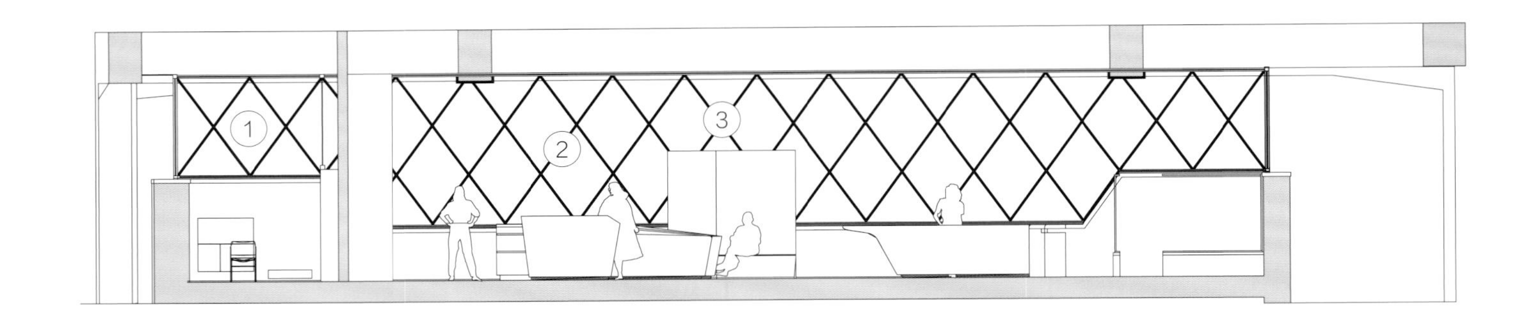

剖面图

Section

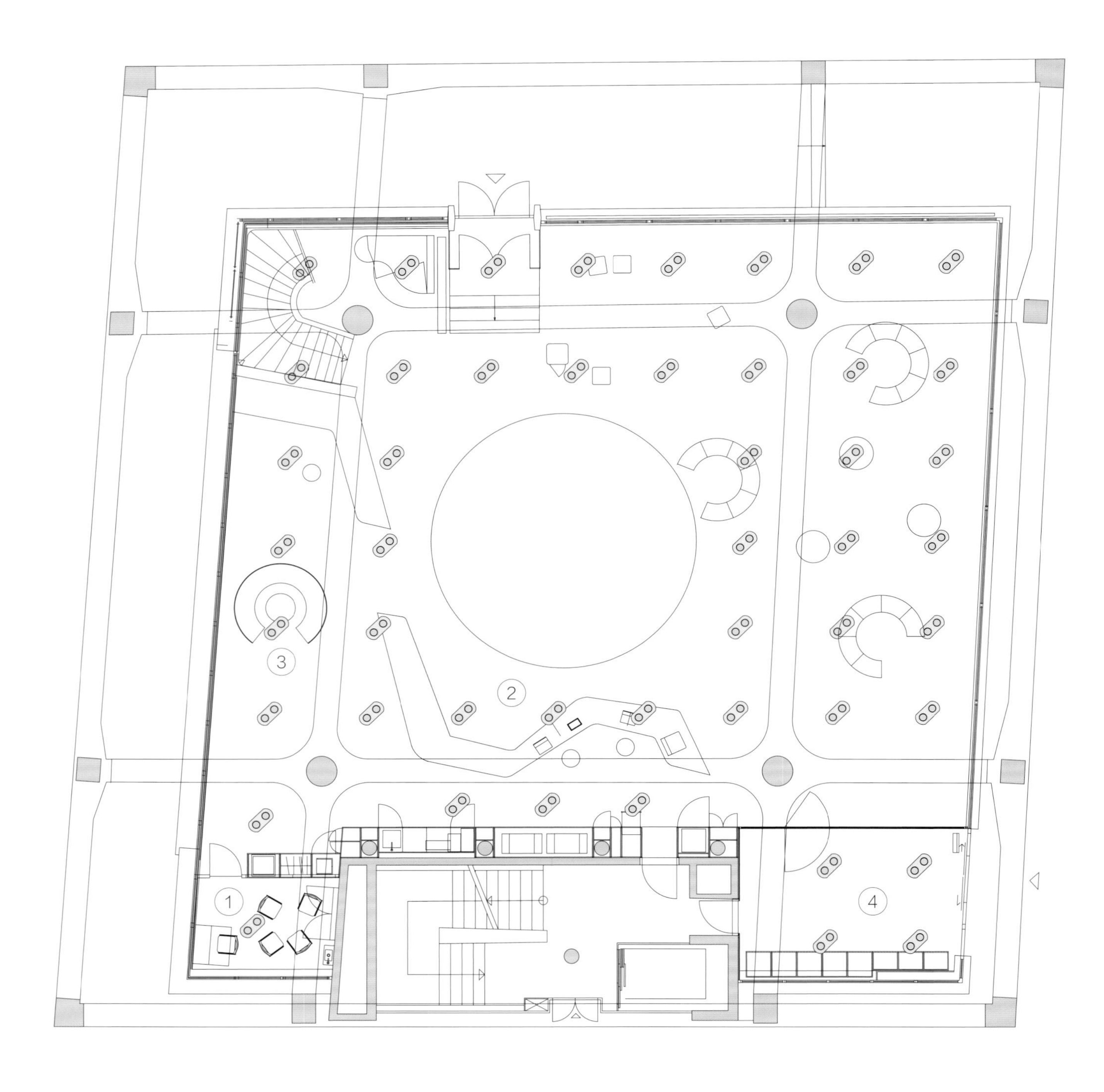

平面图

Floor plan

1.会议室、精神关怀室
2.问询处
3.有屏风的座位区
4.主入口

1. Meeting room / Pastoral care
2. Information desk
3. Screened seating area
4. Main entrance

The Domforum, acting as an information, meeting and event centre of the Catholic Church, is situated in a prominent location. Located opposite the Cologne Cathedral, it presents itself with a special quality of light and space after extensive remodelling and refurbishment by Martini Architekten. The designers at Licht Kunst Licht succeeded at creating a functional, yet atmospheric illumination for the new spaces.

300 years ago, the French abbot Suger of Saint-Denis, one of the first architectural theoreticians of the early Middle Ages professed: Light conveys faith!" This holds true today – but the conditions have become more complex and multifaceted. In a world overloaded with visual distraction, places of rest offering the opportunity for introspection and religious contemplation are not easily established. However, the Domforum provides a convincing example; by acting as the Cathedral's visitor centre, it is also the foyer of Germany's most visited church. It clearly shows that even today, it is feasible to add a contemporary expression to places of faith by means of interior and lighting design.

RECEPTION ROOM FOR THE CATHEDRAL – INTERFACE FOR ENCOUNTERS AND CONTEMPLATION

he Domforum Cologne welcomes pilgrims, organizes tours of the Cathedral, and gives information to visitors and groups. imultaneously, it offers spiritual care for individuals and groups. urthermore, the Domforum functions as a venue with a varied rogram, including both clerically and socially relevant topics, ncompassing a multitude of events and content (spiritual impulses nd encounters, panel discussions, religious education, and cultural ctivities).

RANSPARENCY AND OPENNESS

he building now housing the Domforum is a protected historical nonument, originally built between 1951 and 1954 for the Bank für emeinwirtschaft by Fritz Schaller (1904 – 2002). It was converted nto an information and meeting centre of the Catholic Church y his son Christian Schaller in 1995 / 1996. Due to an increasing hortage of space, new structural and building, services related equirements, and above all a desire for a variety of new uses, n architectural competition for the Domforum's restructuring as held, from which Martini Architekten emerged as the winner.

he aim was to create a publically accessible space with an tmosphere conveying openness and transparency. Another arget was a harmonious and respectful treatment of the existing historical building while taking into account the increased equirements for the Domforum. The architectural team of the onn based Martini office defined the main task in opening the paces towards the outside world without denying their religious filiation. Furthermore, the new interior concept was to underne the existing quality of the façade again. As a result, the new omforum has been conceived as a multifunctional meeting ace with a sacral background that becomes evident only upon oser inspection.

he three-sided fully glazed building façade allows for multiple stas within, outside, and across the room. It also connects a large interior space that is characterized by its natural luminosity with the exterior environment of the Cathedral Square in front. It is now possible to visually grasp the entire room at a glance. In spite of its height, the interior space appears to be wide rather than tall and is easily understandable. The design of the individual spatial installations was kept deliberately restrained; white is the prevailing colour. The grey reinforced concrete girders form a structuring element within the space. This creates the necessary neutral backdrop of an open, bright space that is to be stage, audience, and lounge simultaneously. By means of its natural wood surface, a stylized crucifix contrasts with the remaining wall cladding joint pattern behind the reception desk, thus reminding the visitor of the room's religious association.

MULTIFUNCTIONAL SPATIAL ORGANIZATION AND SEATING

The interior design encompasses three main elements: the reception desk, an appropriately dimensioned multifunctional wall cabinet, and a consultancy island. As a stationary piece of furniture, the new main desk is a free-standing object. Its curved shape is the main contact point for visitors of the Domforum; it is subdivided into an information and counselling section and an area where group tours are coordinated. Made of Corian it presents itself as a space-defining sculpture with a seamless surface.

Another characteristic element of the interior design is a central wall element behind the counter that includes various functions. The wall cabinet includes seat storage with a material lift, a kitchenette, a computer work station with copier, and a cloakroom. It also offers additional storage possibilities for brochures and literature. The wall element allows for an uncluttered foyer room and creates a new spatial zoning of the working area behind the counter. The existing passage to the stair well has been preserved and integrated with the wall cabinet design. Located on the forum's western side, behind the reception desk, is the counselling island. It creates a separate area for the foyer team's pastoral care, offering visual and acoustic privacy. The space can be fitted with seats for events with a frontal orientation or a decentralized configuration. The forum's daytime furnishings, curved seating islands, and the chairs for evening events alike, make use of walnut wood.

FUNCTIONAL ILLUMINATION – ATMOSPHERIC LIGHT

The idea of the lighting concept is based on the requirement of an extensive usage programme: Apart from temporary exhibitions, the foyer houses panel discussions, music and theatre events, as well as lectures. At the same time, the room's clear, formally reduced architectural language is to be emphasized. The task at hand was to introduce a lighting element that would fulfil both the general illumination and accent lighting, while responding to the existing structure of the space. Therefore, the floor's diagonal joint pattern, which is continued on the exterior Cathedral Square, is repeated in the light fixture layout.

The custom designed ceiling luminaire consists of a diffuse frosted acrylic cover, rimmed by a surrounding shadow gap, which forms an integral part of the ceiling. This cover is fitted with two cylindrical recesses which incorporate the adjustable downlights for the direct component. Thus, the directional LED spotlights remain concealed from most viewing angles. In order to realize various distinct lighting scenarios the two lighting components can be controlled and dimmed separately.

The Corian counter with its expressive formal language is the main space-defining design element and is strongly emphasized by the directional spotlights above. Due to these adjustable spotlights, the display furniture and the freely arranged seating islands can be flexibly arranged. The diffuse light guarantees a uniform illumination of the stage. As coordinated with the media designer, remote controlled adjustable LED-projectors are integrated within the ceiling, thus allowing events in this area to be accompanied by dynamic and coloured light.

During the day, an evenly illuminated, friendly space invites the visitor to stay and linger. At dusk, the exterior appearance is inverted as the foyer starts to emanate light from within, thus attracting attention. The control of the lighting system occurs through a touch panel which has been integrated with the main counter. Depending on the spatial requirements, the preset scenes can be individually tuned retroactively.

CLIENT:
Archbishopric Cologne
OCCUPANT:
Archbishopric Cologne
ARCHITECTS:
Martini Architekten, Bonn
TEAM LEADER LICHT KUNST LICHT:
Laura Sudbrock
Isabel Ehm
COMPLETION:
2010
PROJECT SIZE:
326 m^2
OVERALL BUILDING BUDGET:
0.6 million euros
LIGHTING BUDGET:
0.1 million euros

入口区域的玻璃幕墙反射了室内环境和光线。空间在视觉上仿佛有了延伸，人们可以感受其完整的深度。

Light and space are reflected in the entrance area's glass façade. The room appears to be extended and becomes tangible in its entire extent.

柔光照明组件位于灯具磨砂亚克力灯罩的后方，灯罩很好地嵌入天花中。除此之外，定制灯具还包含了可调式LED聚光灯组件，以提供强有力的直接照明。

The diffuse lighting component is located behind a frosted acrylic cover that blends with the ceiling slab. Adjustable LED spotlights are integrated to provide a direct light component.

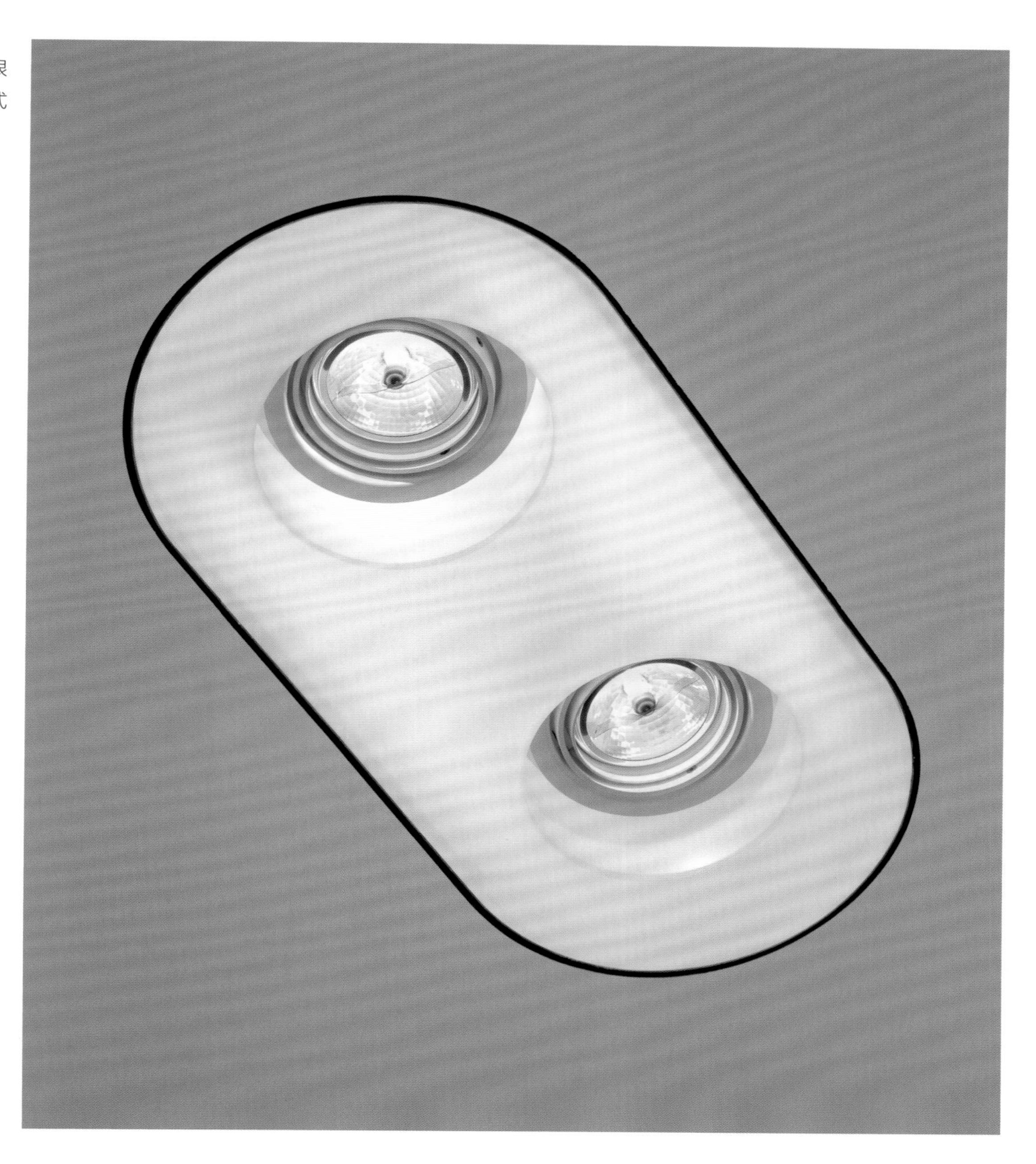

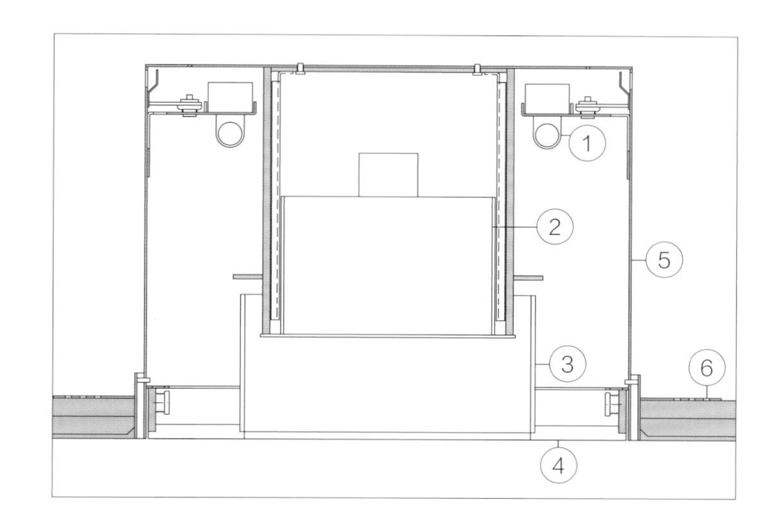

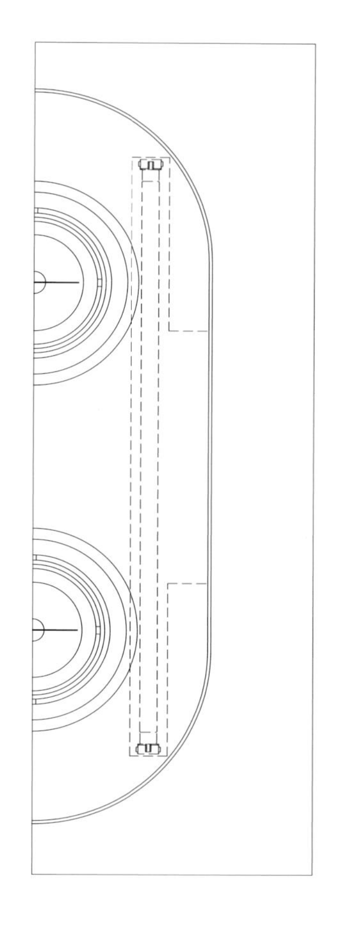

天花嵌入式灯具的细节剖面大样图：
1.24瓦T16荧光灯管
2.带万向环LED光源的照明模块
3.缎面亚克力圆柱
4.磨砂亚克力柔光灯罩
5.灯壳
6.石膏环

Detail section of the ceiling-recessed luminaire
1. 24 W T16 fluorescent lamp
2. Luminaire module with gimbal ring LED lamps
3. Satin acrylic cylinder
4. Frosted acrylic diffuser
5. Luminaire housing
6. Plaster ring

在雕塑般的接待台上方，定制照明的排列形式呼应了地板的铺装造型，这种铺装方式也延续到了相邻的室外大教堂广场。

The layout pattern of the bespoke luminaires in the sculptural reception area reflects the joint alignment of the flooring, which continues on the adjacent Domplatte (Cathedral Square).

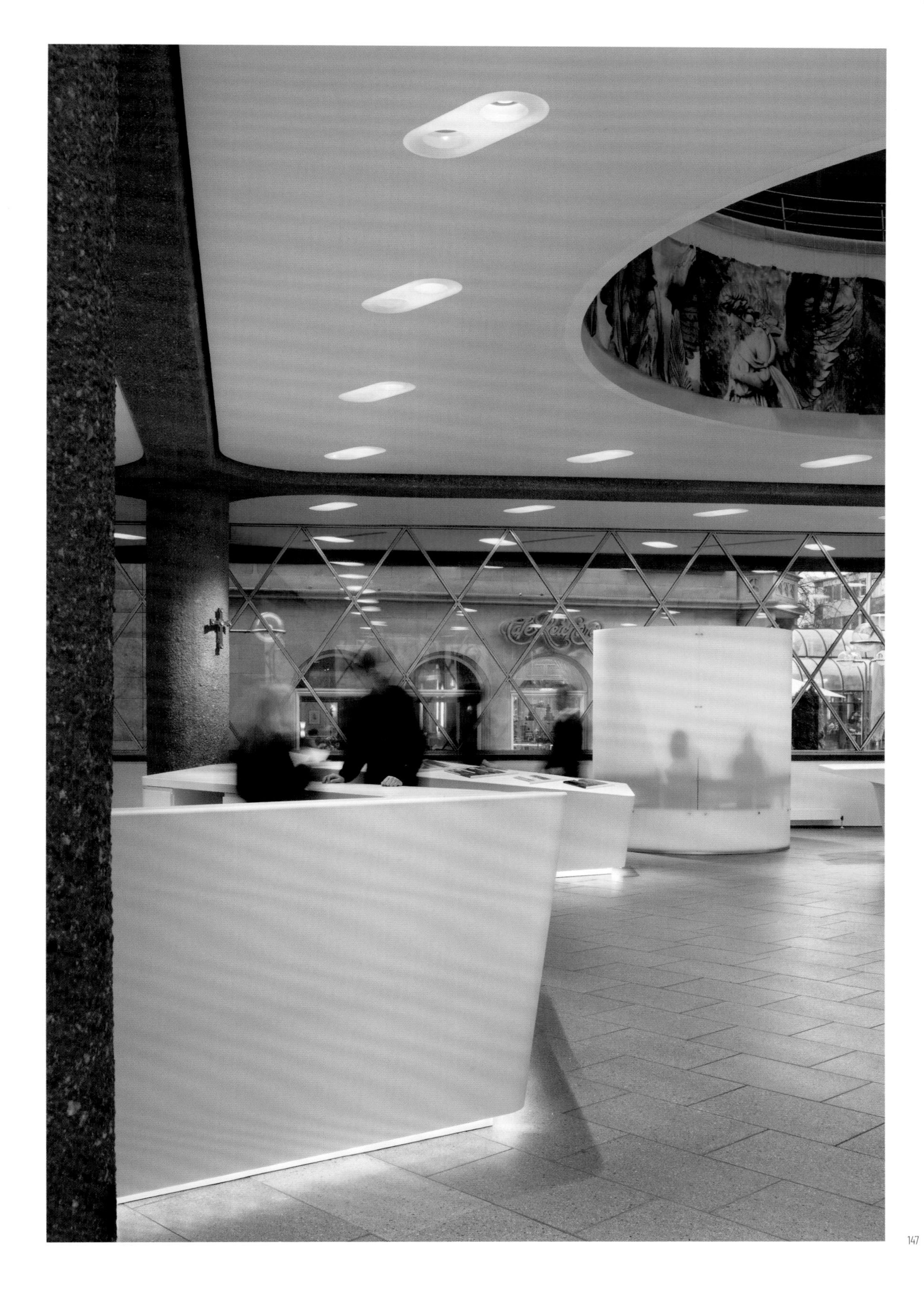

凭借着新设的建筑和照明理念，大教堂文化交流中心向路人展现出更加热情好客的姿态。

With its new architecture and lighting concept, the Domforum extends an inviting gesture to passers-by.

AWARD
WINNER

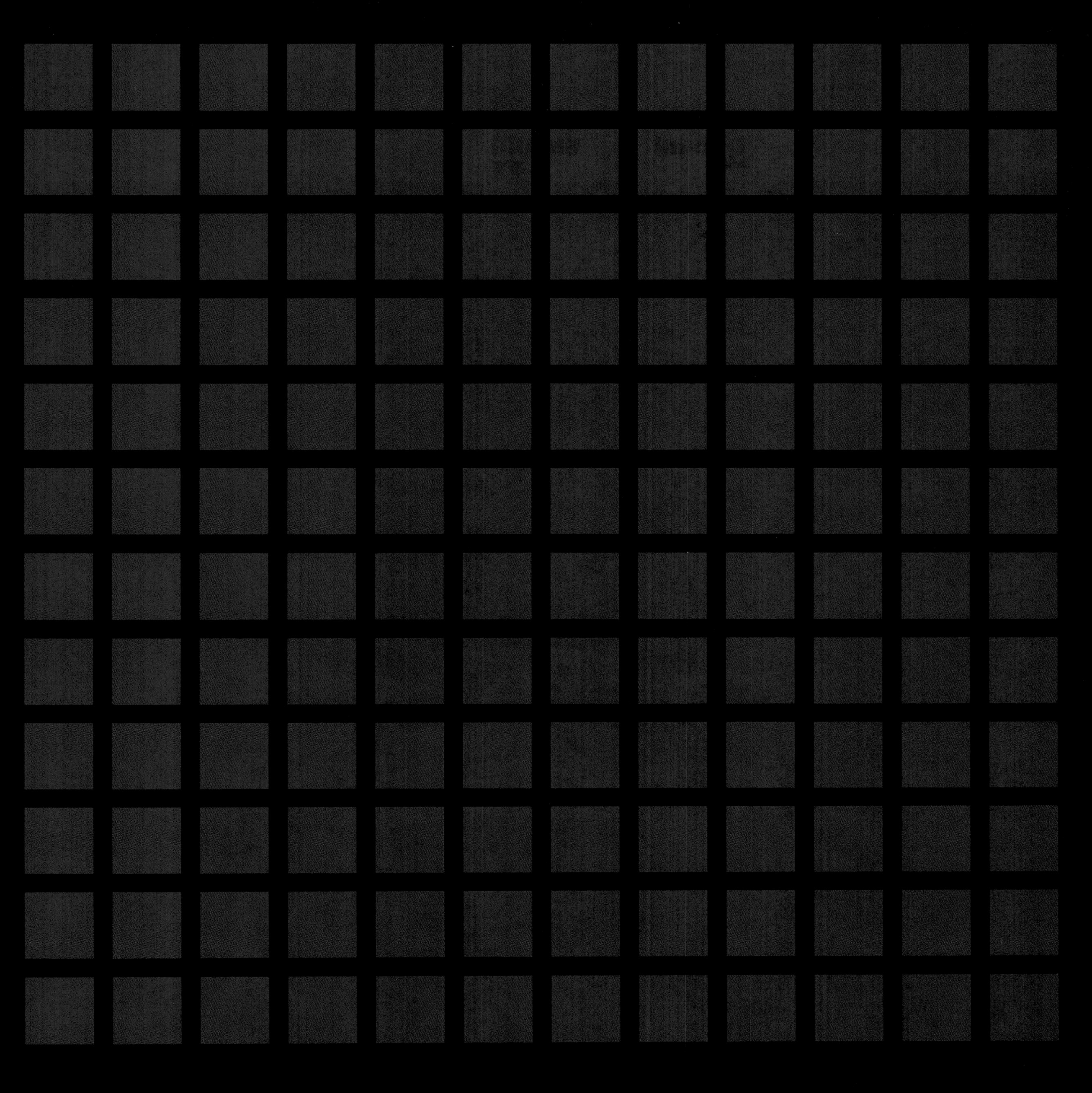

Abadia Retuerta LeDomaine酒店和水疗中心

西班牙萨尔东德杜埃罗

HOTEL & SPA ABADIA RETUERTA LEDOMAINE

SARDÓN DE DUERO, SPAIN

这座以罗马式和巴洛克式建筑风格建造的修道院，如今已经按照新的功能需求进行了翻新并重新投入使用，成为一家五星级酒店和水疗中心。自2015年7月竣工后，这座位于巴利亚多利德市附近，坐落在浪漫的杜埃罗河岸葡萄酒产区的建筑，已成为一处适合修养身心的著名场所。其中，精致的酒店设施是由来自瑞士的意大利建筑师马可·塞拉和主持室内设计的玛琳·德瑞共同设计的。地下水疗馆由瑞士迪纳与迪纳建筑设计事务所和室内设计师米歇尔·隆德利合作完成。项目的建筑总体规划设计为卢尔奇哈特建筑设计事务所。

从马德里市向北约两小时车程，毗邻著名的葡萄酒产区杜埃罗河岸，坐落着始建于12世纪的修道院和Abadia Retuerta LeDomaine酒店。经过长达10年的修建，这座建筑于2012年完成翻新，其中包含了一座五星级酒店，翻新后，酒店呈现出优雅而奢华的空间氛围。2015年7月水疗中心开业，与老建筑的地下空间相连通。建筑师塞拉有效地将原建筑中世纪严谨的建筑风格转化为属于21世纪的建筑形式语言，成功将这个罗马式修道院和酒庄的翻新工程打造为一个卓越的项目。通过简洁的线条、柔和的色彩以及对原有建筑细节的保留，设计师赋予了该建筑清新、和谐、现代化的外观。与此同时，人们仍然可以感知这个场所蕴含的近千年的历史气息，使人们沉浸在幸福愉悦的僻静世界。

在这座老普雷蒙特雷修会修道院的历史建筑中，建筑设计师设计了包括27间客房、3间套房、1间餐厅、1间酒吧、1间葡萄酒窖、1间带有壁炉的休息室、多间多功能会议室以及健身和瑜伽室的五星级酒店。通过被称为“Caballerizas”的酒店外扩空间，客人们可以进入全新的地下水疗中心。

室内设计的连贯性和由此产生的整体化空间结构均体现出业主和建筑设计师的高品质设计追求。对照明设计师来说，老修道院的建筑和空间设计仅允许稀疏的日光射入，这给他们带来了特殊的挑战。照明设计既要强有力地诠释这座古老建筑的复杂结构，又要利用灯光充分展现其历史背景。为了将增设的照明设备对受重点保护的古老建筑结构的干扰降至最低，LKL的设计团队特别研发定制了一套由坚实的青铜制成的系列灯具。

在老修道院餐厅用餐

僧侣们曾经共餐的地方现在被改造成高级餐厅。在保留原有空间比例和结构的同时，设计师根据新用途的要求对空间进行了改造。青铜落地灯分布在餐厅两侧，高度介于客人用餐平均高度和10米高的十字形拱顶之间。通过可单独开关控制的直接和间接光源，这些灯具向古老的石墙和拱形表面投射出温暖的光。墙上大型壁画《最后的晚餐》由一盏精心设置在檐口的聚光灯进行直接照明。

围绕葡萄酒的整体环境体验

葡萄酒吧为客人提供了品尝葡萄酒和用餐的场所。灯具采用的青铜材质与室内设计中的青铜元素巧妙融合在一起。酒吧中的一面墙由青铜金属板覆盖，其上带有一体化设计的青铜支架，用于存放葡萄酒瓶。酒吧下方是葡萄酒窖，在拱形穹顶和两侧的酒瓶架上均设有灯光进行重点照明。空间中的灯具造型经过特殊的设计，由一个弧度与酒窖拱顶半径相协调的青铜拱形支架和一个圆柱形青铜灯座组成。

客房内舒适的照明氛围

出于日常维护和设计的考虑，客房的照明设计要应用于所有标准房间。客房内的整体照明统一为较小的青铜壁灯搭配青铜落地灯和台灯。

历史设施在新设灯光的衬托下显得光彩耀人

老修道院走廊和回廊的照明方案遵循整体连贯的设计原则。兼具直接和间接照明的壁灯沿着细长的空间走向勾勒出有序的视觉节奏感，同时也照亮了回廊拱顶和浅色的石材地面。

老修道院里原先的教堂空间现在可以用来组织各种各样的活动。更新后的照明系统可以通过调取独立设置的灯光场景来灵活地响应不同的空间用途。近12米高的十字形拱顶由可调光的贴面安装式LED聚光灯轻柔地照亮。该设计平衡了室内亮度，同时确保参观者仍可以感知室内天花的高度。

沉浸在幸福美好的世界里

通过“Caballerizas”楼梯间，客人可来到地下空间。这条路径由兼具直接和间接照明分布的青铜壁灯照亮。从酒店大堂开始，人们就可以通过墙壁开口看到室内泳池。泳池正上方设有一个天窗，可以让充足的日光投射到地下空间。此外，无论是泳池区还是大堂，设计师都富有趣味性地布置了吊灯。吊灯的悬挂高度各异，仿佛在空中起舞。在泳池区域，尤其是在夜间，来自不同方向的光线反射和明暗变化会产生令人愉悦的光影效果。在泳池尽头，被水帘遮掩着的地方有一个非常窄的窗口通向接待区。线形轮廓光使水帘柔和地闪烁着光芒。

神奇的户外倒影

两个庭院中心均设有浅水池，也即倒影池。设计师刻意未使用灯光凸显这两个倒影池，意在使其能够生动地反射建筑、景观和周围环境的光线，在水面上显现出倒影。

北部庭院的立面通过地埋式上照灯进行修饰。院子里的树木在明亮的墙面对比下呈现出剪影效果，由此在夜间展示出自身的形态。

在宜人的氛围中漫步于广阔的天地间

总体来看，由于酒店和水疗中心位于乡间，稍有不慎就会造成光污染，因此室外区域仅布置了极少的人工光照明。作为主题灯光，柔和的光线从地面向上渗入建筑和立面开口。具有历史感的定制壁灯与传统灯笼的造型语汇相呼应，标记着建筑的主入口和侧入口。

同样是作为定制灯具系列的一部分来进行研发的短柱灯，顺着葡萄藤沿车道分布，一直延伸到停车场。这些短柱灯由带有菲涅尔透镜的标准光源搭配青铜材质的灯罩和石材底座构成。灯罩的4个开口中配有激光切割的精致装饰网格。在黑暗中，灯光为车道带来富有韵律感的亮光，同时也引导了方向。在酒店的庭院里，设计师采用了同一系列较小版本的短柱灯，沿着道路方向增添了地面照明。

水疗中心的室外倒影池仅由一侧的水下灯光照亮。这样的照明分布避免了光源直接暴露在躺椅一侧客人的视线范围内。由此一来，客人在躺椅上就可以不受干扰地欣赏水池另一侧边缘流水的婆娑光影。

使用人：

Abadia Retuerta LeDomaine酒店

建筑设计：

马可·塞拉建筑设计事务所，巴塞尔

迪纳与迪纳建筑设计事务所，巴塞尔

LKL项目经理：

玛蒂娜·维斯

竣工时间：

2015年

项目规模：

11 000平方米

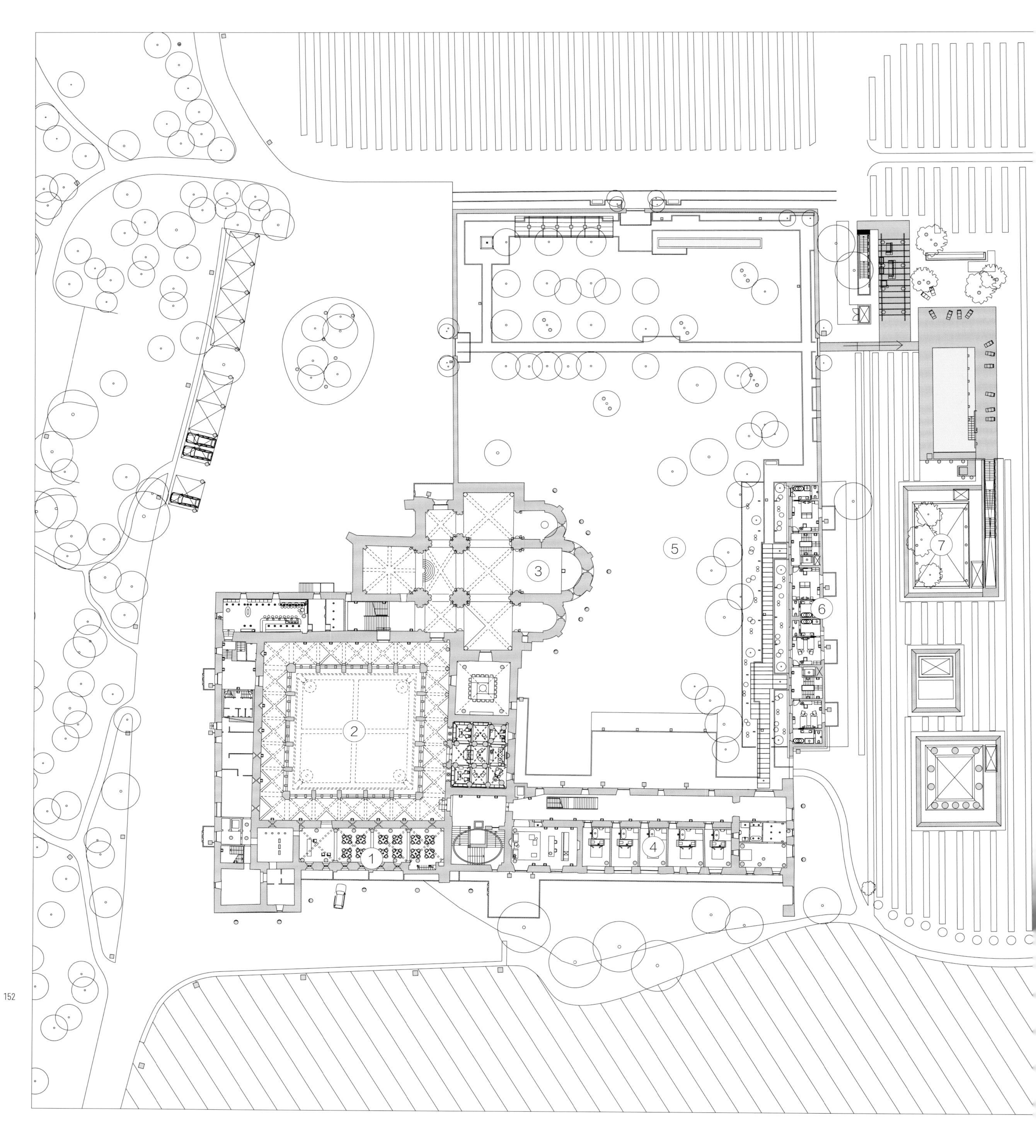

酒店总平面图

Hotel Abadia Retuerta LeDomaine site plan

1.餐厅
2.回廊庭院
3.教堂
4.客房
5.中庭
6. Caballerizas客房
7.水疗区

1. Refectory restaurant
2. Cloister courtyard
3. Church
4. Guestrooms
5. Courtyard
6. Caballerizas – guestrooms
7. Spa area

A former monastery, built with Romanesque and Baroque architecture, has been reintroduced with new purpose, as a Relais & Châteaux five-star hotel and spa. Completed in July 2015, it presents itself as an impressive venue for peace and contemplation in the midst of a romantic landscape in the wine region Ribera del Duero, near Valladolid. The Swiss-Italian architect Marco Serra designed the sophisticated hotel facilities together with interior designer Marlene Dörrie. The underground spa premises, however, were designed by Swiss architects Diener & Diener, in collaboration with interior designer Michele Rondelli. The architect of record was Burckhardt + Partner.

An approximate two-hour drive north of Madrid, adjacent to the famous wine region of Ribera del Duero, sits the former 12th century monastery and winery, Abadia Santa Maria de Retuerta. After a 10-year renovation process completed in 2012, the facility now houses the five-star hotel, 'Abadia Retuerta LeDomaine' in an elegant and luxurious ambience. In July 2015, the spa was opened, exclusively connecting to the old building structure underground. The architect Marco Serra was successful in making the renovations of the Romanesque convent building and winery into an exceptional project, in that he effectively translated the rigor of the medieval building into the formal language of the 21st century. With clean lines, muted colours, and preservation of architectural details, the building ensemble has been bestowed with a fresh and harmonious, modernized appearance. At the same time, the 1,000-year history of the location can still be sensed, thus plunging the guest into a secluded world of well-being and delight.

n the historical building of the former Premonstratensian Monastery, Serra has included the five-star hotel with 27 guest rooms, hree suites, a restaurant, a wine bar and wine cellar, a fireplace ounge, conference and function rooms, as well as fitness and oga rooms. Through the hotel's extension, the so-called 'Caballerzas,' guests enter the new, underground spa.

he high quality design standards of the clients and designers can e seen throughout in the consistency of the interior design and he resulting homogenized spatial structure. For the lighting deigners, the architecture and design of the old monastery building vith only sparse incident daylight, presented a special challenge. he lighting design was to impressively support the architectural omplexity of the prestigious building and at the same time use ghting to adequately represent its historical context. In order for ne lighting equipment installation to interfere only minimally with ne historic, landmark protected building structure, the team at icht Kunst Licht specially developed a diverse custom luminaire amily made of solid bronze.

INING IN THE FORMER REFECTORY

here the monks once broke bread, now houses a restaurant of ne dining. While preserving its original spatial proportions and tructure, the room has been equipped according to the requireents of its new use. Bronze floor luminaires line the sides and nediate between the level of the guest and the ten-meter-high royne vaults. From their separately switchable direct / indirect ght sources, the fixtures emanate warm-toned light onto the ancient stone walls and arched surfaces. The large-scale fresco, which depicts the Last Supper, is illuminated directly with a discreetly placed spotlight on the cornice.

HOLISTIC EXPERIENCE REVOLVING AROUND WINE

The wine bar offers the guests a chance to taste wines and dine. The bronze material used in the luminaires was cleverly introduced into the interior design here. One of the walls is lined in bronze sheet metal panels with integral bronze brackets, holding wine bottles. Underneath the wine bar lies the wine cellar, where the barrel vault ceiling and wine bottle shelves are featured with light. The specially designed fixture is made of an arched bronze bracket adapted to the radius of the vault, and a cylindrical bronze socket.

COMFORTABLE LIGHTING ATMOSPHERE IN THE GUEST ROOMS

For maintenance and design reasons, the lighting concept for the guest rooms was to be applied to all standard rooms. The general lighting in the room is generated from a smaller version of the bronze wall sconce, as well as from bronze floor and table luminaires.

HISTORIC FACILITIES SHINE IN A NEW LIGHT

The lighting of the long monastic corridors and the cloister follows a consistent principle. Wall luminaires with direct / indirect light distribution track and provide rhythm to the elongated spaces, while illuminating the cloister vault and the light-coloured stone floors.

The former church of the monastery now accommodates various events. Its lighting system can respond flexibly to the different uses with individually programmed light scenes that can be called up. The nearly 12-meter-tall groyne vault is gently illuminated from the capitals with dimmable surface-mounted LED spotlights. This creates a well-balanced luminance ratio in the room, while allowing the visitor to still experience the height of the ceiling.

IMMERSION IN A WORLD OF WELLBEING

Through the 'Caballerizas' stairwell, the hotel guest enters the basement. Its path is illuminated with bronze wall-mounted direct / indirect fixtures. Already from the lobby, one has a view into the indoor pool through an opening in the wall. Directly above the pool is a skylight opening that allows an abundant flow of vital daylight into the underground space. In addition, both the pool area and the lobby are decorated with playfully arranged pendant lights, which dance from varying suspension heights. In the pool area, especially at night, a delightful play of light occurs from reflections and shadows. At the end of the pool, a very narrow window to the reception is obscured by a waterfall. Linear light profiles make the water curtain softly shimmer.

MAGICAL REFLECTIONS OUTDOORS

At the heart of both patios there are shallow basins, the reflecting pools. These are deliberately not accented with light, so as to allow them to vividly reflect the architecture, landscaping, and surrounding light.

The façades of the northern patio are displayed from grade recessed uplights. The trees in the courtyard emerge as silhouettes against the bright walls and therefore have a presence of their own in the night.

STROLL IN THE OPEN AIR IN A PLEASANT ATMOSPHERE

In general, little lighting is added to the exterior areas, as the Hotel and Spa's rural location creates a high sensitivity to light pollution. As a theme, soft light penetrates through building and terrain openings from below. Historic wall lights, custom-designed to reflect the formal language of classical lanterns, mark the main and side entrances of the building.

Bollard lights, also developed as part of the custom light family, trace the driveway along the vines to the affiliated parking lot. They are comprised of a standard fixture with a Fresnel lens, which has received a bronze hood and a stone pedestal. The four openings in the hood are fitted with a delicate, lasered ornament grid. The light brings rhythm to the path in the dark while also providing orientation. In the courtyard of the hotel, a small version of the bollard is repeated, adding ground-level lighting along the walkways.

The outdoor pool of the spa is illuminated from one side with underwater lights. These are arranged so that the light sources cannot be seen from the pool chairs. Only the reflections from moving water on the opposite edge of the pool can be observed from the chairs.

OCCUPANT:
Hotel Abadia Retuerta LeDomaine
ARCHITECTS:
Marco Serra Architekt, Basel
Diener & Diener Architekten, Basel
TEAM LEADER LICHT KUNST LICHT:
Martina Weiss
COMPLETION:
2015
PROJECT SIZE:
11,000 m^2

这座罗马式修道院位于著名的杜埃罗河岸的葡萄酒产区，与其附属酒庄如今已被改建为一家典雅的罗莱夏朵五星级酒店和水疗中心。

Located in the famous wine-growing region Ribera del Duero, the Romanesque monastery building with its affiliated wine estate has been converted into an elegant Relais & Châteaux five-star hotel and spa.

考虑到周边的乡野环境，项目采用了非常克制的夜间照明手法。灯光仅柔和地投射在老修道院的外墙上。具有历史感的高杆灯和较低的短柱灯诗意地强调了车道和小径。

Taking the site's rural setting into consideration, the nocturnal illumination has been designed with great restraint. Only the historical monastery walls have been softly illuminated. Historic style pole luminaires and low-level light objects poetically emphasize the paths.

青铜支架上摆放的葡萄酒瓶在墙面上构成了舒缓解压的视觉图案。固定在横梁云台上的聚光灯来自青铜灯具系列，作用是重点强调这些葡萄酒瓶。

A pattern of wine bottles on bronze brackets creates a relief on the wall. They are accentuated by gimbal-mounted spotlights from the bronze luminaire family.

]萄酒窖的灯光由定制灯具提供，灯具的外观造型由弧形
青铜支架和圆柱形灯座构成。灯具光源采用漫反射光分布
]碳丝灯泡，让人联想到原始的壁挂式烛台照明。

The wine cellar is illuminated by custom luminaires, consisting of curved bronze brackets and a cylindrical socket. The luminaire design with its omnidirectional carbon filament lamp is reminiscent of the original wall-mounted candle vault illumination.

客人通过壮观的巴洛克风格楼梯进入客房。楼梯由侧面的青铜壁灯照明，楼梯上方复古精致的灰泥装饰穹顶由凹槽灯提供均匀的间接照明。

The guestrooms are reached via an imposing baroque staircase, that is flanked by bronze wall sconces. The elaborate stucco-adorned oval dome above the stair is indirectly and evenly illuminated by a luminous cove.

曾经的圣器收藏室中，设计师在天花檐口安装了环形
ED凹槽灯，通过偏配光的光线分布照亮上方装饰丰富
穹棱拱顶。现在这里是会议室，由落地灯在视觉上调
较高的空间与参会者之间的比例关系。

In the former Sacristy, the framing cornice incorporates a ribbon of LED profiles that illuminate the abundantly ornamented groyne vault by means of their asymmetric light distribution. Additional floor luminaires mediate between the guests and the monumental scale of this tall space, which now serves as a meeting room.

曾经的牧师会礼堂现在是访客休息室和酒吧。带有半透明纹理的装饰性落地灯和台灯创造出不同的光区，营造了适当的私密氛围。

The former chapter house (Sala Capitular) is now used as a visitor lounge and bar. Decorative floor and table luminaires with opal textile screens create luminous zones for an appropriately intimate atmosphere.

为该项目定制设计的青铜灯具组成了一个家具系列。床的旁边是铰链式铜制阅读灯，客人可以躺着对其单独调节和控制。

The bronze luminaires, specifically designed for the project, form a design family. Adjacent to the hotel beds are hinged bronze reading lights that can be individually adjusted and controlled from a lying position.

老修道院的立面由地埋式上照灯柔和地照亮。钟楼在柔和的灯光下脱颖而出。此外，室内柔和的光线从建筑的开窗中射出，展示出好客的姿态。

The monastery's facade is gently and uniformly illuminated via grade-recessed uplights. The bell tower is emphasized by a slightly stronger brightness. Additionally, soft light emerges from the building apertures and extends an inviting gesture.

过去的避难所、洗礼池和后殿均通过安装在立柱上方檐口处的聚光灯来重点强调。间接照明灯具上的遮光鼻和直接照明灯具上的蜂窝遮光罩有效防止了眩光。

The sanctuary, baptistery, and the apses are all accentuated by means of accent lights mounted on the capital ledges. Snoots for the indirect fixtures and honeycomb louvers for the direct sources prevent glare.

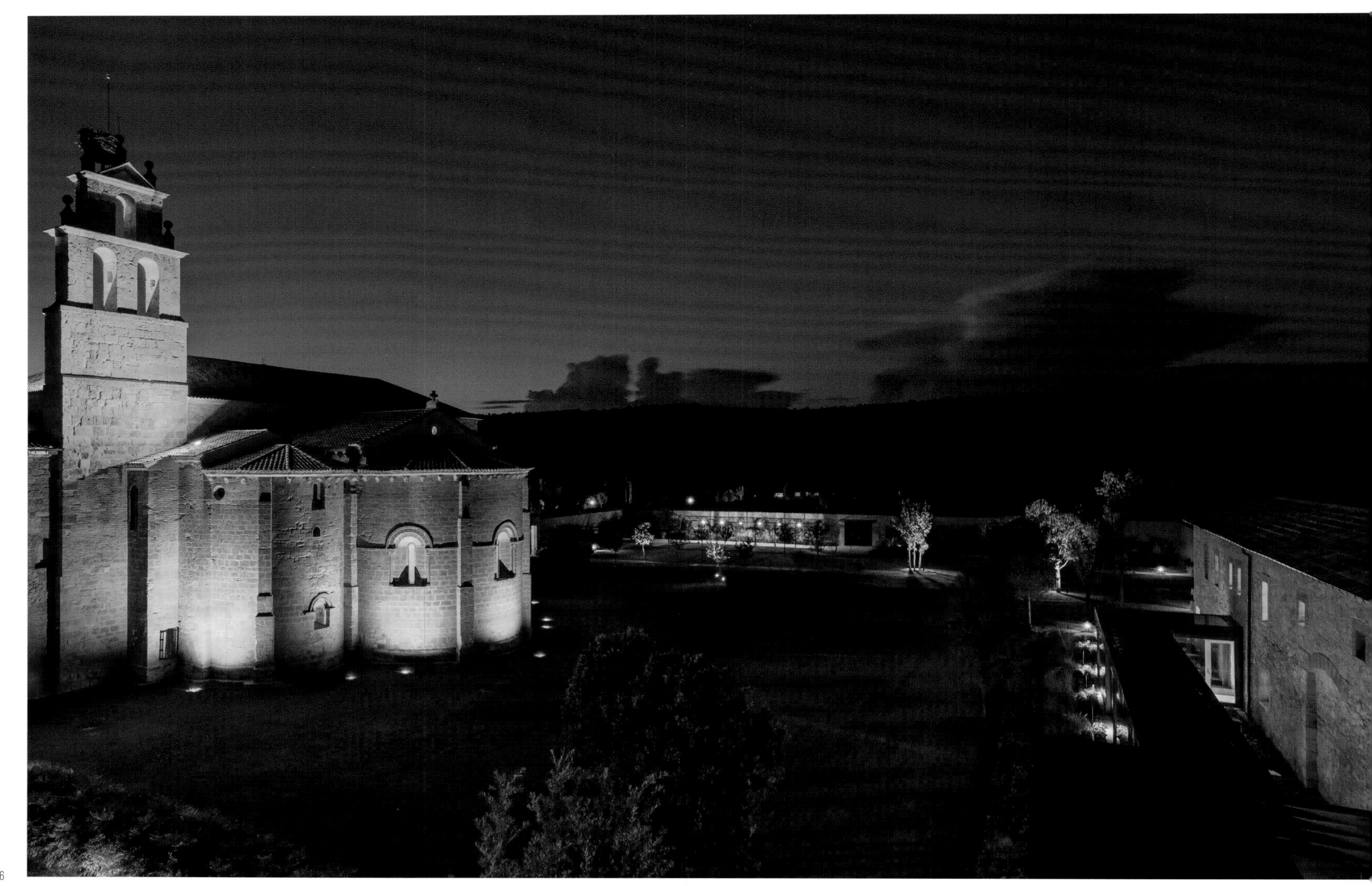

为了补充老修道院有限的立面照明，设计师精心挑选了一些树木和树篱，由地插式灯具进行照明。

Complementing the restrained facade illumination of the monastery, stake mounted luminaires orchestrate selected trees and hedges.

在连接老修道院和楼梯间的空间中，各个连接点都由铜制天花灯具进行重点照明。看起来像自然植物的小型杆式照明装置在为户外绿植区域提供照明的同时，也透过宽大的落地玻璃在走廊中营造出斑驳的光影效果。

In the connecting building between the former monastery and the Caballerizas, the junctions are emphasized by directional light from bronze ceiling luminaires. Small rod light fittings, reminiscent of plants, illuminate the outdoor vegetation and create zoned light in the corridors through the generous facade glazing.

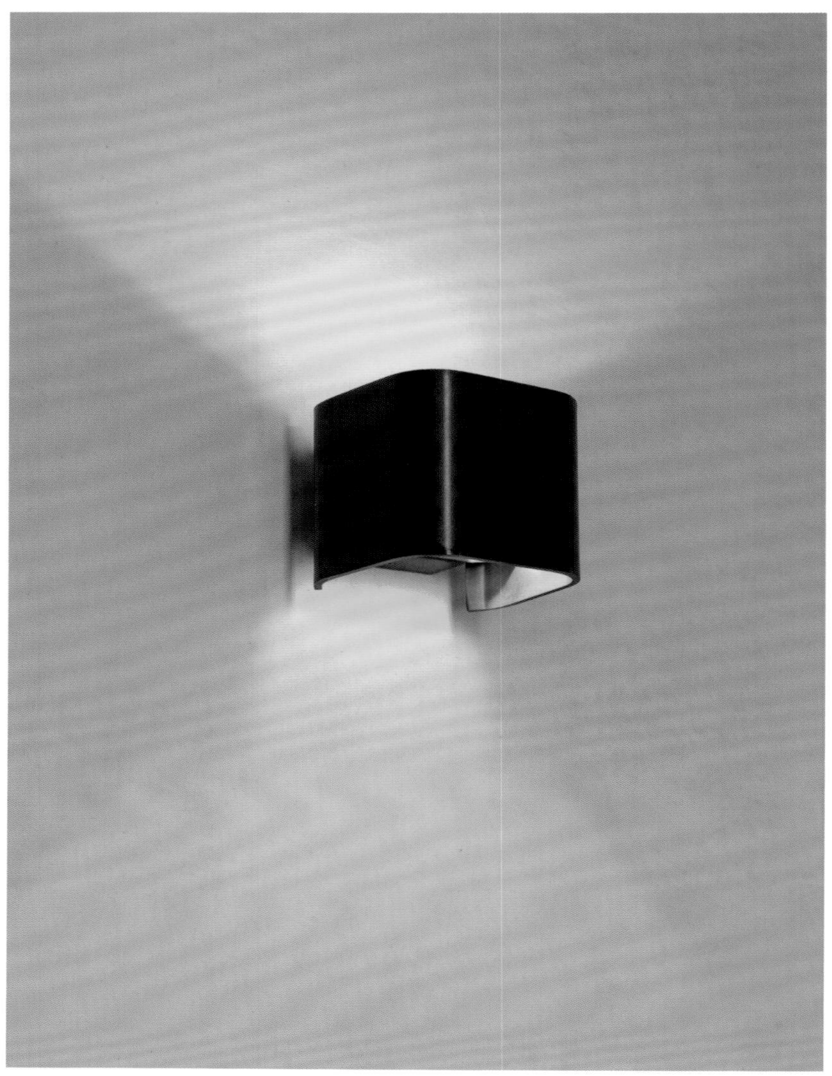

定制系列灯具由德国巴伐利亚州的一家青铜厂生产，灯具的形状和质感均受到老修道院氛围的启发，也彰显了现代设计风格，并将现代化的LED照明技术应用其中。灯具底部精致的青铜网眼为空间增添了温暖的金色色调，同时也将光源隐藏了起来。

Produced by a Bavarian bronze factory, the shape and feel of the luminaires were inspired by the monastical environment, yet reveal through their design the contemporary origins and modern LED lighting technology employed. A delicate bronze mesh at the bottom side lends a warm golden hue to the light while also concealing the light source.

落地灯、台灯和壁灯共同组成了这个灯具系列。这些灯具的灯罩均由四个青铜扇片组成。经过抛光的灯罩内表面将金色的光线反射到空间中。

Forming a luminaire family, the wall sconces are supplemented by floor and table light fittings. Their screens are composed of four fanned bronze shields. Their polished interior surfaces reflect the golden light into the space.

在庭院中心，深色倒影池如镜面一样，将周围的建筑和绿植的景象倒映其中。在南部庭院中，四周的树篱包裹着浅水池，在夜间被地插式灯具温柔地照亮。

In the heart of the patios, dark reflecting pools serve as a mirror for the surrounding architecture and vegetation. At the southern patio, a hedge engulfs the water basin and is softly illuminated by stake-mounted fixtures at night.

璃吊灯悬挂在室内泳池上方，既作为泳池的基础照明，
照亮了接待区。吊灯的悬挂高度不一，为空间增添了
亮的光影反射效果。

Pendant glass luminaires have been designed for the indoor pool, but also illuminate the reception area. They are playfully arranged at varying heights and add brilliant luminous reflections to the space.

特别是在夜间，玻璃吊灯与泳池水面的互动产生了迷人的光影效果。池壁上内嵌了水下灯具，让水中充盈着光亮。

Particularly at night, the interaction of pendant glass luminaires and the pool's water surface creates enticing effects of light and shadow. Recessed in the pool walls are submersible fittings that lustrously fill the water with light.

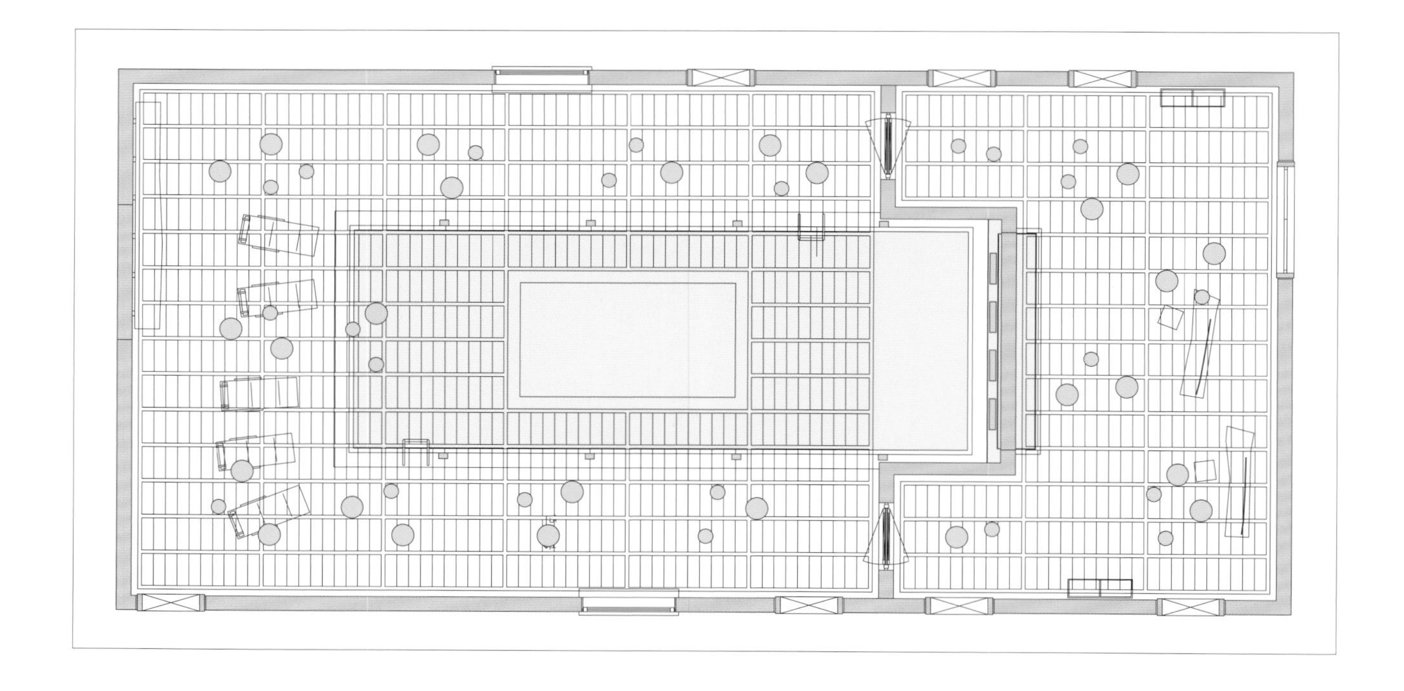

带反向天花的室内泳池平面图

Floor plan with reflected ceiling plan of indoor pool

定制灯具的灯体由吹制玻璃和抛光青铜材质制成。

The body of the specifically designed custom luminaires is made of blown glass and burnished bronze.

AWARD
WINNER

LICHT KUNST LICHT 4

深圳湾壹号 深圳

ONE SHENZHEN BAY SHENZHEN

深圳湾海岸线的灯光旋律

从商业潜力的增长、工程项目的井喷，到国际巨头的青睐，深圳这座承载着中国自1978年改革开放以来快速发展成果的城市，聚焦了来自全球各地投资者的目光。2018年底，在这片交织着光荣与梦想的湾畔上，深圳湾壹号这座涵盖8座综合性塔楼，同时配套了多样化商业裙楼与精致景观花园的城市综合体终于落成。从项目的筹划至落成，近十年时间的打磨使其成为深圳的地标性建筑群。

城市天际线中的新亮点

直到1980年，深圳的总人口仅有约3万，周围被稻田和牡蛎养殖场包围着。随后，与香港毗邻的深圳被宣布规划为经济特区，深圳的繁荣时代自此开始。如今，这个超级大城拥有超过1700万居民，而且正如不断更迭变化的城市天际线，仍在继续发展与壮大。

这座欣欣向荣的城市综合体位于深圳迅速崛起的后海区，项目两侧分别是繁华喧嚣的市区与宁静的深圳湾。深圳湾壹号伫立在这两种截然不同的城市环境之间，绝佳的地理位置条件为其提供了成为两者“接口”的机会。这个沿海的大型地产项目包含2座办公和酒店塔楼、6座公寓塔楼、7栋豪华别墅、1座商业裙楼以及文化与运动场馆。不同于大多数城市综合体，该项目所有的塔楼都向外扩散，直至场地边缘，在地块中央为使用者提供了极具私密性的公共和私人花园。这些塔楼通过具有不同视觉效果与空间层级的裙楼基本设施连接，包括人行通道、休闲长廊、花园和水景等。

环境友好型建筑

尽管该项目的体量庞大，功能复杂，但这并不影响照明设计师对照明质量的严格把控，以及对使用者身心舒适度的高度重视。早在规划项目整体朝向和各座塔楼的位置前，科恩 · 皮德森 · 福克斯建筑设计事务所的建筑师们就特别重视深圳湾壹号未来使用者的视域。一方面，规划设计需要保证从室内欣赏到深圳湾海岸以及周围山峦的景色，并突出场地与水域、绿地之间的关联。另一方面，私家花园、泳池以及大型露台的设计也鼓励使用者亲近自然。建筑运行的高能效性体现在对太阳能的有效收集、用水量的控制设计和绿色屋顶设计等方面。与此同时，塔楼的朝向和排列也有利于遮阳和自然通风，从而将空调的运行成本降至最低。该项目从设计之初就旨在达到LEED金级认证的严格要求，同时获得了美国绿色建筑WELL金级认证以及中国绿色建筑三星级设计认证。

采用“少即是多”的照明设计理念

在深圳湾鳞次栉比、五光十色的高科技商业建筑群中，建筑照明的密度有增无减，但即便如此，照明设计师仍然在本项目大部分的建筑立面上采用了“少即是多”的照明理念。塔楼的高度和朝向各不相同，照明设计师采用整体连续的照明方式，希望通过有选择性的灯光线条来凸显建筑简洁但充满力量的存在。水平方向的建筑幕墙线条灯光恰到好处地彰显出建筑群有凝聚力的身份象征，照明设计师在突出每一个建筑体块特征的同时，采用较为节制的照明表达方式，避免对幕墙过度照明。照明设计师精心挑选了部分建筑元素，将纤细水平方向的灯光线条与塔楼拐角、裙楼及景观体块无缝衔接，勾勒出精巧而细腻的建筑轮廓。在建筑裙楼部分，连续的水平线条灯光勾勒出裙楼的边界，而在公寓塔楼中，灯光仅对部分露台的边缘形态进行突出。直接照明的LED线条灯具内嵌在建筑幕墙指定的露台凹槽中，与石材包边的表面齐平，隐蔽式的灯具不会干扰建筑的日间视觉效果。主楼立面上的照明设计也沿用了相似的设计手法，通过更为紧密的灯具排布，使主楼在建筑群中脱颖而出。远而观之，整个项目富有韵律感的光线分布组合方式仿佛温柔的海浪，在深圳的天际线中呈现出文雅精致的姿态。

在近人尺度上，设计师有意减少了私家花园区域的灯光元素，以营造出宁静祥和的光环境氛围。雕塑和景观小品被光线稍加强调，人行道区域由扶手嵌入式灯光勾勒其轮廓。这些视觉层级中的照明元素均以静态的白光呈现，与上方塔楼的光线效果形成和谐的视觉对比。

注重细节

照明设计团队十分重视项目使用者的居住体验，因此一直强调采用防止眩光的照明解决方案，以确保最大化塔楼中用户的视觉舒适度。对灯具安装细节的严格把控，确保了建筑外立面的灯光无论从安装位置还是发光强度上均不会对附近的居民造成干扰。为了避免眩光以及溢散光进入室内空间，照明设计师不断与项目设计团队讨论研究，通过视觉样板模型来验证灯具在幕墙立面上的安装细节和排布方式。这样的深入协作旨在更好地明确灯具的安装位置，同时简化灯具的安装方式。此外，幕墙石材外侧略微倾斜的遮光石板结构，可以实现精准的截光角度，同时确保灯具的隐蔽安装。所有灯具均采用RGB+白光的LED光源，最大限度地提高能源效率和项目整体的运行能力。

面朝城区和海岸的媒体立面

业主要求在高达341米的T7塔楼的完整东西建筑立面上设置媒体立面，这给照明设计师提出了严峻的挑战。灯具的安装位置仅限于楼层与楼层之间的玻璃幕墙空间，以保证室内用户的视野不受阻碍。在这种情况下，楼层之间的横向灯具有3.5米高的垂直空间间隔，如此之大的间隔很容易导致媒体立面上的内容无法被清晰地表达出来。因此，照明设计师模拟了多种观看距离，并针对人们不同的视觉感知和媒体立面图像分辨率进行研究，得出了较高的灯具像素分辨率可以对较大的垂直像素间距进行补偿的结论。最终，每个立面都实现了312×1996像素的多媒体矩阵。与此同时，对灯具精准的像素控制使游客可以从千米之外欣赏到壮丽的图像，且塔楼内的用户也可以不受干扰地欣赏户外的美景，完全不会注意到自己所处空间的窗户上方与下方存在灯具。

T7媒体立面演示内容的概念是与当地环境相呼应的，并与城市文脉形成互动。通过安放在塔顶的精密传感器系统，当地实时的风向、风速、湿度、温度以及潮汐状态都可以被感知和记录下来，并进行诗意多彩的媒体可视化表达。项目首次媒体立面内容由光说故事（北京）文化创意有限公司构思制作。四个灯光场景寓意着自然界形态的多样性、生命的活力以及时刻发生着的变化。此外，灯具的控制硬件设备能够让媒体立面在特殊节日和活动中呈现主题性内容表达，甚至可以与深圳湾区域其他楼宇的媒体立面进行联动。

借由灯光表达的企业形象

T7塔楼较窄的南北建筑立面以大型水平建筑遮阳板为特征。从远处看，它们可以被识别成一个整体的图形，与该项目开发商的企业标识相似。夜晚，窄光束角的上照灯有效地突出了建筑元素的底面，让建筑翼角变为一个发光的符号，从附近的香港过境点清晰可见。

致敬湾区时代的照明设计

各个塔楼上的水平灯条共同奏响了光线的旋律。在近距离观看时，用户可以获得无眩光的舒适视觉体验，而整体观赏时，则可以感受到其呈现出的强烈的象征意义。从更广的视角眺望，深圳湾壹号项目的媒体立面和动态灯光元素与这座年轻、充满活力的城市气质相呼应，而抽象、雅致的媒体立面内容也凸显出该项目在深圳湾岸边的建筑形象与气质。

业主:
深圳市鹏瑞地产开发有限公司
建筑设计:
科恩 · 皮德森 · 福克斯建筑设计事务所，纽约/香港
LKL项目负责人:
卢卡斯 · 金
LKL团队成员:
江怡辰、斯蒂芬 · 锡勒
竣工时间:
2018年
项目规模:
460 000平方米
整体项目预算:
6.3亿欧元
照明预算:
758万欧元（不含室内装饰性照明）

T7塔楼立面图 Elevations T7 Tower

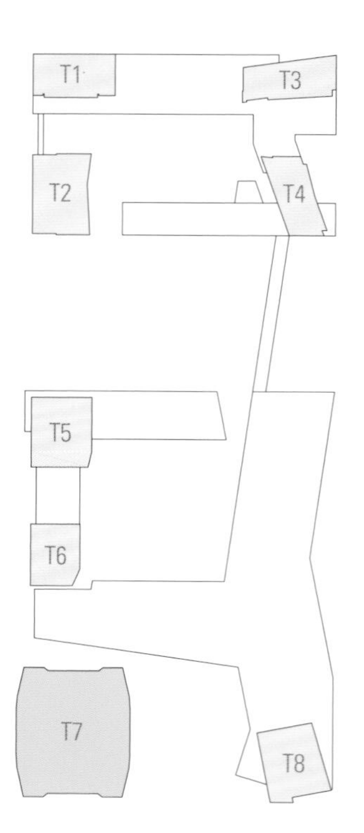

场地平面图 Site plan

From an explosion in construction and commercial potential, to attracting significant international investments, Shenzhen has gained international attention due to its rapid development since China's 1979 Reform and Opening Up policy. Furthering this momentum, One Shenzhen Bay held its grand opening at the end of 2018. The project consists of eight towers rising above a diverse podium, in which sophisticated landscape gardens are integrated. Spanning nearly ten years from the beginning of project planning to the final close-out, One Shenzhen Bay has become a landmark of the city.

NEW HIGHLIGHT IN THE SKYLINE

As late as 1980, Shenzhen had a population of approximately 30,000 inhabitants and was surrounded by rice fields and oyster farms. Then, the region across the border from Hong Kong was declared a Special Economic Zone and the boom began. Today, the megacity has over 17 million inhabitants and, as the ever-changing skyline proves, continues to grow.

The exceptional plot of land within the Houhai district provides One Shenzhen Bay the opportunity to serve as an "interface" between the bustling city and tranquil Shenzhen bay. The ambitious real estate development near the coast includes two office/hotel towers, six residential towers, seven exclusive villas, a retail arcade, as well as sport and cultural facilities. Pushed to the project's perimeters, the unique layout of the architecture creates intimate public and private parks within the center of the plot. The towers are linked by a remarkably versatile podium infrastructure, with walkways, promenades, gardens, and water features.

ENVIRONMENTALLY FRIENDLY ARCHITECTURE

Despite the project's large scale and mixed-use program, great attention was given to quality and the well-being of the individual user. When planning the orientation and alignment of the eight towers, the architects at Kohn Pedersen Fox gave emphasis to the view corridors towards Shenzhen Bay and the surrounding mountains, emphasizing the site's relationship to nature. The private gardens, swimming pools, and large balconies encourage a lifestyle that is lived both inside and out. Building operation efficiency is reflected in the project's solar collectors, water use reduction, and green roof designs, while the towers' orientation and sectional stacking promotes shading and natural ventilation, minimizing air conditioning costs. The entire project has been designed to achieve the stringent requirements of the LEED Gold rating, and has achieved precertification of WELL Gold rating, as well as Three Star Green Building Design Label in China.

COURAGE TO APPLY "LESS IS MORE"

In a high-tech megacity where the density of light is ever increasing, a "Less is More" concept was intentionally pursued for the majority of the facades. With various tower heights and orientations, a consistent lighting approach was envisioned to utilize a condensed quantity of light lines to provide a simplified, yet powerful presence. Horizontal light lines construct a cohesive identity by revealing each architectural mass without overpowering the facades themselves. Architectural elements carefully chosen for the integration of lighting house thin horizontal light lines outlining the angled towers, podium, and landscape volumes. For the podium area, they run continuously around the entire perimeter, while only the outer edges of alternating balconies are traced with light on the residential towers. Direct view linear LED profiles were selected and integrated into the balcony reveals, aligned to the surface of the stone cladding as to not be visible during the day. The feature tower boasts similar horizontal light lines on its facade; differentiating itself with a much higher density of products. From afar, this grouping of rhythmic arrangements combines like gentle waves into a sophisticated distinction along Shenzhen's skyline.

At a human scale, lighting elements were intentionally reduced in the private garden areas to promote a calm and tranquil atmosphere. Sculptures and landscape elements are gently highlighted, while walkways are delineated by handrail recessed lighting. All lighting at these levels remain static and white, establishing a peaceful contrast from the tower lighting above.

ATTENTION TO DETAIL

Following the design team's focus on the user's well-being, anti-glare solutions have been emphasized to ensure maximum visual comfort for the inhabitants of the development's residential towers. Mounting details ensure neither the location nor the intensity of the exterior lighting products would disturb residents. To achieve this, the mounting detail and arrangement at facing facades were constantly modeled and studied. This process was used to determine the lighting locations and further helped to refine installation details. A slight inclination of the stone facade panels further serves to provide sufficient cut-off angles to shield the luminaires from view. All luminaires were specified with RGB+White LED light sources, maximizing both energy efficiency and operation capabilities.

MEDIA FACADES ON BOTH CITY AND SEASIDE

The client request to integrate media facades over the entire height of the east and west facades of the 341-meter-high T7 tower presented a particular challenge. The installation locations were limited to the space in between the glazing of the floors to allow unobstructed views to the outside. This large spacing of 3.5 m between horizontal luminaires had the potential to negatively affect the readability of the media facades. Various perception and image resolution studies at a number of viewing distances were conducted, which concluded that a higher pixel density could compensate for the large vertical pixel pitch. As a result, a media matrix with a 312 × 1,996 pixel density was realized for each facade. Precise pixel control allows to display spectacular graphics visible from many miles away, while occupants of the towers enjoy vast views to the exterior, unaware of the luminaires mounted directly above and below their windows.

The content of the T7 media facade was conceptualized to echo with local environment and to interact with urban context. Through the use of sophisticated sensors on its rooftop, data of the real-time wind, humidity, temperature, and oceanic tides were recorded and transferred to the poetic and colourful media visualizations. The first content of the media facade is a dynamic installation conceived by Studio Lighting Stories in Beijing. Four light scenes stand as metaphors for the diversity, vitality and constant change in nature. Furthermore, the control hardware enables themed content for special occasions and even the ability to synchronize One Shenzhen Bay's facade lighting with other nearby media facades.

IDENTITY FEATURED WITH LIGHT

The narrower north and south facades of T7 tower are characterized by large horizontal sun-shading fins. From a distance, these can be read as a graphic structure, which resembles the development's logo. The undersides of these architectural elements are effectively highlighted with a narrow beam uplight, which turn the fins into a glowing symbol at night, distinctly visible from the nearby Hong Kong border crossing.

LIGHTING PAYS TRIBUTE TO THE AURA OF THE BAY

Together, the horizontal light lines across the various towers form a rhythmic ensemble of light which remains unobtrusive at close proximities, yet strongly symbolic when viewed collectively. At a broader scale, the media facade and dynamic lighting components provide One Shenzhen Bay the ability to relate to the active atmosphere of the city, while the purposefully gentle content is designed to reinforce its identity on the banks of Shenzhen Bay.

CLIENT:
Shenzhen Parkland Real Estate Development Co., Ltd
ARCHITECTS:
Kohn Pedersen Fox Associates, New York / Hong Kong
TEAM LEADER LICHT KUNST LICHT:
Lucas King
PROJECT TEAM LICHT KUNST LICHT:
Yichen Jiang
Stephan Thiele
COMPLETION:
2018
PROJECT SIZE:
460,000 m^2
OVERALL BUILDING BUDGET:
630 million euros
LIGHTING BUDGET:
7.58 million euros excluding interior decorative lighting

两侧建筑立面都实现了312×1996像素的多媒体矩阵，精准的像素控制确保了在建筑立面上可以呈现出壮观精美的图像。

A media matrix with a 312 × 1,996 pixel resolution was realized for each facade. Precise pixel control allows to display spectacular graphics.

水平条形灯与精心挑选的建筑元素完美融合，在夜间重新定义了塔楼拐角、裙楼和景观元素。

Horizontal light lines are carefully integrated into select architectural elements, giving definition to the angled towers, podiums, and landscape volumes.

这些塔楼由功能丰富的裙楼基本设施连接，人行通道、休闲长廊、花园和水景在各个目的地之间起连接作用。

The towers are linked by a remarkably versatile podium infrastructure, with walkways, promenades, gardens, and water features mediating between destinations.

水平方向的灯光线条勾勒出每个建筑体块的特征，同时不影响建筑立面本身的表达，打造建筑群极具凝聚力的夜间形象。

The horizontal light lines construct a cohesive identity by revealing each architectural mass without overpowering the facades themselves.

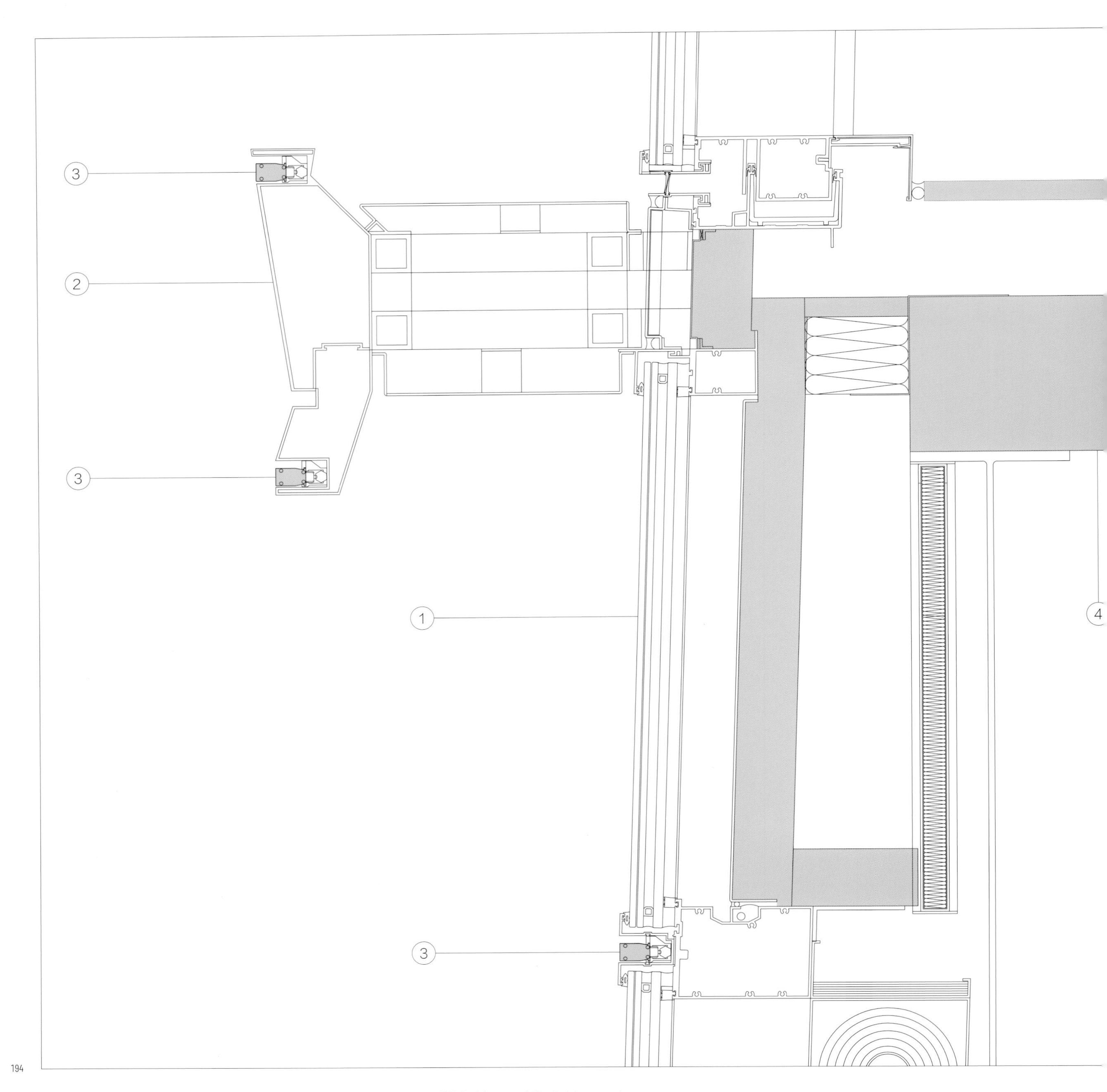

T7塔楼建筑翼灯光剖面图

T7 Architectural fin lighting section

1.玻璃
2.建筑翼角
3.条形 RGB+白光灯具
4.层板

1. Glazing
2. Architectural fin
3. RGBW light line
4. Floor slab

在深圳湾壹号塔楼样板的现场，深圳楼宇顶部的样子。

Impressions above the roofs of Shenzhen during a mockup at the tower of One Shenzhen Bay.

水平方向的条形灯的现场模型。

On-site mockup of the horizontal light lines.

落客区雨篷吊顶天花上采用背光照明，穿孔铝板的图形组合模拟了邻近的深圳湾海浪图案。落客区的灯光气氛与酒店宴会厅的装饰性吊灯相得益彰，呈现出欢迎宾客的极佳姿态。

The backlit perforations in the drop-off ceiling are patterned to mimic the nearby waves of Shenzhen Bay. Coupled with the decorative chandeliers in the hotel and banquet foyers, the drop-off area forms a brilliant ensemble to welcome guests.

穿项目各种建筑体块和景观特征的水平条形灯共同形
了有韵律感的灯光合奏，近距离观看时并不引人注
，但当整体观赏时，则可以感受到强烈的象征意义。

Together, the horizontal light lines across the various architectural volumes and landscape features form a rhythmic ensemble of light which remains unobtrusive at close proximities, yet strongly symbolic when viewed collectively.

AWARD
WINNER

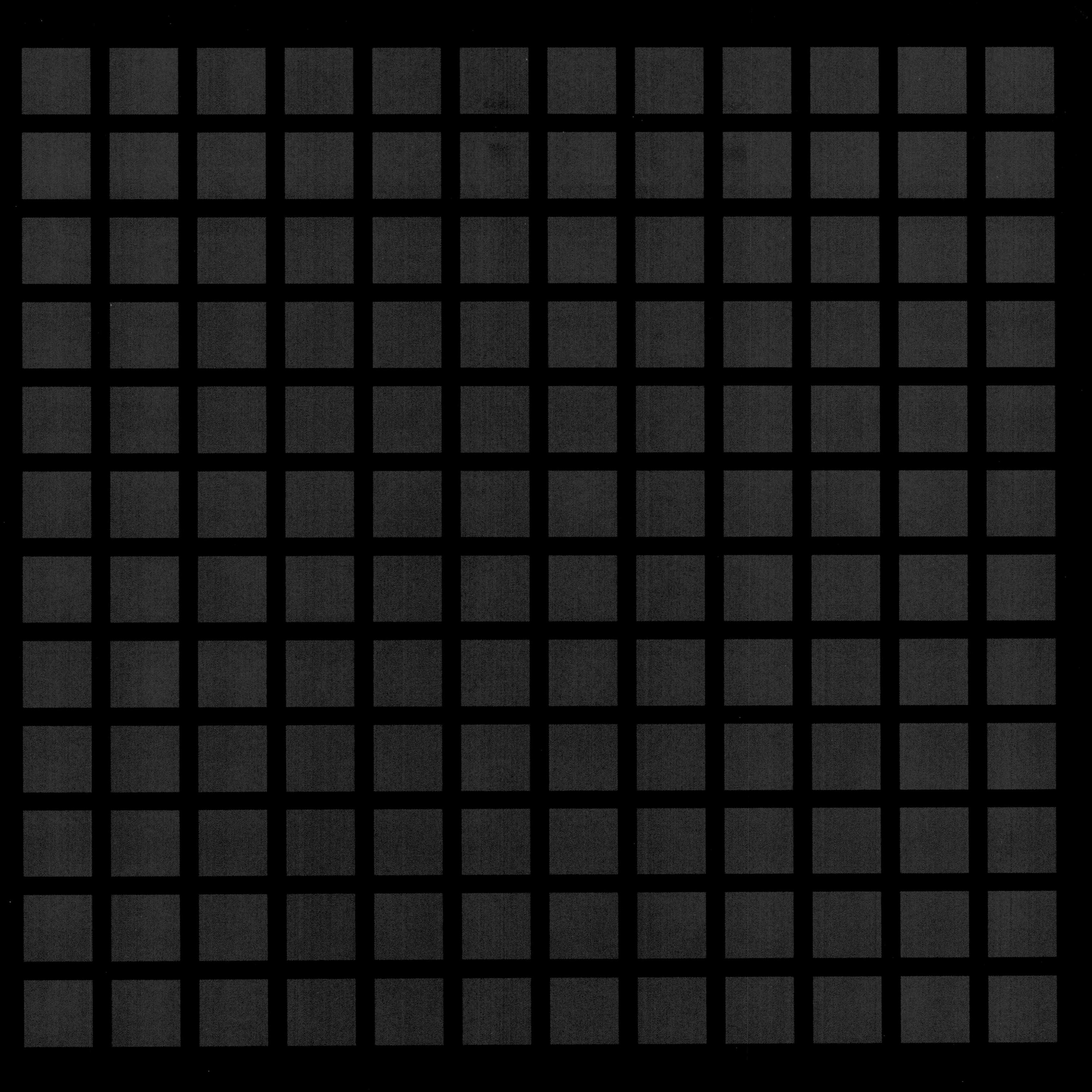

LICHT KUNST LICHT 4

巴伐利亚国王博物馆 霍恩施万高

MUSEUM OF THE BAVARIAN KINGS HOHENSCHWANGAU

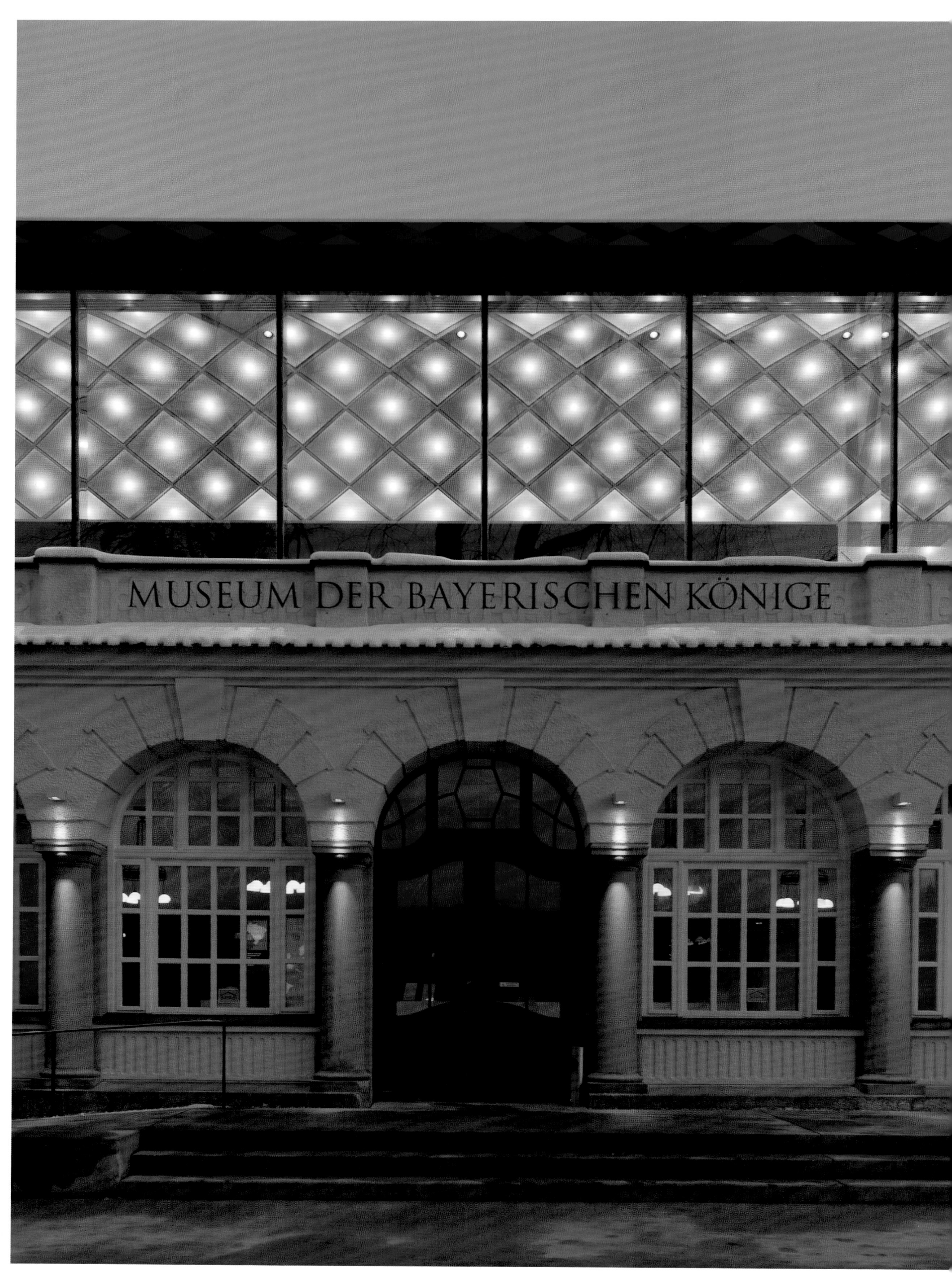
MUSEUM DER BAYERISCHEN KÖNIGE

展现维特尔斯巴赫王朝的历史

巴伐利亚国王博物馆展现了维特尔斯巴赫王朝至今的发展历程，旨在让参观者对这一家族的历史有所了解。展厅约为1000平方米，设置在新旧交叠的错层空间中十分醒目的位置。针对这一博物馆，LKL拟定了一个能够满足多种功能需求的照明方案，以令人印象深刻的方式凸显展品，同时强化建筑本身的形象标识。

巴伐利亚国王博物馆的所在地正是之前的阿尔卑斯玫瑰大酒店。这是一座拥有100多年历史的综合体建筑，位于阿尔卑斯湖畔，新天鹅堡和旧天鹅堡脚下。建筑设计师精心处理了这个历史悠久且被列入保护名录的建筑结构，并大胆地增添了覆盖三个长廊的穹顶，并将其打造为博物馆的展陈空间，其独具一格的空间特征让参观者耳目一新。

以光辉四溢的门厅来迎接宾客

参观者可通过位于建筑中部的入口大厅进入博物馆，该大厅之前是阿尔卑斯玫瑰大酒店的餐厅。白天，充足的日光会透过大厅的大窗户洒入室内。浅色的水磨石地板、方格天花板和白色的墙壁营造出明亮的空间印象。傍晚时分，大厅内光辉四溢的灯光气氛吸引着访客进入参观。为了达到这样的视觉效果，照明设计师升级了目前仍在使用的老式吊灯，将光源换成卤素灯泡。从中央大堂的楼梯向上走可通往二层的永久性展厅，该展厅设立在新建的带有三个长廊的穹窿之下。

优雅、静谧而又耀眼夺目的光效

对于施塔布建筑设计事务所来说，通过在原餐厅位置之上加建一层来进行博物馆的扩建设计，如何在人字墙之间建造一个支撑性结构是非常关键的，如此才可以改善门厅结构元素缺失的现状。这一设计目标最终通过建造三个带有钢结构的穹顶来实现，这些穹顶仅在四个支撑点处予以支撑。建筑方案为构建三个大型无柱空间，它们与立面走势相平行，其形式与带有两个侧廊的中殿结构相似。

中央展览空间被策展人定义为“皇家城堡大厅”。展览以大型摄影作品和华丽的展品为主，专门围绕着旧天鹅堡和新天鹅堡这两座皇家城堡及其建造者国王马克西米利安二世和国王路德维希二世展开。建筑一体化的照明解决方案将整个穹顶设计为发光天花，为展览营造出恰到好处的灯光氛围。

经过一系列的模型实验和样板模拟测试后，照明设计师们研发了一套照明解决方案，既可以保留菱形的方格钢结构，又能使天花完成面保持简洁，且不受技术设备的干扰。为了实现这一效果，设计师将带有LED背光照明的曲面亚克力玻璃面板安装到每个天花镶板内。当LED光源设备安装于钢拱架之间的每一格菱形镶板中时，会在亚克力玻璃盖板上呈现出微妙的光晕。此外，设计师挑选了部分天花面板，在其面板后方设置了投光灯，通过定向的重点照明来突出核心展品。

面向城堡、山脉和湖畔的开阔视野

发光天花的设计也被使用在另外两个较窄拱顶的侧廊空间中。在这两个空间里，发光天花仅从立墙一侧的天花结构开始，延伸至穹顶的顶部——因为从穹顶顺着外侧幕墙方向延伸至地面，一面巨大的玻璃幕墙为参观者提供了绝美的观赏视角，可将周边壮丽的风景尽收眼底。两个侧廊的宽窄各异，其中靠近建筑后方的侧廊较宽，而朝向旧天鹅堡和室外湖畔的侧廊较窄。但无论空间是宽还是窄，它们所展现的绝世美景总能让人流连忘返。

宗谱展示

在宗谱展室中，后墙的夹层玻璃是展示维特尔斯巴赫宗谱的陈列墙。为了清晰地呈现玻璃上的图像和文字，照明设计师通过在窗格玻璃底部边缘的槽道中嵌入线形LED擦墙灯，营造出均匀的背光照明。窗格之间还有一层日光防护的特殊材料，为LED照明提供额外的反射面，进而使光线更加柔和均匀。

较小展厅中的信息传递和互动

针对原有建筑的多个较小空间，设计师通过鲜明的对比方案，使参观者对展陈信息和展品产生更加直观和生动的印象。在展厅中，设计师在整体深棕色可丽耐材质的展陈家具环境中插入了背光照明的大面幅照片、明亮的陈列柜以及交互式屏幕和投影装置。为了衬托深色的空间背景和丰富展陈色彩之间的内在联系，环境照明以低调、质朴的灯光予以回应。窗户上的遮光屏对入射日光进行限制，以满足室内环境照明的最低功能性需求。

在这样的场景中，整体照明的光源应保持隐蔽。因此，照明设计师在展厅之间的通道过梁上少量嵌入了设计严谨的灯具。每一盏灯具均由两种照明光源构成：向地面投射定向光的LED光源，以及可以发出强烈漫射光的日光灯。操作人员可分别对这两种光源进行开关和调光控制，并且根据展览的需求调节光输出量。在非营业时间，日光灯管还可以用作清洁灯和安保灯。展厅中唯一没有完全隐藏在建筑细节中的灯具，是位于天花单独点位的可调式投光灯。当展览需要强调额外的场景，或展柜一体化照明无法满足观展需要时，这些投光灯就可以派上用场。

为沿用至今的楼梯量身定制灯具

在这座古老的建筑中，展陈区域的顶层和底层由老楼梯相连。对于该区域的楼梯和走廊，照明设计师采用无论从光线效果还是造型设计上都能够与老建筑的历史产生共鸣的灯具来进行照明。由坚实青铜制成的、包括直接和间接照明元素的定制壁灯很好地满足了上述需求。借助这些灯具，设计师不仅可以为通行区域提供充足的照明，并且还能在墙壁表面添加掠射光效，同时还能够为较长的走廊空间营造舒适的视觉感受。

具有远观效果的“场刊”

巴伐利亚国王博物馆的展览方案是由巴伐利亚历史博物馆制定的。受托的管理委员会主席、巴伐利亚档案馆前馆长赫尔曼·朗姆舍特尔教授在国王博物馆开幕典礼上对其功能进行了说明。他做了一个恰当的比喻：“……用戏剧的语言来说：该博物馆首先应被视为关于旧天鹅堡和新天鹅堡这两个作品及其创作者的说明性‘场刊’。”

他的比喻恰如其分，并且该建筑具有象征意义，因为这一“场刊”在参观者游览这两座城堡的过程中十分显眼。从新、旧天鹅堡两个视角中均可以看到国王博物馆的新建拱形屋顶。建筑设计师十分清楚该建筑物可以达到的视觉效果，并将屋顶视为该建筑的“第五立面”，因此在屋顶拱形结构上覆盖了金属板，使之在色彩视觉效果上与现存的砖瓦屋顶相匹配。此外，博物馆的照明方案并不仅仅考虑了室内的视觉效果。通过国王博物馆走廊的落地窗，照明设计师将带有矩阵状光点的发光天花呈现在室外游客的视线中，给人留下惊艳的视觉印象。作为一大设计亮点，它们为这片充满浪漫色彩的土地锦上添花，并向所有参观皇家城堡的游客“发出邀请”，欢迎他们进一步了解欧洲最古老的王朝之一——维特尔斯巴赫王朝的历史。

业主：
维特尔斯巴赫补偿基金会
使用人：
利德尔城堡酒店
建筑设计：
施塔布建筑设计事务所，柏林
LKL项目经理：
马耳他·西蒙
竣工时间：
2011年
项目规模：
1000平方米
照明预算：
70万欧元

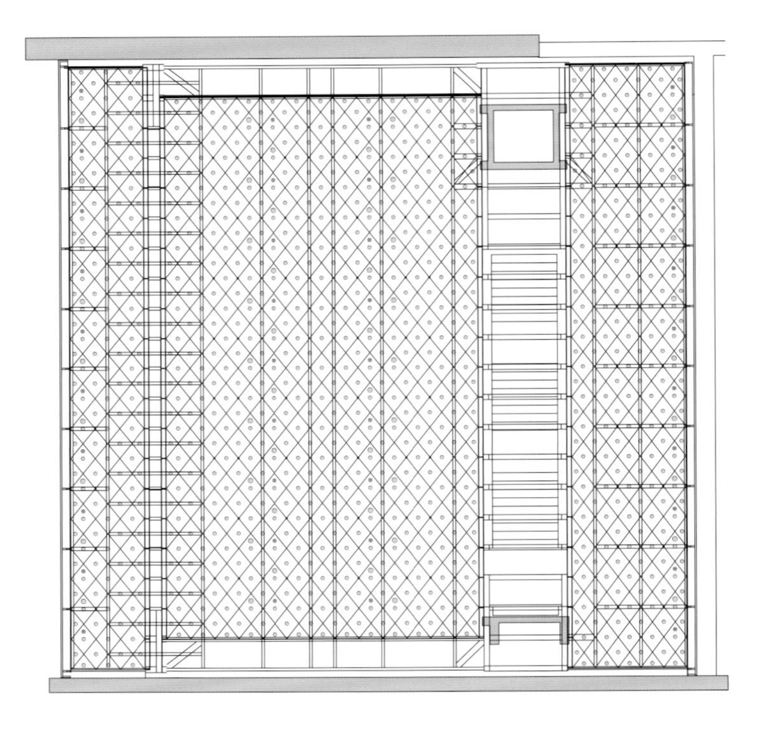

三个拱形穹隆的反向天花平面图

Reflected ceiling plan of the three barrel vaults

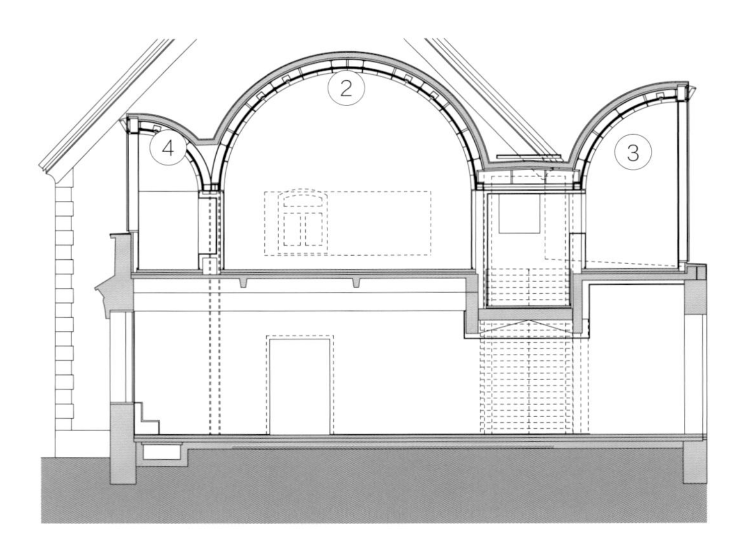

皇家城堡大厅剖面图

Section through Hall of the Royal Castles

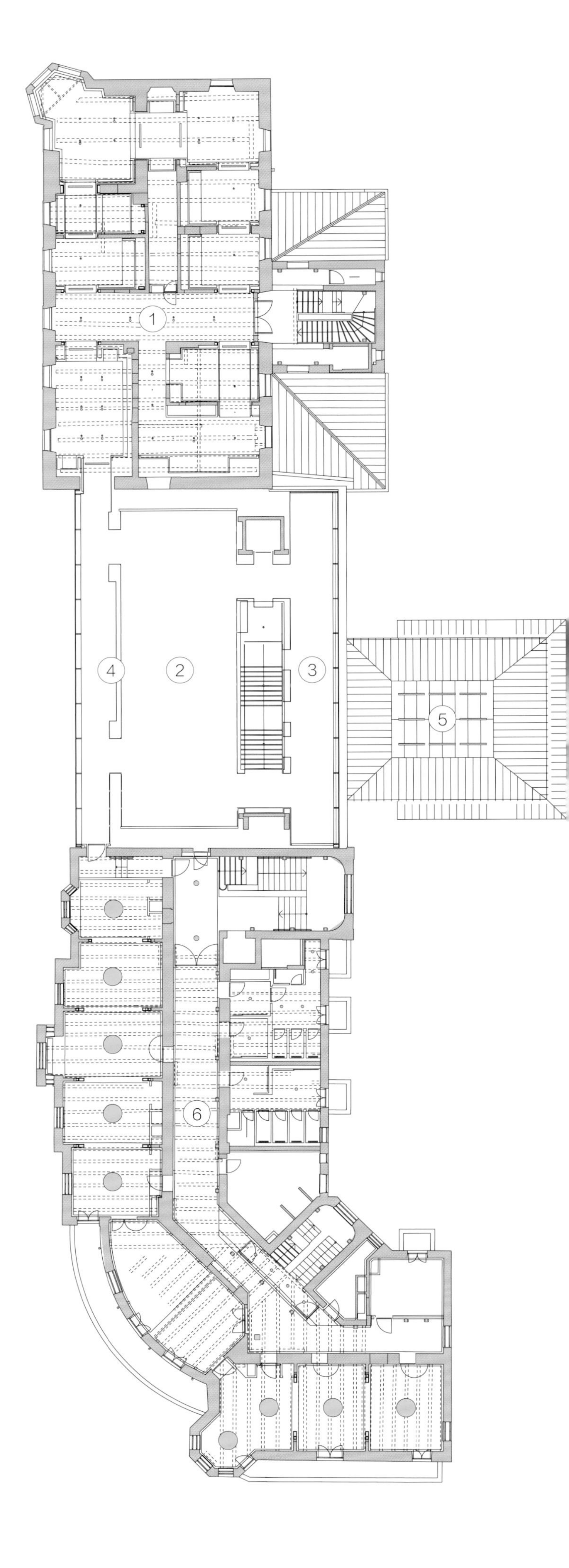

1.Jägerhaus展室
2.皇家城堡大厅
3.宗谱展示区
4.全景座位区
5.棕榈屋
6.阿尔卑斯玫瑰大酒店餐厅

1. Exhibition rooms in Jägerhaus
2. Hall of the Royal Castles
3. Genealogy display
4. Panorama seat
5. Palm House
6. Dining at the Alpenrose

二层平面图

First floor plan

e Museum of the Bavarian Kings provides an insight into the history of the House of Wittelsbach from its beginnings to the present y. 1,000 m² of exhibition space extend in a prominent setting over multifaceted spatial setup consisting of both old and new. For this useum, Licht Kunst Licht has planned a lighting concept that satfies the diverse functional requirements, impressively stages the hibited objects, and essentially shapes the identity of the building.

e home of the Museum of the Bavarian Kings is the former Grand otel Alpenrose, a more than 100-year-old building complex on e banks of Lake Alpsee and at the foot of the Neuschwanstein d Hohenschwangau castles. A careful handling of the historic sted building structure and the bold addition of a barrel vault with ree naves have created museum spaces that take the visitor by rprise by using very distinct characteristics.

BRILLIANT WELCOME IN THE FOYER

cess to the museum is via the centrally located entrance hall, the rmer dining hall of the Hotel Alpenrose. Through its large windows, e entrance is supplied with ample daylight. The light-coloured rrazzo flooring, the coffered ceiling, and the white walls create a minous overall impression. During the evening hours, the visitor is elcomed by a brilliant lighting atmosphere. To accomplish this, the isting historical chandeliers were technically revised and fitted th halogen lamps. A staircase located in the main foyer provides cess to the permanent exhibition on the first floor, which resides nder the newly built barrel vault with its three naves.

LEGANT STATIC, FULMINANT EFFECT

designing the museum extension by adding a new level above e historical dining room, it was important for Staab Architekten Berlin to create a supporting structure between the gable walls at preserves the absence of structural elements in the foyer hall. is was achieved by implementing three vaults with steel structure, hich are supported at only four bearing points. This solution ows for three large, column-free spaces running parallel to the çade, like a central nave with two aisles.

e central exhibition space has been declared the 'Hall of the oyal Castles' by its curators. With large photographic displays and agnificent exhibits, the presentation is dedicated to the Royal astles of Hohenschwangau and Neuschwanstein as well as to eir builders, King Maximilian II and King Ludwig II. The architecre-integrated lighting solution creates an appropriate setting for e exhibition, as the entire vault is realized as a luminous ceiling.

ter a series of model experiments and mock-ups, the lighting esigners at Licht Kunst Licht found a solution that both preserves e diamond-shaped coffered steel construction and maintains a eiling surface free from technical equipment. For this purpose, n LED backlit curved acrylic glass panel is fit into each coffer. e LED is mounted in each diamond between the steel ribs and stributes a subtle corona on the acrylic glass cover. In selected eiling panels, additional projectors have been placed behind the over in order to emphasize particular exhibits with directional ccent lighting.

OPEN VIEWS OF CASTLES, MOUNTAINS AND LAKESIDE

The luminous ceiling is continued in the two narrower vaults in the aisles. Here, however, it only stretches from the rearward ceiling boundary to the crown of the vault; from the crown down to the floor, a glass façade offers magnificent views onto the surrounding landscape. The galleries of varying widths – wider towards the back of the building and more narrow towards the lakeside and Hohenschwangau Castle – invite you to stay and linger, while offering incomparable panoramas.

GENEALOGY DISPLAY

The laminated glass of the rear wall serves as a display for the Wittelsbach genealogy room. In order to properly present the images and texts printed onto the glass, a linear LED graze light is recessed in a channel at the bottom edge of the glass. A sun protection material, applied between the panes, offers an additional reflective surface for the LED light.

INFORMATION AND INTERACTION IN THE SMALLER ROOMS

The exhibition concept in the smaller rooms of the old building provides palpable access to information and exhibits by means of contrast. Large-format, backlit images, brilliantly illuminated showcases, interactive screens and projections are embedded in exhibition furniture made of dark brown Corian. The concept for the ambient lighting replies to this correlate of colourfulness against the dark background with unobtrusive modesty. Daylight intake is curbed by blackout screens at the windows, the ambient lighting is limited to a functional minimum.

In this context, the light sources for the general lighting should remain hidden. Consequently, discreet luminaires have been slightly recessed in the lintels of the passages between the spaces. Each of these luminaires consists of two components: An LED which casts directional light onto the floor and a fluorescent lamp, which radiates an intense diffuse light distribution. Both light sources can be switched and dimmed separately and their light output can be adjusted according to the requirements of the exhibition. Outside business hours, the fluorescent sources serve as cleaning and security light. The only luminaires not fully integrated in the room details are adjustable projectors located at the ceiling monopoints. They are employed when an additional scenographic accentuation is desired to benefit the exhibition, or in the rooms where the furniture integrated illumination is not sufficient.

A CUSTOM DESIGNED LUMINAIRE FOR THE OLD STAIRCASE

The top and ground floor of this section of the historic building are connected via an existing staircase. For the stairs and hallways in this area the lighting designers wanted a luminaire with a lighting effect and design in tune with the formal language of the old building. A direct / indirect wall luminaire made of solid bronze met the above mentioned requirements. It provides sufficient illuminance levels in the circulation zones, adds grazing light to these surfaces, and pleasantly structures longer corridors.

PROGRAMME BOOKLET WITH LONG-DISTANCE EFFECT

The exhibition concept for the Museum of the Bavarian Kings has been developed by the House of Bavarian History. The chairman of the Board of Trustees, Prof. Dr. Hermann Rumschöttel, former Director General of the Bavarian Archives, described the museum's function at its opening, with an apt parable: "... to put it in the language of the theatre: This museum is first and foremost a type of explanatory programme booklet for the Hohenschwangau and Neuschwanstein productions and their creators."

He was spot-on with his statement, also in a figurative sense, because this programme booklet is quite present during a visit of the two castles; from both viewpoints the new barrel roof of the museum can be seen. The architects were quite aware of the visual effect of the building and treated its roof as 'fifth façade.' Therefore, the barrel structure is covered with metal shingles in a colour scheme matching the existing tiled roof. The lighting concept does not limit itself to the building interior. The luminous ceilings with their geometric matrix of luminous dots display their stunning effect through the glazing of the galleries. They enrich the romantically charged landscape as a highlight and extend the invitation to all who wish to learn more about the history of one of Europe's oldest dynasties after visiting the Royal Castles.

CLIENT:
Wittelsbacher Ausgleichsfonds
OCCUPANT:
Schlosshotel Lisl GmbH & Co KG
ARCHITECTS:
Staab Architekten GmbH, Berlin
TEAM LEADER LICHT KUNST LICHT:
Malte Simon
COMPLETION:
2011
PROJECT SIZE:
1,000 m²
LIGHTING BUDGET:
0.7 million euros

巴伐利亚国王博物馆坐落在阿尔卑斯湖畔，新天鹅堡脚下。

The Museum of the Bavarian Kings is located below the Royal Neuschwanstein Castle on the banks of Lake Alpsee.

夜晚，建筑立面的光晕好似一个个发光的灯笼。

At night, the facade apertures are transformed into luminous lanterns.

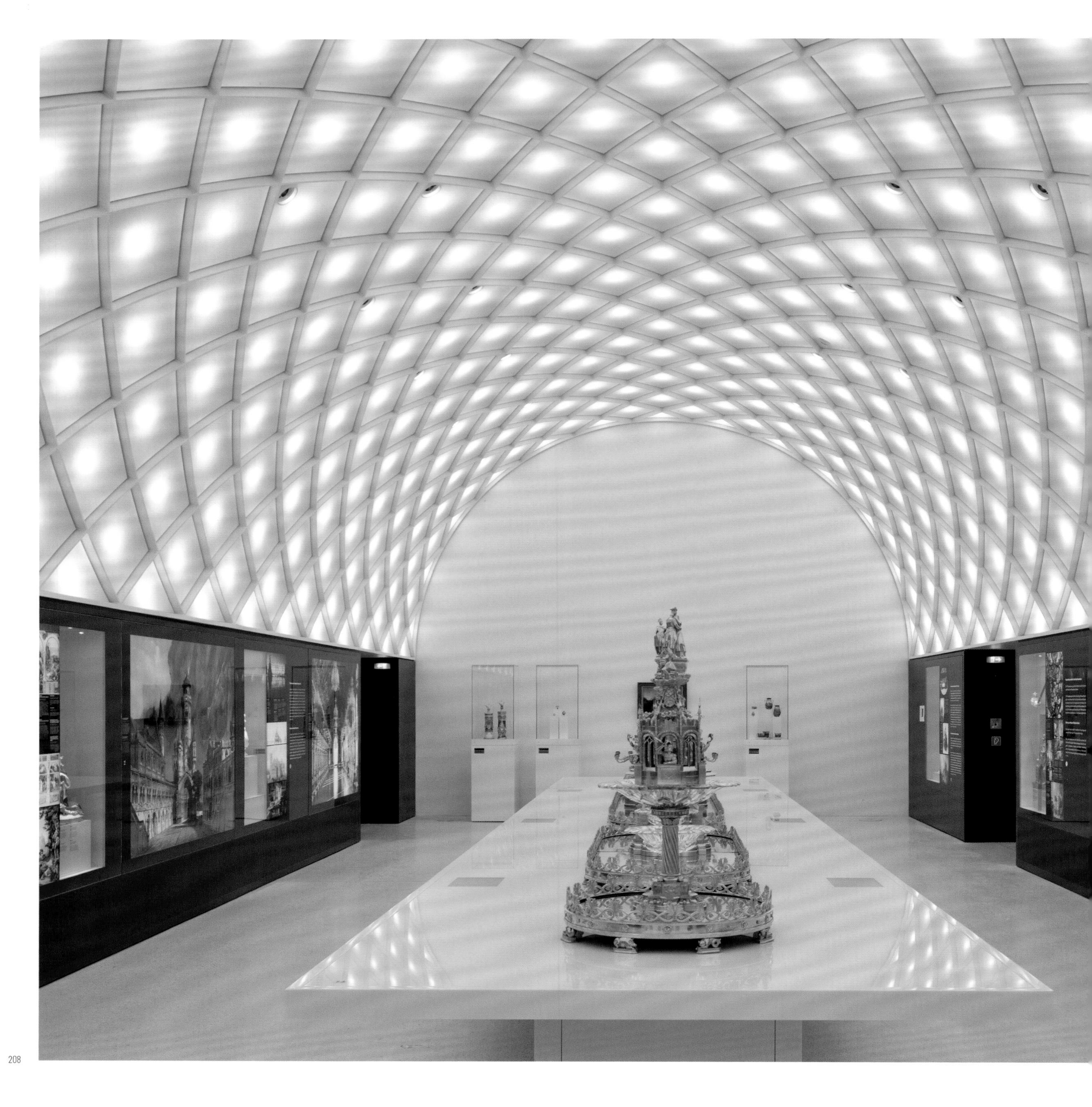

皇家城堡大厅陈列了尼伯龙根氏族的核心展品。大厅及其两个侧廊空间的拱形穹顶由带有背光照明的凹形亚克力花格镶板构成。

The 'Hall of the Royal Castles' exhibits the Nibelungen centerpiece. The barrel vault of the hall and its two auxiliary rooms are composed of coffers made of backlit, concave acrylic panels.

每片花格镶板中均设有一个LED背光照明组件，将天花上一个个明亮的发光点以及周边柔和的光晕结合在一起。拱顶表面内嵌了可调节聚光灯，用于展品的重点照明。

Each coffer is fitted with a LED backlight that combines a brilliant luminous point in the ceiling with a gentle corona. Fully adjustable spotlights for the accentuation of exhibits are recessed into the vault surface.

从落地窗向外望去，可以看到广阔的阿尔卑斯湖，湖面上映着旧天鹅堡的倒影。发光天花的照明手法也被使用在了大厅两侧较窄的侧廊中。

Panoramic view across Lake Alpsee with a reflection of Hohenschwangau Castle. The luminous ceiling is continued in the two narrower aisle vaults.

独立式玻璃陈列柜由安装在天花上的可调节聚光灯提供精准的重点照明。

Freestanding glass showcases are discreetly emphasized through ceiling-mounted adjustable accent lights.

随着先进LED技术的应用，即使光敏度较高的展品也能受到很好地保护，并通过灯光对展品进行视觉突出。

With the implementation of advanced LED technology, even sensitive exhibits are visually enhanced while remaining considerate to their conservation.

精心布置的灯光显著强调了扩展展柜和走廊通道的空间秩序。

The spatial sequence of the extended exhibition cabinets and the corridor shaped passages are strikingly accentuated by the light orchestration.

在展厅和展厅之间，照明灯具被内嵌在走廊门楣中。它们通过两种照明光源为空间提供精准的整体照明。其中，LED光源模块发出的定向光恰到好处地补充了荧光灯管的柔光照明。

Between the individual exhibition spaces, luminaires are recessed in the door lintels of the connecting passageways. It provides a discreet general illumination by means of two light components. Soft fluorescent light is complemented by directional light from LED modules.

设计师巧妙地将照明设备隐藏在展柜中，在对展品进行重点照明的同时，避免其暴露在参观者的视线之内。电子显示屏和采用背光照明的图像以生动的方式向参观者传达信息。

The lighting equipment integral to the showcases is concealed from view and accentuates various exhibits. Display screens and backlit images convey information in a lively fashion.

在这座古老建筑的楼梯和走廊空间，设计师布置了包括直接和间接照明元素的坚实铜制壁灯。

A direct / indirect wall sconce made of solid bronze is used in the staircase and hallways of the historical building section.

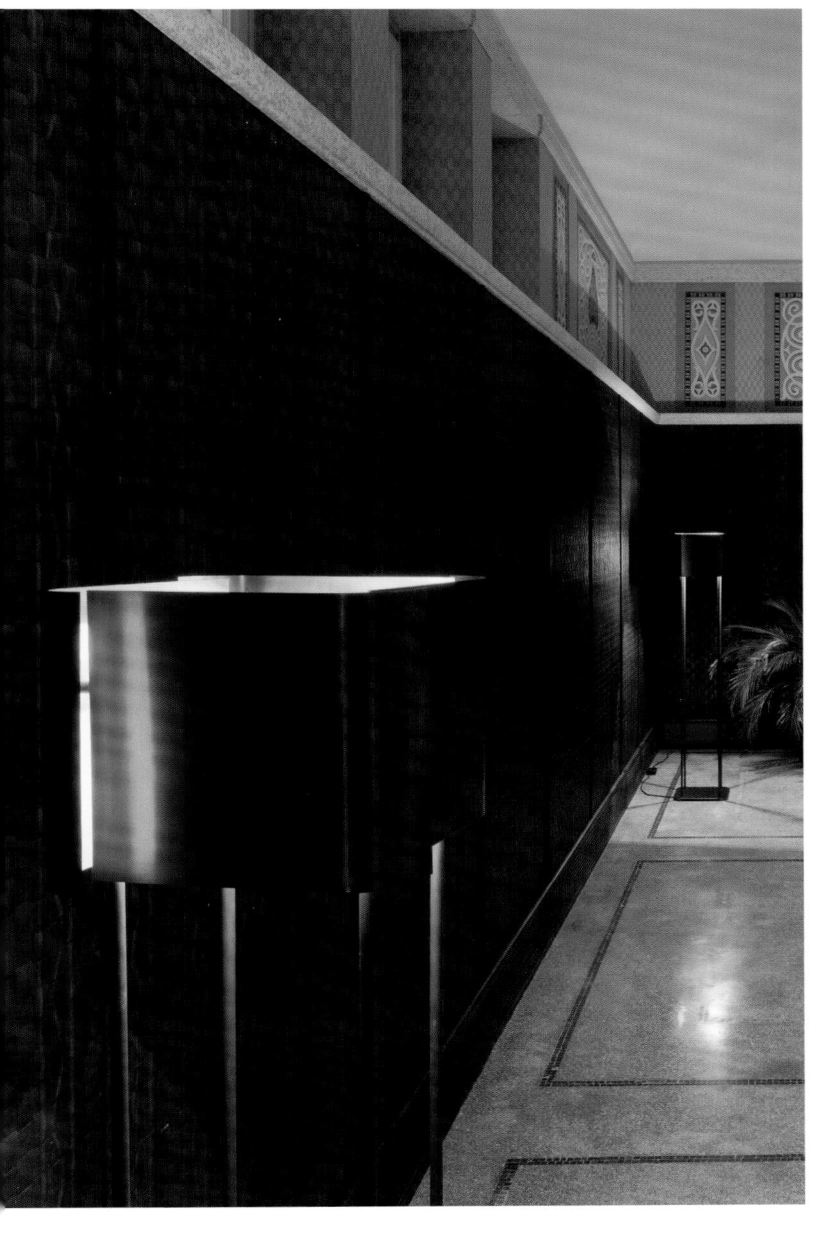

落地灯与定制壁灯的设计相呼应，共同组成这个现代化灯具系列。

In conjunction with the wall sconces, matching floor luminaires form a contemporary luminaire family.

在灯具出光口配置铜制遮光网，可以避免游客直接看到光源，同时为光线披上了一层金色的衣裳。

A bronze mesh prevents direct views of the light source and gives a golden luster to the illumination.

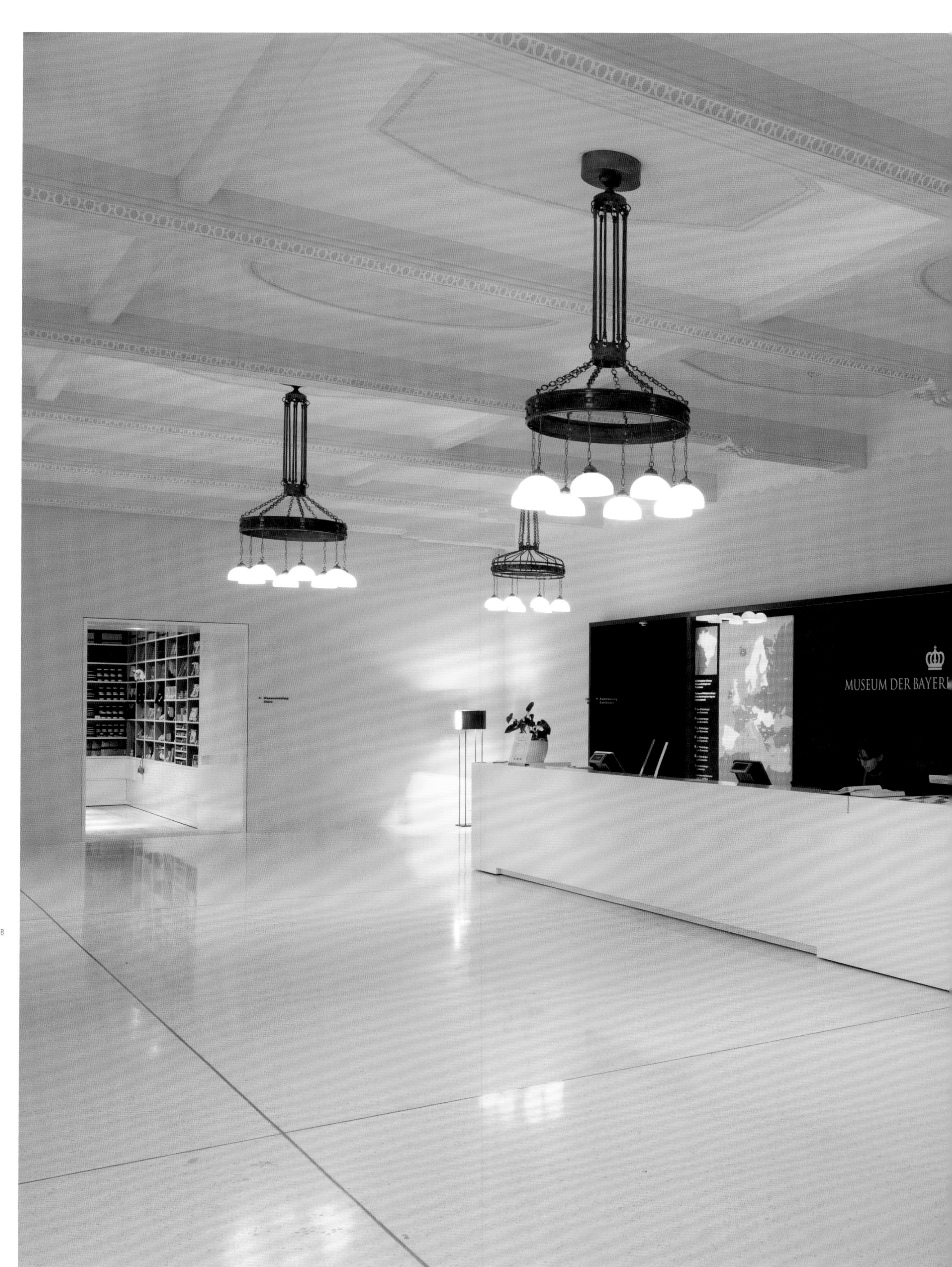
MUSEUM DER BAYER

← WC
← Garderobe
Wardrobe

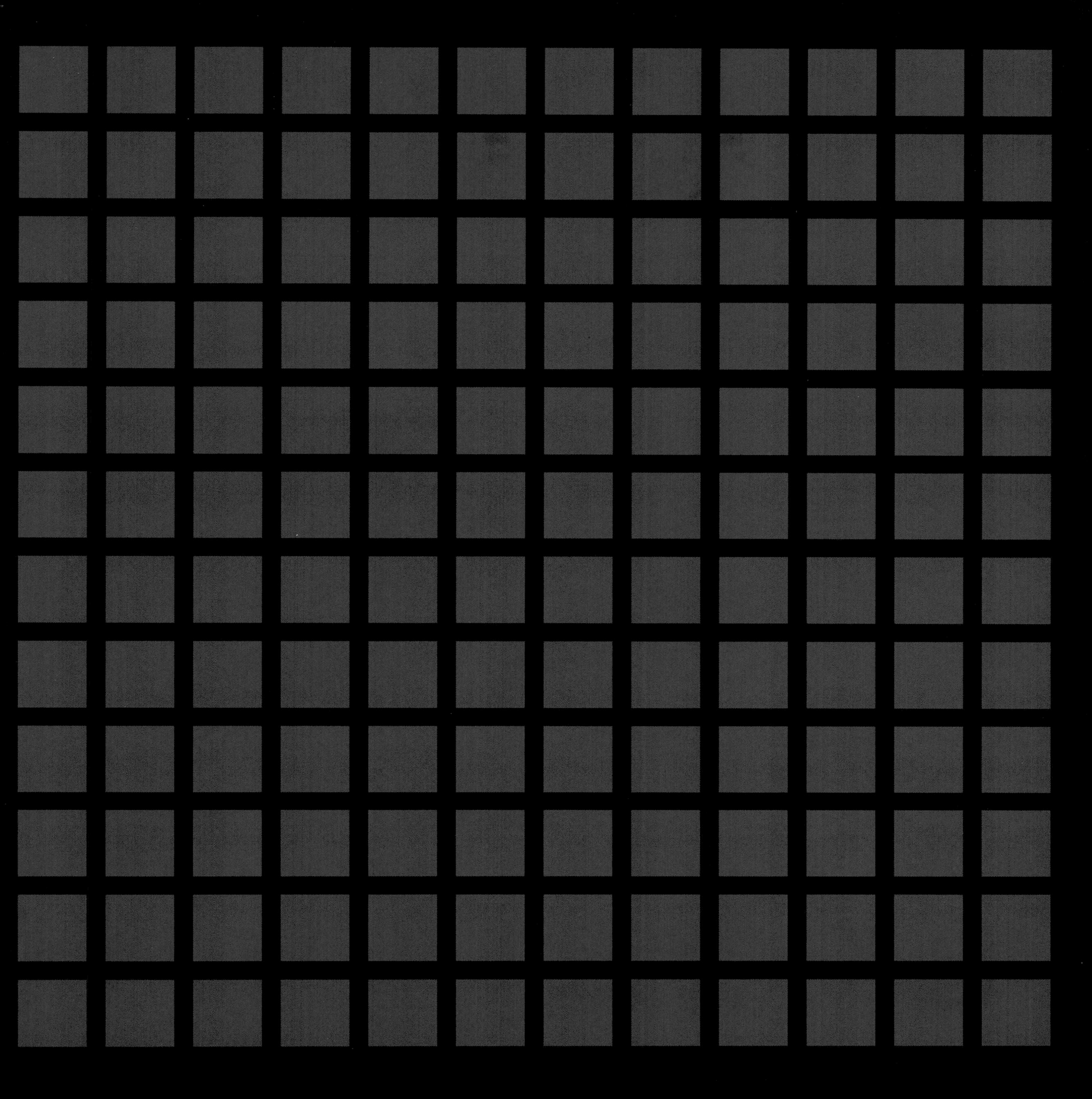

布雷斯劳尔广场地铁站 科隆

BRESLAUER PLATZ SUBWAY STATION COLOGNE

Severinstraße
Kartäuserhof

新建筑与新秩序——科隆布雷斯劳尔广场地铁站

雷斯劳尔广场位于德国北莱茵-威斯特法伦州人最多的城市科隆，是科隆最著名的广场之一，邻中央火车站。布雷斯劳尔广场地铁站西侧的前广直接通向科隆老城中心极具代表性的城市立面及享誉全球的科隆大教堂。此外，布雷斯劳尔广面向莱茵河的一侧主要分布着一些“二战”后建的交通基础设施，还有部分已经老化了的20世纪0年代德国“经济奇迹”时期的建筑物。

的城市发展规划致力于开发城市旅游交通，向游更好地展现优美的莱茵河畔风光，并力求让该广的更新设计满足21世纪的城市发展需求。因此，为城市更新计划的一部分，布雷斯劳尔广场逐步变为更加开放的现代化交通枢纽。其中最为显著改造工程要属新地铁站建设项目。建成后，该地站可以连接多条以科隆中央火车站为中心的城市共交通线路，每天可运送280 000余名乘客。

雷斯劳尔广场地铁站是科隆城市轨道新南北交通线的第一站，也是该市精细化地下隧道系统的起。在整个项目中，每一个地下站点均由不同的建事务所来负责。总部位于科隆的布德和门泽尔建设计事务所最终赢得了布雷斯劳尔广场地铁站的计竞赛，并获得了该地铁站的设计合同。

步走进清晰明亮的地下深处

筑设计师在广场的三个位置规划了地铁站的进出，引导游客从广场进入地铁站。其中最明显的一进出口直接贯通了广场中心区域与科隆中央火车的中心通道。

铁站的建筑设计理念为“当代大教堂”，为了契这一理念，建筑师将地铁站地上入口部分设计为天空伸展的风亭造型，由外围的立柱支撑，空间得宽敞通透。入口区域三角形状的建筑造型确保中有一侧始终向乘客敞开，而面向广场的立面则用透明的金属网覆盖，从而保护游客免受广场上机动车交通影响。

客从广场上的风亭进入地铁站后，便可对长25米、宽30米的地铁站全貌一览无余。经过入坡道和自动扶梯，以多条长廊通道形式排开的接层映入眼帘。长廊优雅地缓缓延伸，确保乘在各个方向上都可以看到空间的全景。经由这连接层，乘客可以通往三个站台。站台在地铁中沿着一条长长的曲线分布，相应的轨道则限了其最深处的标高。

了使乘客不受遮挡地看到地铁站的全貌，建筑师建筑所有的结构支撑性立柱尽可能设计为轻巧美的样式。此外，建筑师还对部分墙壁进行穿孔，时采用透明材料制作固定装置和栏栅。总体而言，建筑材料凸显了地铁站大方、友好且明亮的感观印象。深色金属格栅形成了整齐的天花造型，墙壁则由浅色抛光混凝土制成，同时地板衬有高级水磨石。

一个通行和人流系统

一方面，通过地铁站入口处的大型开口以及地下空间明亮表面的相互反射，日光基本能自然且均匀地洒落到车站内。另一方面，地铁站内的人工光照明，特别是在与中央火车站相连接的区域，则需要通过设计来满足一系列复杂的视觉要求。因此，包括地铁站的入口、坡道、通行区域、等候区域和站台邻近区域等空间均需要采取不同的照明方案，以及对照明均匀度、照度水平和眩光防控的高标准要求使得该项目无法仅依照设计美学来规划照明方案。在追求设计美感的同时，照明方案还必须考虑地下空间不同的安装高度条件、设备维护策略，以及方位引导和疏散系统的要求。

最终，照明设计师将这些看上去由技术需求与美学追求所产生的设计冲突通过活用建筑自身的结构得以解决。在现有通行区域的整体形状、造型曲线和高度条件下，在视觉上已经颇具技术感的金属格栅天花为设计师提供了可以在天花中排布高功率嵌入式筒灯的美学基础，以及根据功能性要求在特定位置提高或降低灯具分布密度的技术条件。设计师精心研发了一套对设备维护要求较低且照明设备能在外观上保持统一的照明系统，这样一来，不同功率等级甚至部分安装高度较低的灯具就可以统一集成到天花系统中。为此，设计师完善了天花嵌入式圆环中的内嵌细节，使灯具光源、灯体和定制改装配件均可嵌入一个标准单元之中。该标准单元的外观与天花面板的颜色和材质相匹配。

地铁站内的人流交汇处、台阶和坡道区域由灯光进行重点突出，而道床则大部分保持无照明状态。这种视觉对比更加凸显了由光线强调的建筑表面。乘客在地铁站行走时，视线由近及远打量整个空间，可以感受到鲜明多样且动态变化的灯光气氛。

在广场上营造闪耀的光华

到了夜晚，广场上几处入口风亭在灯光的映衬下熠熠生辉。它们不仅向游客展现出广场的方位指向和通行方面的安全感，还拥有独特的艺术氛围。得益于强有力的天花照明，风亭的金属网立面接收并反射了精心布置在天花上的灯具的光线，而立柱则以剪影形式映衬在这些金属网立面上。风亭立柱在夜间的亮度正好与其白天的状态相反，但仍然可以呈现建筑的通透感。

连接层的通透感和深度感

主要投向地面区域的照明方案增强了连接层走廊空间的通透感，设计师未对单层空间中的垂直表面做特别的光线处理，这使地铁站内的多层空间，特别是楼梯和站台区域，成了乘客的关注焦点。

站台上的空间感和安全感

地铁站站台通过两种照明方式提供光照。在站台上方，下照灯系统沿着细长的隧道弯道，为地面和墙面提供相对较高的照度水平。在较低区域，连接层地面下方的天花上则安装了外露的荧光灯带。通过这些柔和的光线，能够均匀地照亮沿道床建造的展示墙。

这种差异化的照明手法使中央区域始终保持着清晰的空间尺度感，也使空间设计中的高度重叠和路径交错形式易于识别，从而准确传达出方位信息，让乘客获得通行方面的安全感。

作为城市名片的地铁站

布雷斯劳尔广场地铁站是连接科隆公共交通系统与科隆中央火车站的枢纽。因此，对游客来说，该地铁站也是一张城市名片。从技术角度来看，定制的照明解决方案十分节能，并且满足了这种交通运输设施规模下高度复杂的功能性需求。此外，灯光与建筑有效地结合在一起，设计师在严谨的灯具设计方案的基础上，通过光线语言确保了建筑物的通透感和开阔性，并且将复杂的空间结构变得易于理解。

业主：
KVB-科隆运输服务股份公司
使用人：
KVB-科隆运输服务股份公司
建筑设计：
布德和门泽尔建筑设计事务所，科隆
LKL项目经理：
LKL团队
竣工时间：
2011年
项目规模：
4900平方米
照明预算：
40万欧元

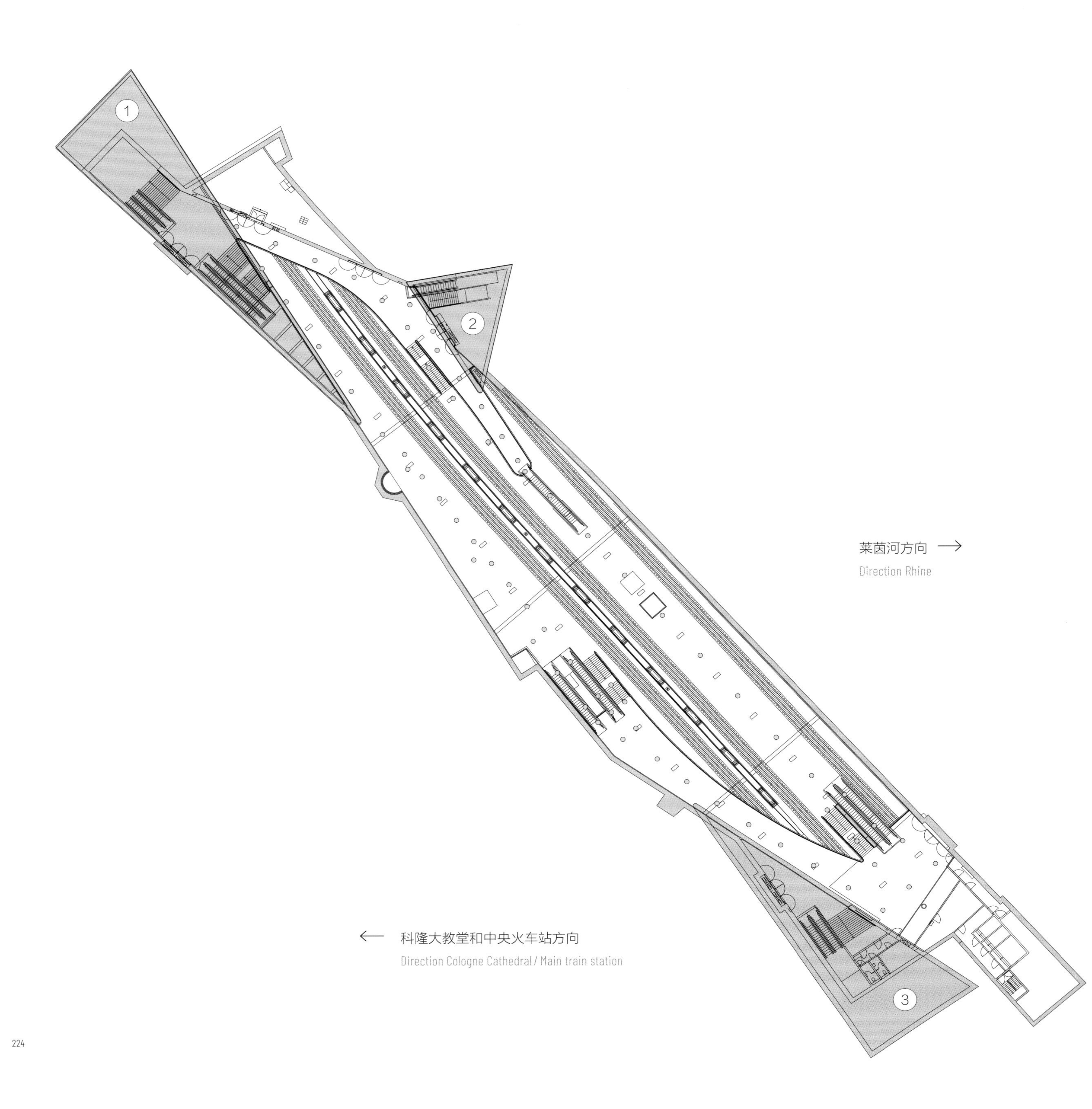

连接层总平面图

Overview plan of the junction floors

1.艾根史坦出口
2.库尼伯特区出口
3.布雷斯劳尔广场出口

1. Eigelstein exit
2. Kunibert Quarter exit
3. Breslauer Platz exit

Breslauer Platz in Cologne is one of the most prominent squares flanking the central station of North Rhine-Westphalia's most populous city. The station's forecourt on the western side forms a direct connection to the representative façades of Cologne's historical city centre, and the world-famous Cologne Cathedral. Meanwhile, the Rhine-facing side of Breslauer Platz was primarily known for its mundane postwar transportation infrastructure and dated buildings constructed during the 60's economic miracle era.

Endeavoring to open up the city and its Rhine views to travelers, and to adapt the square's design to the requirements of a 21st century city, the new urban development plan called for Breslauer Platz to undergo a gradual conversion into a more overt and contemporary transportation hub. The most obvious update is the construction of the new underground rail station, which now smoothly links Cologne's mass transit lines with the Cologne Central Station, transporting 280,000 passengers per day.

At the same time, the Breslauer Platz subway station is the first stop on the new north-south axis of the Cologne city railway and the starting point of the sophisticated tunnel system underneath the city. During the course of the entire project, each station was awarded to a different architectural office. The design competition and resulting design contract for the Breslauer Platz station was won by the Cologne based architecture firm, Büder + Menzel.

GENTLE SWEEP TOWARDS LUCID DEPTHS

Guiding visitors into the underground station, the architects have penetrated the square's surface in three strategically chosen locations. The most prominent one directly connects the middle of the square to the central passage of Cologne's Central Station.

Alluding to the notion of railway stations as 'Contemporary Cathedrals', the above ground entrances are built as skyward reaching pavilions, supported by perimeter columns and containing a spacious limpidity. Their triangular shape ensures that one side is always open towards passengers, while façades facing the square are shrouded with a transparent metal mesh to shield the visitors from motorized traffic.

When accessing the station through one of the pavilions, its 25 meter long and 30 meter wide expanse unfolds immediately. Across ramps and escalators, the entrances provide access to junction floors arranged as galleries, which perform elegant sweeps to ensure passengers maintain an overview from all directions. From these junction floors, you have access to three platforms that trace a long curve through the station and whose tracks define its deepest elevation.

To allow for an unobstructed view across the entire station, all structurally required columns are designed to be as light and elegant as possible, walls are partially cut away, and fixtures and railings are made of transparent materials. As a whole, the material concept underlines the generous, friendly and bright impression evoked by this train station. While the dark metal grating forms a homogenous ceiling, the walls are made of light-coloured polished concrete and the floors are lined with high-grade terrazzo.

A SYSTEM OF PATHS AND MOVEMENT

Almost automatically, daylight floods into the station with an even distribution via the large entrance openings and inter-reflections of the bright surfaces within. The electric lighting in an underground station, particularly if it involves a connection to a central station, entails a complex set of tasks. Entrances, ramps, circulation areas, waiting areas, and platform edges all call for different lighting concepts. Their high standards regarding uniformity, illuminance levels and glare control counteract a purely design-driven planning approach. Beyond that, the concept must account for different mounting heights, maintenance strategies, orientation and egress systems as well.

What appears to be a conflict between design demands on one hand and strict technical requirements on the other, ended up being resolved through the architecture itself. The already technical look and feel of the metal grating ceiling allowed for a layout of powerful recessed downlights that follow the shapes, curves, and levels of the circulation areas while condensing or spreading out according to requirements. Through the sophisticated development of a low-maintenance lighting system with a consistent design appearance, different power levels and even some lowerable fixtures could be integrated into the ceiling. For this purpose, a recess detail within a regressed circular plate has been designed. This allows for all luminaires, details, and custom modifications to be embedded in a standard unit, which has been adapted to match the ceiling colour and material.

Junctions, stairs, and ramps are emphasized by light, while the track beds remain largely unlit. The emerging contrasts make the surfaces underlined by light stand out. Through the views and vistas within the station, and by means of passenger progression, a differentiated and dynamic light atmosphere emerges.

INVITING A PLAY OF LIGHT ON THE SQUARE

At night, the entrance pavilions shine brightly and clearly across the square, their inviting allure conveying more than just orientation and a sense of safety. Through the resonant illumination from the ceiling, their columns appear in silhouette against the metal mesh façade, which receives and brilliantly reflects light from strategically placed luminaires. The luminance distribution during the day is practically reversed, yet the transparent effect of the pavilions remains.

TRANSPARENCY AND DEPTH ON THE JUNCTION FLOORS

The transparency of the junction floors' galleries is supported by the lighting concept, which involves primarily illuminating the floor area. Vertical surfaces in single-storey spaces are not specifically treated with light, which makes the multi-storey areas of the station, particularly the stairs and platforms, emerge as centers of attention.

A SENSE OF SPACE AND SECURITY ON THE PLATFORM

On the platforms, two styles of lighting are being used. Above the platform, the downlight system follows the elongated bend of the tunnel and applies relatively high illuminance levels to the floor and walls. In the lower zones, below the junction floors, bare fluorescent strips are mounted to the ceiling. By means of their diffuse light distribution, they uniformly illuminate the display walls along the track bed.

Through this differentiated illumination technique, the height and width of the central space remain impressively noticeable, while the staggering of heights and paths in the spatial arrangement are made readable, thus conveying orientation and a sense of safety.

A SUBWAY STATION AS THE CITY'S BUSINESS CARD

Breslauer Platz station is the pivotal connection between Cologne's mass transit system and the Cologne Central Station. It can therefore be understood as the city's business card for visitors. The bespoke lighting solution is very efficient from a technical standpoint, and fulfils the highly complex demands for a transportation facility of this size. At the same time, the lighting effect has been effectively combined with the architecture, making the building's transparency, expanse and complexity perceptible while maintaining a discreet luminaire design.

CLIENT:
KVB – Kölner Verkehrs-Betriebe AG
OCCUPANT:
KVB – Kölner Verkehrs-Betriebe AG
ARCHITECTS:
Büder + Menzel Architekten BDA, Cologne
PROJECT LEAD LICHT KUNST LICHT:
Team Licht Kunst Licht
COMPLETION:
2011
PROJECT SIZE:
4,900 m^2
LIGHTING BUDGET:
0.4 million euros

地铁站的入口风亭坐落在著名的布雷斯劳尔广场上，紧邻科隆中央火车站和科隆大教堂，它向游客展示出热情、好客的姿态。

Located on the prominent Breslauer Platz square, in immediate vicinity of the Cologne Central Station and the Cologne Cathedral, the entrance pavilion of the subway station displays its inviting aura.

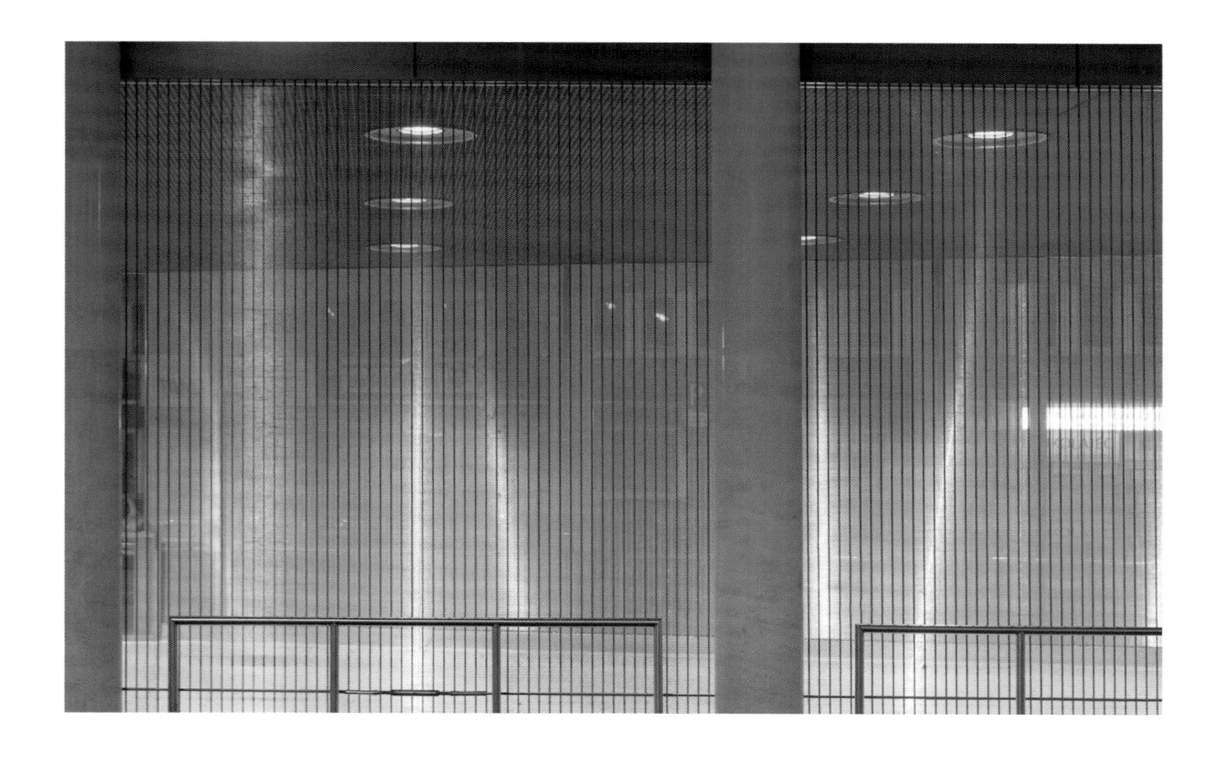

风亭的金属网立面接收定向照明光线，同时也保持着通透的观感。

The pavilion's metal mesh facade picks up the directional light, yet remains transparent.

立柱的排布方式与金属网立面相呼应，为乘客提供了广阔的视野，也将乘客与广场的机动车交通分隔开来。

The column layout and metal mesh façade allow for vistas while screening the passengers from the square's motorized traffic.

地铁站的入口风亭与车站的主要通道相连，引导乘客通往明亮的地下空间。

Connecting to the central passage of the main station, the entrance pavilion guides the visitor towards the light-filled underground.

天花上连续的内嵌细节允许不同瓦数、不同安装高度和不同维护方式的设备嵌入其中。

A consistent recess detail permits various wattages, mounting heights, and maintenance mechanisms.

沿着细长的弧形空间，乘客可以轻而易举地看到地铁站空间的全貌。

Following the elongated curvature, one glance reveals the station's entire expanse.

灯具遵循天花面板的形状合理分布，并且根据情况需要，可以在指定区域提高或降低其分布的密度。

The luminaire layout follows the shape of the ceiling slab while condensing and spreading out according to requirements.

透明的建筑材料和灯具的差异化分布相结合，产生了光线反射和空间透视效果。

Transparent materials and a differentiated layout of light sources create reflections and vistas.

AWARD
WINNER

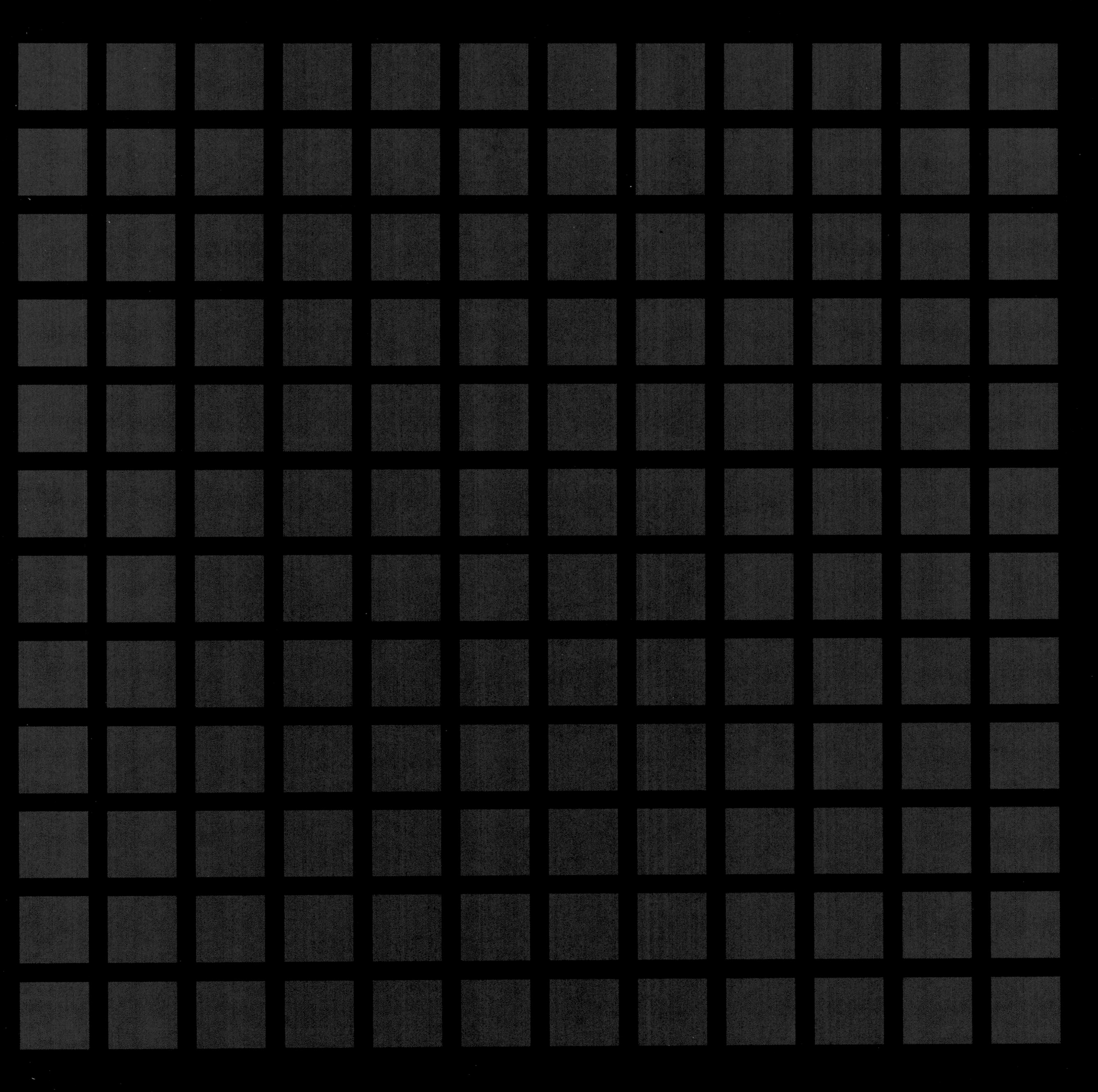

LICHT KUNST LICHT 4

巴登-符腾堡州内政部新大楼

斯图加特

BADEN-WÜRTTEMBERG MINISTRY
OF THE INTERIOR STUTTGART

形结构化特征和令人激动的连接排布——位于斯图加特的巴登-符腾堡州内政部新大楼

008年初，施塔布建筑设计事务所设计的巴登-符堡州内政部新大楼的方案在一次建筑竞赛中脱而出。历时5年多，该项目于2013年竣工，可供00余名职员使用。最后胜出的建筑设计方案让本大赛的评委和雨果-哈林奖评委十分信服。

座位于威利-勃朗特街道总长200米的建筑群，筑设计的指导原则是提出一个能够很好连接一系大小不一的长方体体块的建筑方案。这些长方体块略微交错排布，内部中庭形态各异。这种建筑构赋予政府不同部门以独特的空间特征，同时创了拥有较高利用率的城市楼体轮廓线。建筑位于长地段，朝东北方向逐渐收窄，处于复杂的城市构环境中。该项目的照明方案支持了建筑设计的图，使中庭内侧的立面看起来像散发着光芒的窗。值得一提的是，设计师以相对简单的照明手法现了理想的灯光效果。

使用者和访客角度出发的人本尺度的大型建筑

为细长形的6层高建筑群，该项目临街分布，从市地块层面划分了城堡花园中部街区和东部相邻克纳街区。建筑群的前楼向北依次连接了5个建体块。由于地势的倾斜，地下室在项目最南端的园一侧变成了地面层空间，因此非常适合在南端筑物中设置内政部的公共活动区域。中央大厅和动室在狭长的建筑南端设有专门的入口，从该入进入，可以抵达餐厅、日托中心和图书阅读区。公活动进出口位于威利-勃朗特街道的建筑主入，为内政部与环境、气候和能源部下设的600多办公室提供服务，同时也为与之相邻的农村事务消费者保护部，以及联邦州政府、警察局和民部门的几个安全局势中心提供服务。稳重的外面设计与大面积的箱形窗户形成强烈的视觉对。新建筑的一个主要特点是，从外部看不出这个23 000立方米的建筑“盒子”中有超过19 000平米的室内可用面积。

具有户外观感的中庭

中庭内侧，丝带状的墙体开口可让宾客看到建筑层的路径通道，同时也强调了中庭的视觉特征。庭设计采取严格的正交形式，采光充足，打破了型办公建筑结构常有的单调感，并借助精妙的照设计展现出了令人振奋的空间氛围。建筑设计师采用了种类较少的建筑材料，例如喷砂外露混凝、抛光地板、烟熏橡木、玻璃和水泥基亚光烤漆花板，并通过这些材料的巧妙组合营造出宽敞、净的空间环境。

形建筑结构中的光线

建筑设计相似，在照明规划过程中，设计师有可研发出一套照明方案—— 仅采用几种重复出现灯光元素，对不同形状的空间进行整体把控和设计。5个中庭的结构是室内设计的核心，它们包括横向和纵向通道，中厅和其他房间一样，要与项目整体平面的几何形态融为一体。中央大厅作为重要的人流交汇口显得十分雄伟，在其尽头仅有一个洒满日光的简约中庭空间。这种设计组合营造出巧妙的，甚至带有戏剧感的空间秩序。尽管所有中庭都是封闭的，却让人感觉像是来到了户外庭院或广场。这种视觉感受要归功于中庭穿孔的立面造型以及环绕在其周围的办公区走廊。

为了强调这一特点，照明设计师萌生了一个想法——用线形灯光勾勒出所有的走廊，并在两个区域之间通过灯光引入清晰的分界线。线形灯条内嵌在天花凹槽中，首尾连接的T16荧光灯管，直接照亮了走道空间。同时，周围走廊的墙壁表面由线形灯管发出的杂散光照亮。相比之下，中庭的穿孔立面保持较暗的状态，而地面则由窄光束HIT聚光灯提供功能性照明。在中庭里，人们被走廊中散发光芒的窗口包围；而在走廊上，人们也可以朝中庭望去。光“线”的照明理念顺理成章地演变为应用在所有楼层的灯光主题，甚至延伸到了建筑角落。在屋顶花园旁边，线形灯光介于庭院和周围走廊之间，充当了室内和室外区域的分隔元素。

艺术品增强了中庭的辨识度

设计师在前3个中庭统一设置了由艺术家莱克·埃里亚斯创作的大型金属环艺术装置。其中一个金属环立在墙壁上，另一个被高高悬挂在天花下方。这些圆环巧妙地展示出中庭结构逐渐向一个方向收窄的建筑形态，其轻盈的外观将这个明显的地形劣势转化为一件成功的艺术品。

具有雕塑感的楼梯

如果有人想步行上下楼，那么他会进入长期对外开放的楼梯间。楼梯间藏身于装饰格栅后方，仅能隐隐约约地从缝隙中寻见。遵循整体项目中线形灯光引导路径的照明理念，所有楼梯间统一安装了线形LED扶手灯。楼梯扶手的照明设计与建筑格栅相呼应，形成了清爽的视觉对比。当人们从中庭望向楼梯间时，扶手灯光显得十分内敛且令人愉悦。除了一体化照明，扶手自身的长方形剖面尺寸仅有5厘米×3厘米，这与建筑师最初的设想相契合。之所以能够做到如此小的尺寸，得益于现阶段投入使用的小型高功率LED灯具技术。

完全隐藏在金属天花格栅后方的首层照明

与中庭主要以纵向采光不同，面向建筑水平表面的照明手法更加多元。特别是首层与中庭相邻的区域，呈现出不一样的视觉观感。设计师为门厅和通往会议室的走廊区域单独设计了一种照明方式，由隐藏在金属天花格栅后方的HIT下照灯提供照明。为了强调空间的多样性，设计师还在座椅上方增加了装饰性吊灯，以此来界定空间中的休闲区域。在会议室前方，由紧凑型荧光灯管作为光源的壁灯在墙壁两侧散发出清晰的光斑，营造出具有极高视觉对比度且令人兴趣盎然的入口区域。

业主：
巴登-符腾堡州基金会，由巴登-符腾堡州财产和建设办公室、斯图加特办事处执行
使用人：
巴登-符腾堡州内政部，环境、气候和能源部，农村事务和消费者保护部
建筑设计：
施塔布建筑设计事务所，柏林
LKL项目经理：
埃德温·斯米达
竣工时间：
2013年
项目规模：
33 000 平方米
项目总预算：
6500万欧元
照明预算：
130万欧元

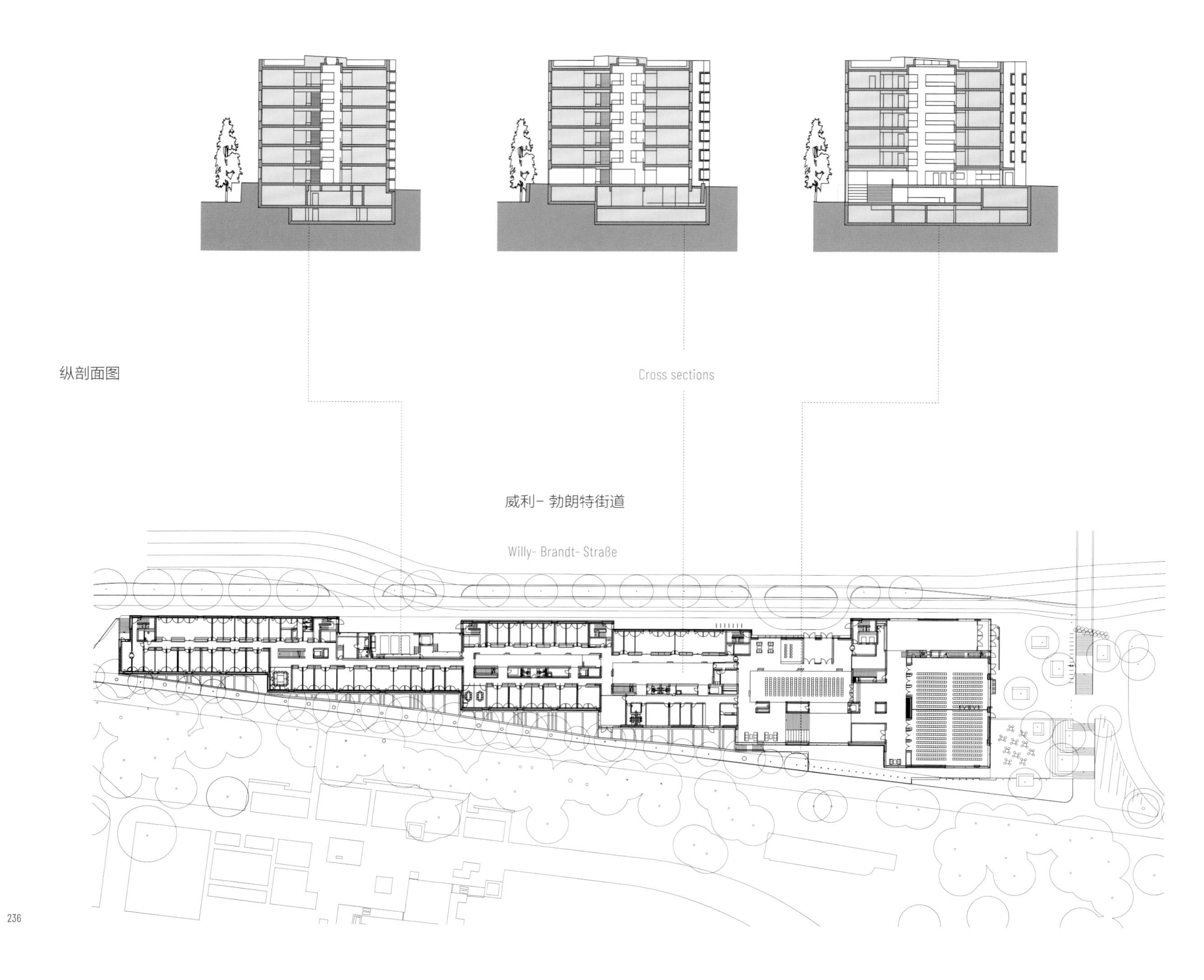

纵剖面图

Cross sections

场地平面图

Site plan

Emerging from the endorsement of an early 2008 architectural competition, the Baden-Württemberg Ministry of the Interior's new building was successively completed in 2013 for the use of its more than 600 employees. The winning design by the Berlin planning office Staab Architekten was convincing, both to the competition's jury, as well as the jury of the Hugo Häring Prize.

The guiding principle for the 200-meter-long building on Willy-Brandt-Strasse is the procession of a series of differently sized and slightly staggered rectangular volumes with various dimensioned internal courtyards. They provide the different Ministry departments a spatial identity while at the same time generating an effective urban building outline. Located on a narrow lot, tapering to the northeast, the building is located in a sophisticated urban setting. The lighting concept supports the intentions of the architecture and makes the inner façades of the courtyards appear as interspersed illuminated windows. The lighting design achieves this effect by relatively simple means.

LARGE BUILDING WITH A HUMAN SCALE FOR USERS AND OBSERVERS

As an elongated complex, the building is six stories tall on the street side and marks the boundary between the Mittlerer Schlossgarten and the subsequent eastern Kerner quarter. The front building is adjoined by five other segments to the north. The sloping topography turns the basement into a ground floor on the park side of the south end of the building. It is therefore ideally suited to accommodate the public areas of the Ministry. In addition to the central foyer and event rooms having a separate entrance on the narrow south side, there is also a dining facility, a daycare center, and a specialized library. The main entrance, located on Willy-Brandt-Strasse, is used for the more than 600 offices of the Interior and Environment departments, as well as the Rural division with its adjoined Customer Protection Agency and several extra secure situation centres of the Federal State Government, the Police and Civil Defense. The calm façade design with large box windows unfolds a great visual impact. A key quality of the new building is, that from the exterior, no perspective exists that would reveal its interior area is over 19,000 m^2 and 123,000 m^3.

COURTYARDS WITH AN EXTERNAL EFFECT

Ribbon-like wall openings, that allow a view of the surrounding access areas on all levels, also shape the character of the courtyards. These atriums, designed strictly orthogonal and generously supplied with daylight, break the potential monotony of the large office structure and unfold an exciting atmosphere with the help of a sophisticated lighting concept. Only a few materials are used, such as sand-blasted exposed concrete, polished floors, smoked oak, glass, and cement based matte painted chipboard, which altogether are used and employ a pleasant visual calm and spaciousness.

LIGHT LINES IN A LINEAR BUILDING STRUCTURE

Similarly, during the planning process, it was possible to develop a lighting concept that serves the various room shapes with only a few recurring elements. The sequence of five atriums forms the centerpiece of the interior design. They include the horizontal and vertical access and – like all other rooms – blend in with the tapered geometry of the project. While the main foyer acts extremely noble in its function as a prestigious junction, a rather straightforward daylit space emerges at the end. This produces an altogether skillful, almost dramatic sequence of spaces. All atria present themselves as exterior courtyards or squares, thanks to their perforated facades and the surrounding gallery corridors of the office areas.

To underline this characteristic, the idea arose to frame all gallery corridors with a linear light channel and introduce another clear break between the two areas. The light lines are embedded in ceiling recesses with overlapping T16 fluorescent lamps and efficiently apply light directly to the walkways. The surrounding corridor walls are also illuminated from the light lines' stray light. The perforated atrium facade, by contrast, remains relatively dark, while the courtyard's floor receives light from narrow-beam HIT spotlights. Within the atrium, one is now surrounded by glowing windows while in the gallery one experiences the feel of looking into a courtyard. The concept of light lines logically develops as a continuous theme on all floors, even to the furthest corner of the building. At the top, however, next to the roof terraces, the light line mediates between the actual courtyards and the surrounding corridors, acting as the separation element between interior and exterior.

ART EMPHASIZES THE ACUTENESS OF THE ATRIA

In each of the first three atria there is a giant metal ring from the artist Raik Elias. One stands on the wall, while another is high above, attached under the ceiling. The rings impressively display that the atria are tapered in one direction, yet with their weightless appearance, they transform this apparent disadvantage into a successful work of art.

STAIRWAY CONNECTING SCULPTURES

If one wants to move from floor to floor by foot, you enter the essentially open stairwell. Hidden behind louvres, it can only be seen hazily from the void. Following the concept of linear transportation path lighting, all stairwells are fitted with a linear LED handrail. In unity with the lamellae, the handrail lighting on the stairs acts as a refreshing contrast and – when seen from the atrium – appears pleasantly restrained. Despite integral lighting, the rectangular handrails are only 50 x 30 mm, just as the architects had originally envisioned. This is due to the currently available small, yet powerful designs of LED technology.

LIGHT ON THE GROUND FLOOR COMPLETELY BEHIND METAL MESH CEILINGS

Specifically on the ground floor, areas adjoining the atria present themselves differently. As a result, the lighting at the foyer and the connection to the conference rooms changes to HIT downlights behind metal mesh ceilings. To add variation, object-like pendant luminaires are used to mark individual seating areas as rest zones. In front of the conference area, there is a particularly high contrast and stimulating entry area resulting from distinct light emanated by wall sconces with compact fluorescent lamps.

CLIENT:
Baden-Württemberg Foundation gGmbH, represented by the Baden-Württemberg State Office of Property and Construction, Stuttgart office

OCCUPANT:
Baden-Württemberg Ministry of the Interior
Baden-Württemberg Ministry of Environment, Climate and Energy
Baden-Württemberg Ministry of Rural Affairs and Customer Protection

ARCHITECTS:
Staab Architekten GmbH, Berlin

TEAM LEADER LICHT KUNST LICHT:
Edwin Smida

COMPLETION:
2013

PROJECT SIZE:
33,000 m^2

OVERALL BUILDING BUDGET:
65 million euros

LIGHTING BUDGET:
1.3 million euros

在走廊中，灯光通道清晰地划分了中庭和走廊区域。线形灯条发出的散溢光也照亮了走廊后墙。

In the gallery corridors, light channels create a clear break between the atria and the corridors. Stray light from the light lines illuminates the gallery rear walls.

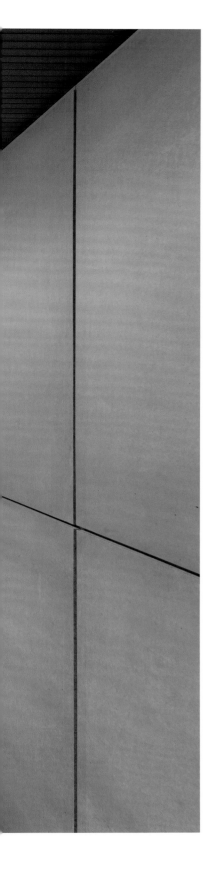

楼梯台阶透过装饰格栅的缝隙依稀可见。LED灯带集成在细长的扶手栏杆内，勾勒出楼梯的走向。

The stair flights can only be seen hazily through the lamellae. LED profiles have been integrated into the slender handrails to illuminate the treads.

走廊和楼梯台阶在中庭最边缘处交会。隐蔽安装在天花夹层的灯带与扶手灯光汇聚于此，形成集中的光区，强调了建筑水平和垂直方向上的连接关系。

At the end of the atria, gallery corridors and flights of stairs overlap. The concurrence of the ceiling light channel and the handrail light engenders a concentration of light that marks the connection of horizontal and vertical progression.

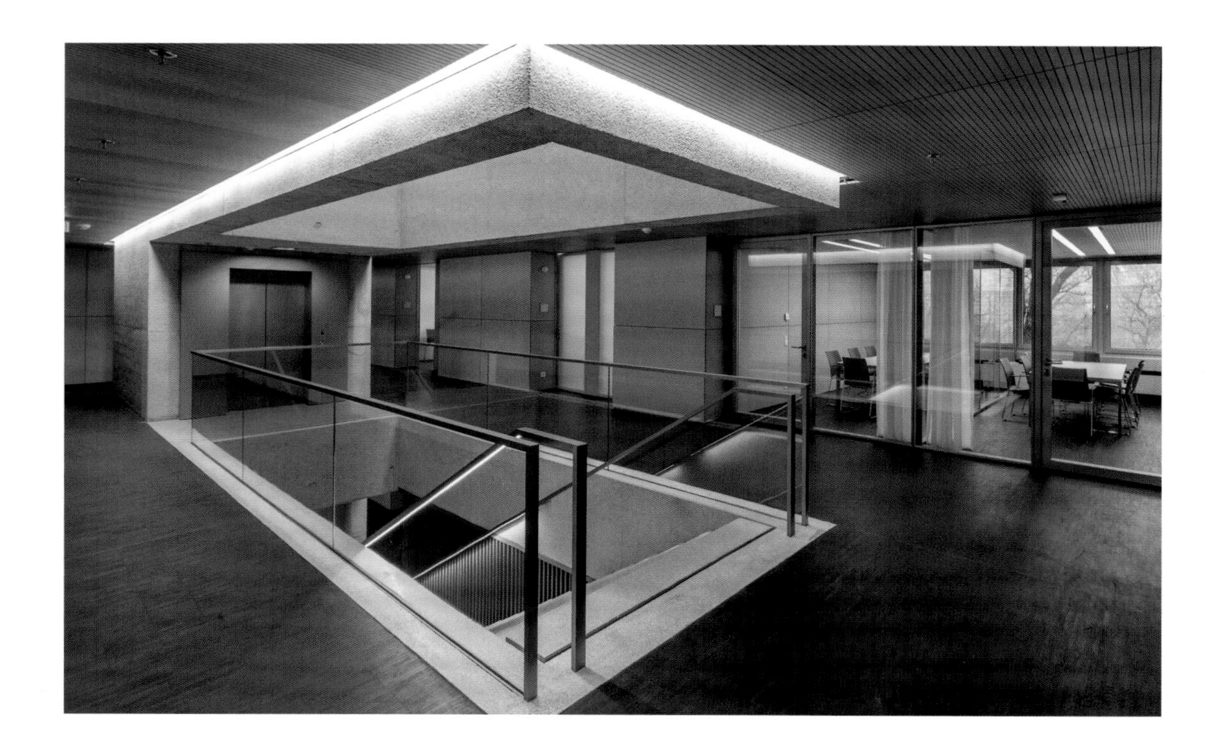

天窗勾勒出楼梯井最顶部的边界。与其他楼层一样，隐藏在天花夹层的灯带界定了室内和室外空间。

Skylights form the uppermost boundary of the staircases. Here too, a light channel marks the boundary between the interior and the exterior.

在被灯光强调的楼梯造型与墙体立面的合力衬托下，竖向格栅仿佛变成了半透明的屏风，朦胧之美油然而生。

The brightly illuminated treads and wall sections transform the louvers into a translucent layer, that allows for diffuse vistas.

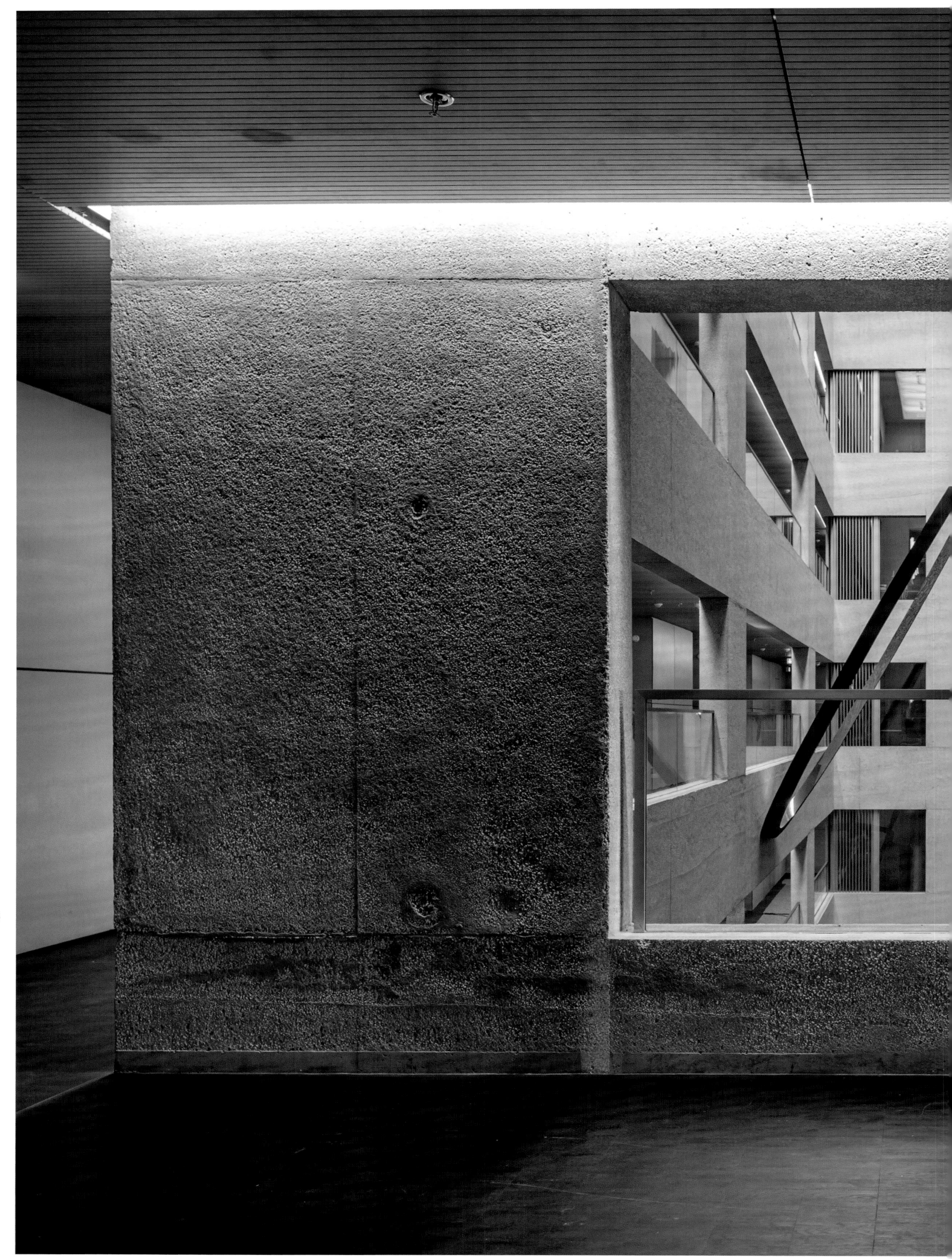

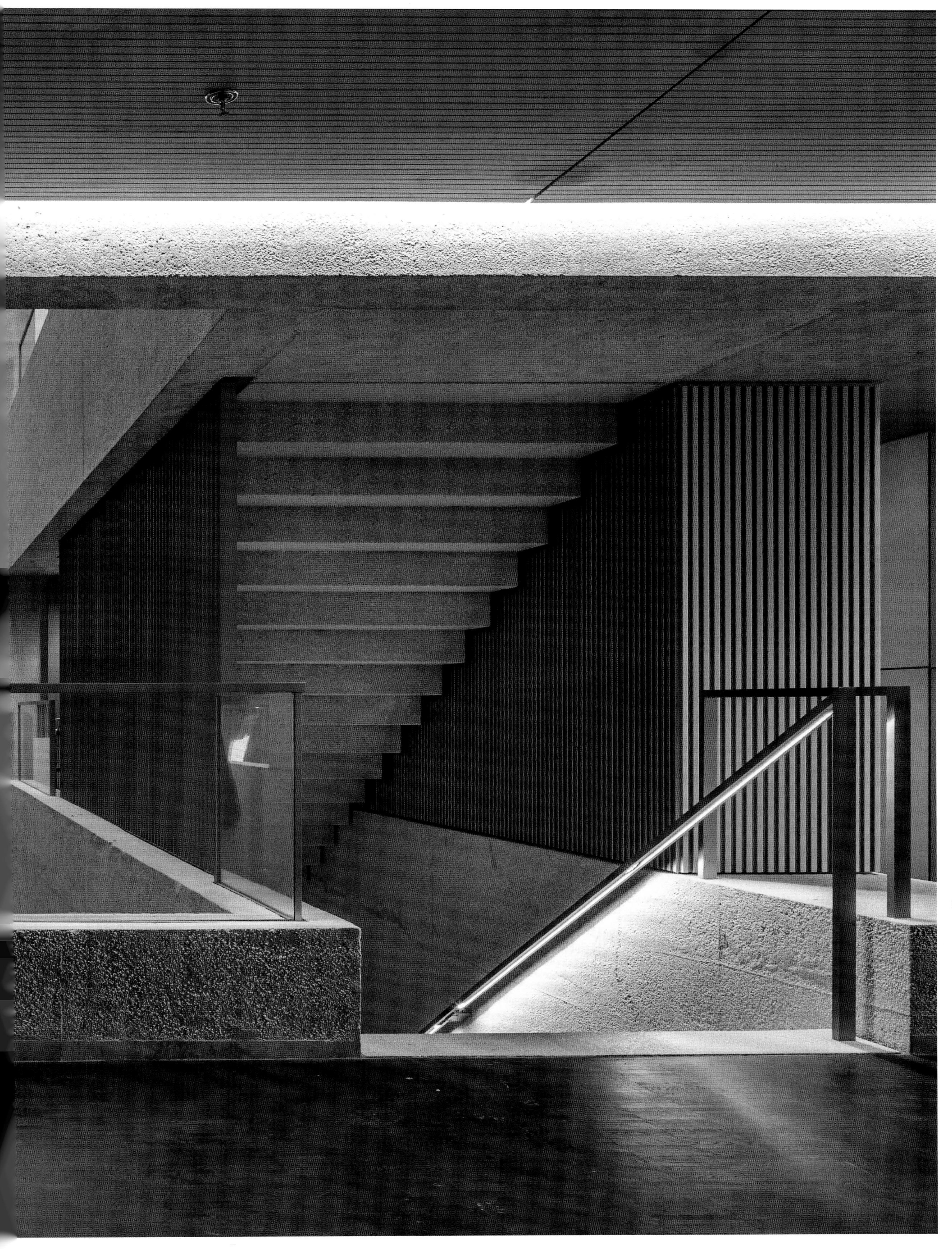

AWARD
WINNER

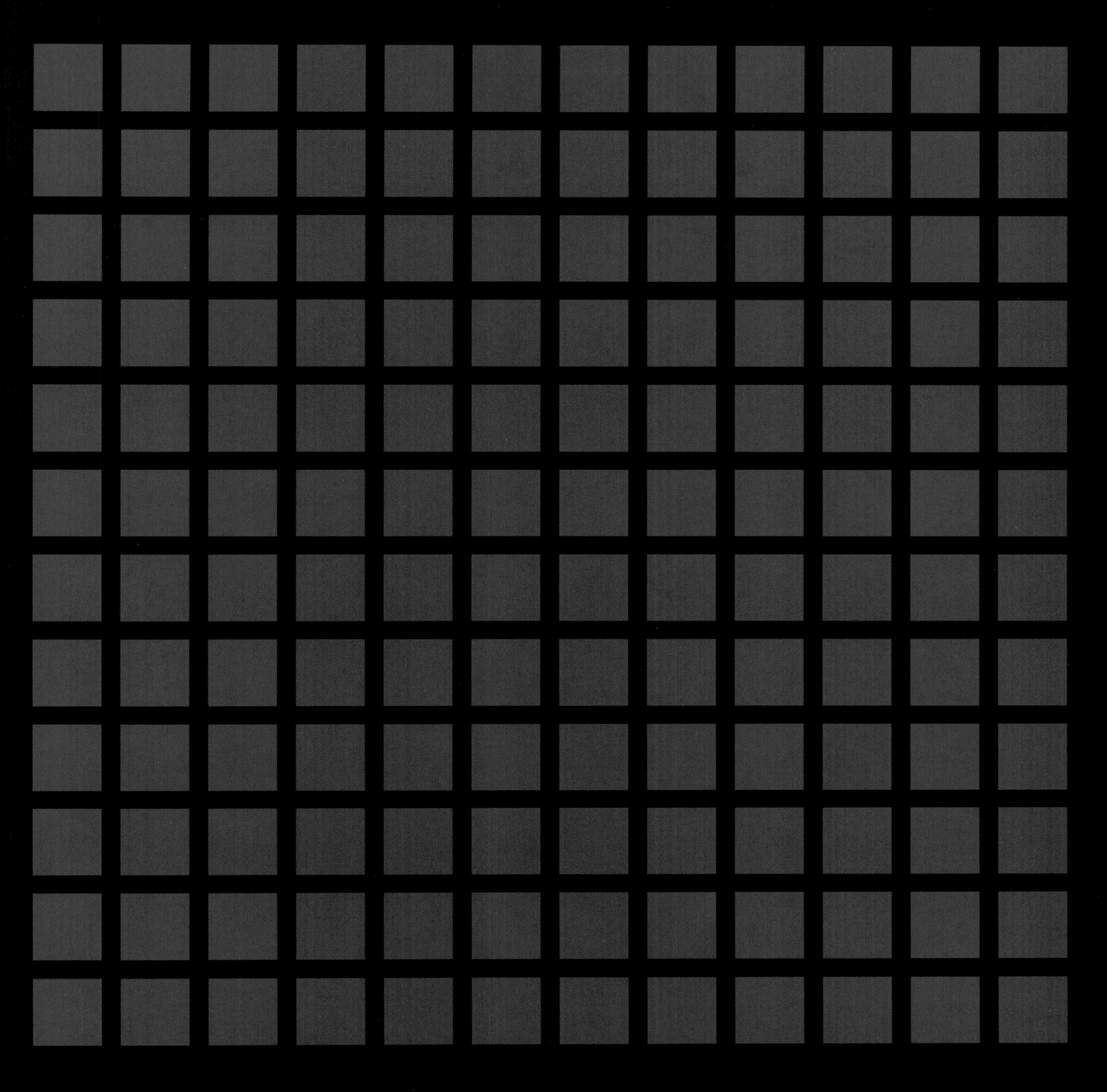

LICHT KUNST LICHT 4

利布弗劳恩圣玛利亚教堂骨灰安置所

多特蒙德

COLUMBARIUM LIEBFRAUENKIRCHE DORTMUND

细腻共情的照明氛围——采用LED氛围照明的多特蒙德利布弗劳恩圣玛利亚教堂

这座19世纪重新翻修的教堂建筑，如今成为多特蒙德市区内一座令人印象深刻的举行骨灰瓮埋葬仪式的地点。细腻共情的照明设计完全由LED技术实现，有效强调了教堂的室内设计理念。

重新规划室内空间的提案并非特别新颖——这在其名上就能体现。基督教早期的骨灰安置所是鸽巢式的、半地下或完全地下的瓮墓。由于城市新建公墓日益紧缺，加之建设成本的增高，德国各地的市政当局都在规划新的墓地地点。由此看来，将目前使用率较低的市区教堂再次利用的方案也是合情合理的。于是，包括多特蒙德在内的德国部分城市，已经将过去的教堂建筑改建为骨灰瓮墓地。1883年建成的新哥特式利布弗劳恩圣玛利亚教堂曾经风光无限，如今的访客却寥寥无几。因此，多特蒙德市政府与帕德伯恩大主教团合作，提出了将其改建为骨灰瓮墓地的方案。

无遮挡的空间通透感

在招标和合同授予的过程中，业主和评审团都展现出了谨慎、稳重的态度。最终，施塔布建筑设计事务所凭借极简化的概念设计，从众多竞标方案中脱颖而出，赢得了第一名，获得了该建筑设计竞赛的框架内合同。与其他竞标方案不同的是，施塔布建筑设计事务所不打算将骨灰瓮墓地设计成迷宫般的紧凑式堆叠布局。相反，建筑师拆除了大量的现有陈设，力求重现空间过去的开放性和通透感。

在该项目中，建筑设计师被要求只能在首层靠近地面的空间设置墓葬场地来安置骨灰瓮。骨灰瓮的低位布置传达了这个墓地与传统墓葬的清晰类比。骨灰瓮沿着中殿的中心线分布，延展到侧廊。它们之间摆放的长凳可以让祭拜者与逝者离得更近一些。沿着建筑轴线，通畅无阻的中央走廊营造了空间的轴向视线。从这里望去，祭拜者可以看到死亡之书和复活节蜡烛基座。墓葬区域的完成面呈方形，用来安置骨灰瓮和铭文牌。室内设计方案的视觉效果与紧凑排列的教堂长椅相类似，使教堂呈现出通透和开放的空间氛围。

教堂半圆形后殿中设有木制祭坛、读经台、骨灰翁石碑和一个木制十字架。这里是举行葬礼的地方。入口处带有天使雕像的圣水钵和中殿尽头的复活节蜡烛象征着教徒从洗礼到葬礼之间的人生之路。

基于适宜性的照明设计

多特蒙德骨灰安置所——利布弗劳恩圣玛利亚教堂的照明设计是基于其空间设计和宗教内涵的，经过仔细推敲得出的适宜性设计方案。照明在满足葬礼和大公服务视觉要求的同时，为哀悼者营造出静谧、沉思的空间氛围。

LKL的照明设计师解释道："将分别进行开关和调光控制的环境照明与重点照明相结合，可以满足多种照明场景的需求。为了避免分散人们对教堂建筑形态的注意力，灯具设备以视觉对象的形式出现，与环境背景融为一体。"

整个空间只在入口处设置了一盏清晰可见的灯具。设计师选取一盏包含直接和间接照明组件的圆环状吊灯，对立有雕像的圣水钵进行重点照明。中殿的照明设计首先将访客的视线沿着中轴线引导到前方的唱诗班区域。与教堂的其他区域相比，唱诗班区域的照度水平更高，因此唱诗班区域成了整个空间的视觉参考点。设计师将投光灯具隐蔽地安装在立柱后方，对祭坛、读经台、骨灰瓮石碑和十字架进行重点照明。唱诗班区域的长椅则由隐隐约约、均匀的光线进行照明。

LED带来的氛围照明和高效照明

中殿的环境照明由包含直接和间接照明组件的LED定制灯具实现，这些灯具安装在中殿立柱的柱头上方，安装高度为12.7米。灯具采用紧凑的块状灯体设计，内设6盏可调节角度的投光照明模块。灯体完成面的外观灵感来自相邻的石材表面，照明模块很好地隐藏在灯体内。

在定制灯具内部，其中4盏可调式投光灯通过暖色调的光线温柔地照亮地面，营造出平静的灯光氛围，同时强调了铜制瓮墓的厚度和材质的色彩，使墓碑上的浮雕内容清晰可见。剩下的2盏投光灯通过宽角度的漫射光照明，表现了拱形天花表面，并实现了空间内良好的光线平衡。与此同时，空间的高度在光线的烘托下变得清晰可见。安装在灯具上的防眩遮光罩可以避免拱门上出现杂散光，此外，所有直接照明的投光灯都带有蜂窝遮光格栅，避免访客受到眩光干扰。

走廊和唱诗班空间也沿用了中殿的照明设计方案，但这些区域立柱的柱头略低，因此灯具的发光强度也相应降低了一些。通过LED可调光技术，使用者可以轻而易举地调节光线，以适应各种使用场景。由于该控制系统中应用了感光传感器，因此人工光照明系统也能对日光的变化做出反馈。在自然光照度水平较高的情况下，中殿和走廊空间的人工光照明组件会自动关闭。

教堂里所有LED灯具都是通过可寻址的DALI系统来进行控制的，这满足了效能优化、高光质和极长维护间隔的照明要求。照明控制系统被整合在建筑相关的控制通道中，使用者可以通过直观式控制面板调用5个预置的照明场景。

光、建筑、功能和谐统一

人工光照明的概念方案支持了建筑设计师对教堂室内空间的重新诠释。灯具的极简化定制设计使其悄然融入了空间环境。借助LED可调光技术，设计师不仅成功建立了对室内日光的补充照明机制，而且通过光线对拱顶及其延展面进行强调，在夜间营造出让人平静和沉思的光环境氛围。

业主：
东鲁尔区天主教会协会
使用人：
多特蒙德利布弗劳恩圣玛利亚教堂
建筑设计：
施塔布建筑设计事务所，柏林
LKL项目经理：
劳拉·苏德布罗克
竣工时间：
2011年
项目规模：
1900 平方米
项目总预算：
330万欧元
照明预算：
10万欧元

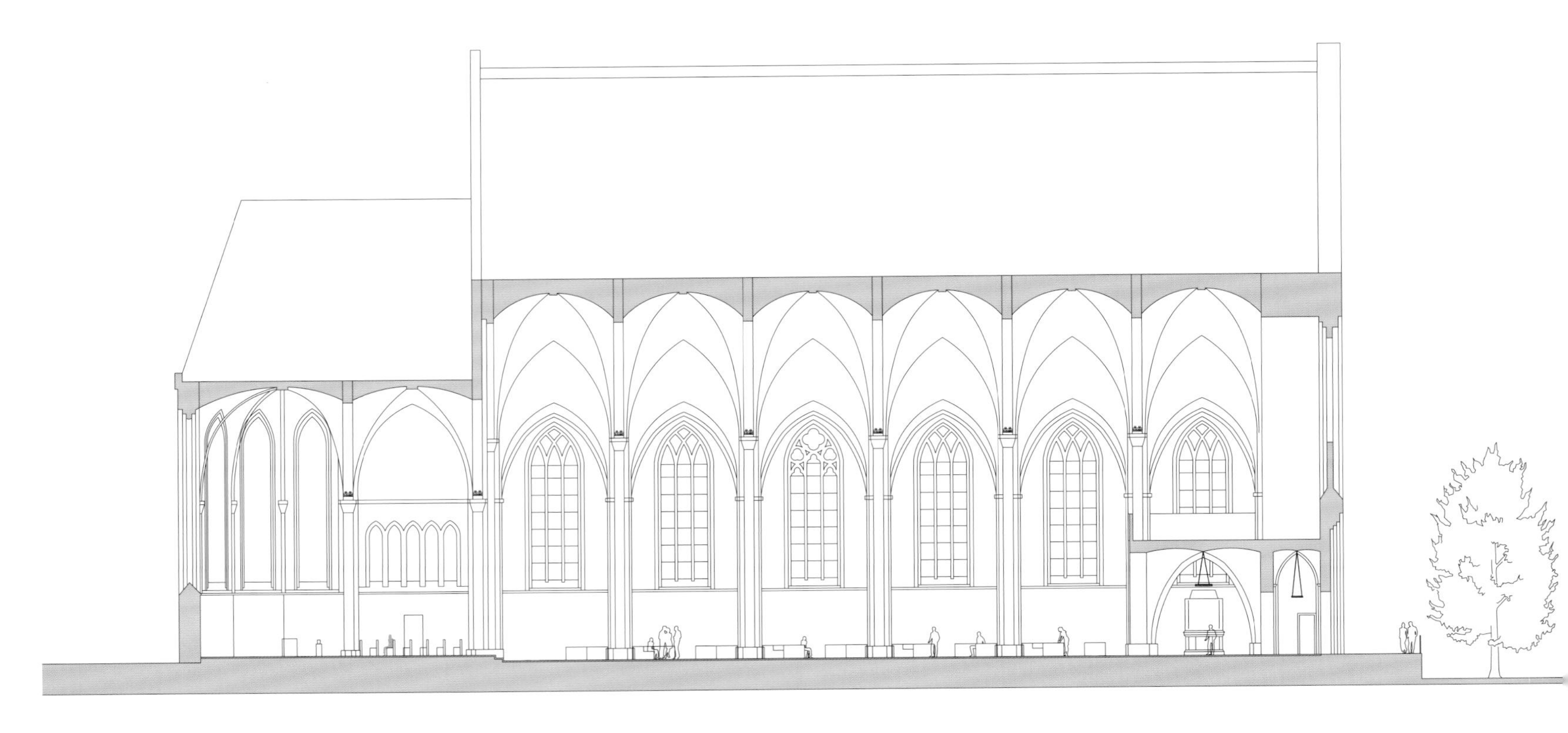

纵剖面图

Longitudinal section

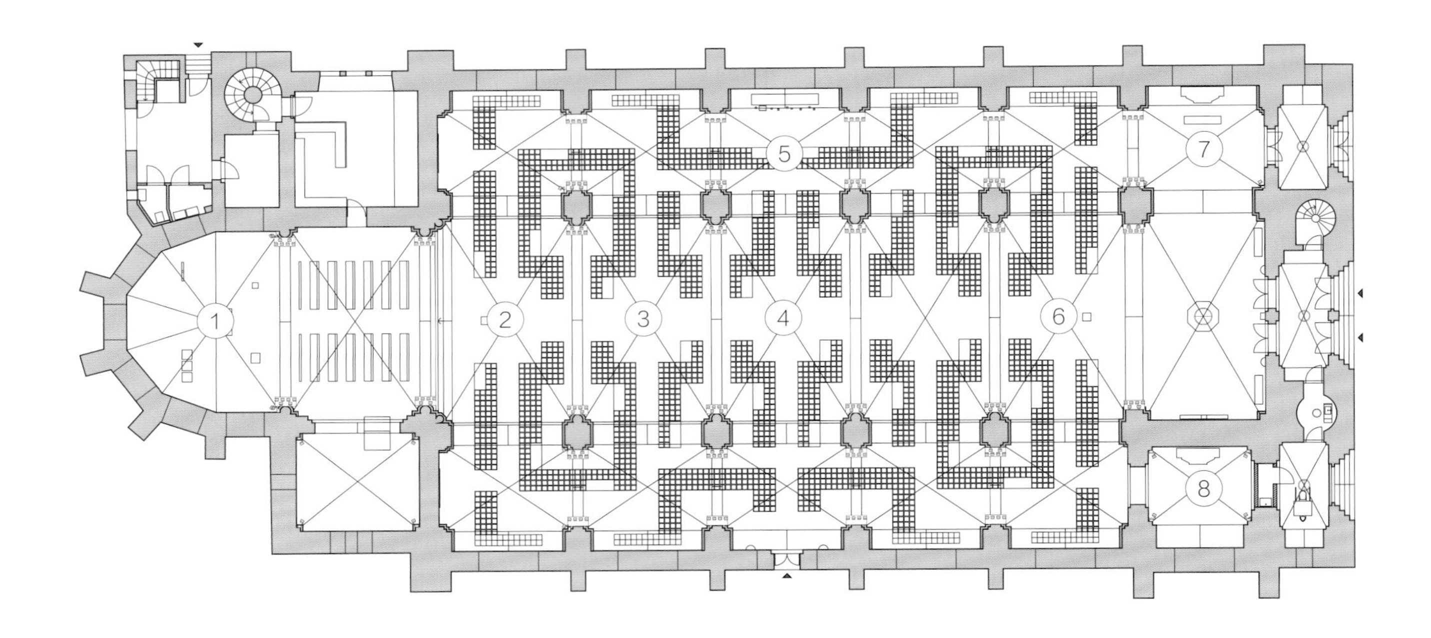

平面图

Floor plan

1.丧葬圣堂
2.复活节蜡烛
3.瓮棺墓地
4.安息之地
5.骨灰堂
6.死亡之书
7.圣母怜子雕像（带烛台）
8.约瑟夫圣堂

1. Funeral Chapel
2. Easter candle
3. Urnfield
4. Final resting place
5. Place of the Forgotten
6. Book of the Dead
7. Pietá with candle holder
8. Joseph Chapel

restored church building from the 19th century presents itself an impressive location for urn burial ceremonies in urban eas. The sensitive lighting design – solely realized with LED chnology – effectively underlines the interior design concept.

e idea is not particularly new – this is apparent in the name. lumbaria' in the early Christian period were dovecote like, mi or fully subterranean urn graves. The growing lack of w urban cemeteries and specifically the cost pressure have used municipal authorities throughout Germany to establish w locations for urn burials. Thus, it makes sense to utilize rdly used urban inner-city churches. In several German cities, so including Dortmund, former church buildings have been modeled for urn cemeteries. The neo-Gothic Liebfrauen church ilding, erected in 1883, had seen better days and was regisring few visitors. Thus, in cooperation with the Archbishopric Paderborn, the municipal administration of Dortmund has omoted the idea of its conversion into an urn burial location.

IOBSTRUCTED OPENNESS

the tendering process and award of the contract, the owner d jurors showed a sure hand in their decision-making. As a sult, the impressively minimal concept by Staab Architekten om Berlin had been awarded the 1st prize and the contract thin the framework of an architectural competition. Unlike the her contributors, Staab Architekten decided against developing labyrinth-like and stacked layout for the urn graves. On the ntrary, the designers strived to restore the former openness d transparency of the space by generously removing existing rnishings.

e architects were confined to the construction of a ground-lev burial site to accommodate the urns. The low-set positioning the urn graves conveys a clear analogy to traditional burial. e urns are distributed along the centre line of the nave and read into the aisles. Benches among the urns allow the visitors be near to the dead. An unobstructed central corridor serves as axial line of sight along the building's axis. Here, the pedestals r the Book of the Dead and the Easter candle are located. The p surface, with its square contour, serves to accommodate e urns and later the inscription plates. This solution, which sembles compact church pews, gives the church a sense of ansparency and openness.

cated in the apse are the wooden altar, the ambo, the urn stele d a wooden crucifix. This is where funeral services occur. The oup with angel statue at the entrance and the Easter candle at e end of the central nave impressively symbolize the passage etween baptism and funeral.

GHTING DESIGN ACCORDING TO THE PRINCIPLE F APPROPRIATENESS

e development of a lighting design for the Columbarium ebfrauenkirche in Dortmund is based on the idea of a lighting oncept which is appropriate in the context of a sacred building in s space and atmosphere. The illumination satisfies the demands to provide an appropriate setting for funeral and ecumenical services as well as to offer the mourners a place of calm and contemplation.

"Combining separately switched and dimmable ambient and accent lighting, we can accommodate the requirements for various lighting scenarios", explain the planners from Licht Kunst Licht. "In order not to distract from the architecture of the church building, the luminaires appear as a visual object, blending into the background."

Only in the entrance area exists a distinctly perceptible luminaire, a direct-indirect circular chandelier which accentuates the bronze basin with statue. The lighting concept in the nave initially directs the visitor's view along the central axis into the choir. Due to its higher light level compared to the rest of the church, the choir serves as visual reference point. Projectors inconspicuously mounted behind columns, accentuate the altar, the ambo, the urn stele, and the cross. The benches in the choir receive a subtle and homogenous illumination.

ATMOSPHERIC AND EFFICIENT LIGHTING WITH LED

The ambient lighting of the nave is achieved by custom designed direct / indirect LED-luminaires mounted above the capitals at a height of 12.7 meter. The luminaires are designed with a compact block-shaped housing, accommodating six adjustable projectors, with a finish inspired by the adjacent stone surfaces. The lighting elements largely disappear in this housing.

Four of the adjustable projector heads subtly illuminate the ground with a warm-toned light and create a calm lighting atmosphere, which emphasizes the depth of the material and colour of the bronzed urn grave surfaces and makes the engraving readable. The two remaining projectors reveal the vaulted ceiling with wide-angled diffuse light and create a good balance of light levels within the space. At the same time, the height of the space becomes palpable. Anti-glare shields prevent the arches from receiving stray light, while the use of honeycomb louvers on all direct projectors prevents glare for the visitor.

The lighting principle of the nave is continued in the aisles as well as in the choir on a slightly lower capital height and accordingly lower light intensities. Thanks to dimmable LED technology, the brightness can be easily adapted to various user scenarios. With the implementation of a light sensor, the artificial illumination also reacts to changing daylight situations. In case of higher natural illuminance levels, the artificial lighting components in the nave and in the aisles are switched off.

All LED luminaires in the church are DALI addressable, which fulfills the requirements of optimized efficiency, high light quality, and extremely long maintenance intervals. Combined into architecture-related switching channels, five programmed lighting scenes can be called up via an intuitively designed control panel.

LIGHT, ARCHITECTURE, AND FUNCTION IN HARMONY

The artificial lighting concept supports the architectural reinterpretation of the church interior. Due to their simplified form, the custom-designed luminaires unimposingly blend into the space. With the dimmable LED technology, the designers not only succeeded at establishing a natural complement to the incoming daylight, but also created an atmosphere of calm and contemplation at night by carefully emphasizing the vaults and their expanse.

CLIENT:
Gemeindeverband Katholischer Kirchengemeinden
Östliches Ruhrgebiet
OCCUPANT:
Grabeskirche Liebfrauen Dortmund
ARCHITECTS:
Staab Architekten GmbH, Berlin
TEAM LEADER LICHT KUNST LICHT:
Laura Sudbrock
COMPLETION:
2011
PROJECT SIZE:
1,900 m^2
OVERALL BUILDING BUDGET:
3.3 million euros
LIGHTING BUDGET:
0.1 million euros

教堂中殿的灯光将访客的视线沿着中轴线引导至前方的唱诗班区域。中殿中央的整体照明由安装在立柱上方的定制 LED 灯具提供，灯具包含了直接和间接照明组件。

The light orchestration in the church nave directs the visitor's view along the central axis into the choir. The central nave's general illumination is provided by direct / indirect custom LED luminaires, mounted above the columns.

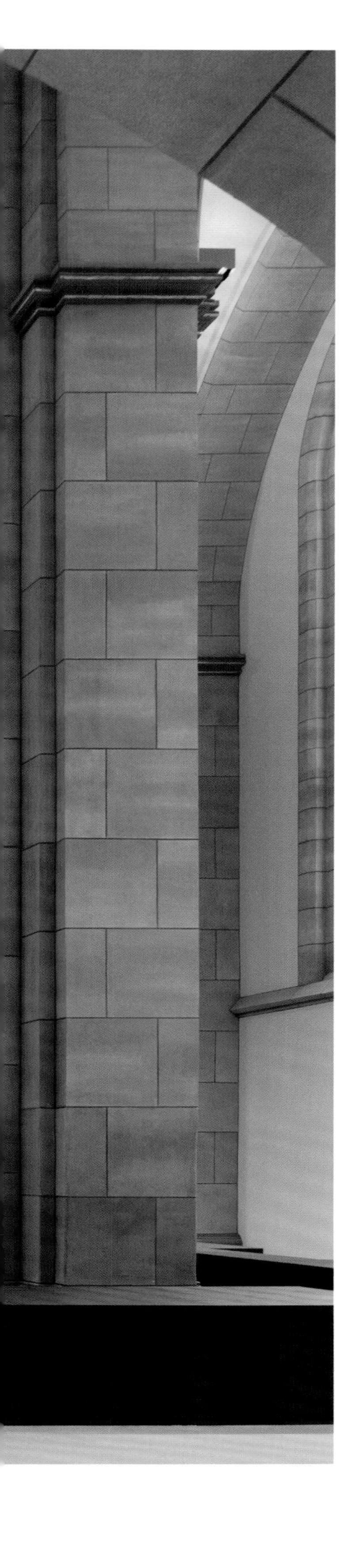

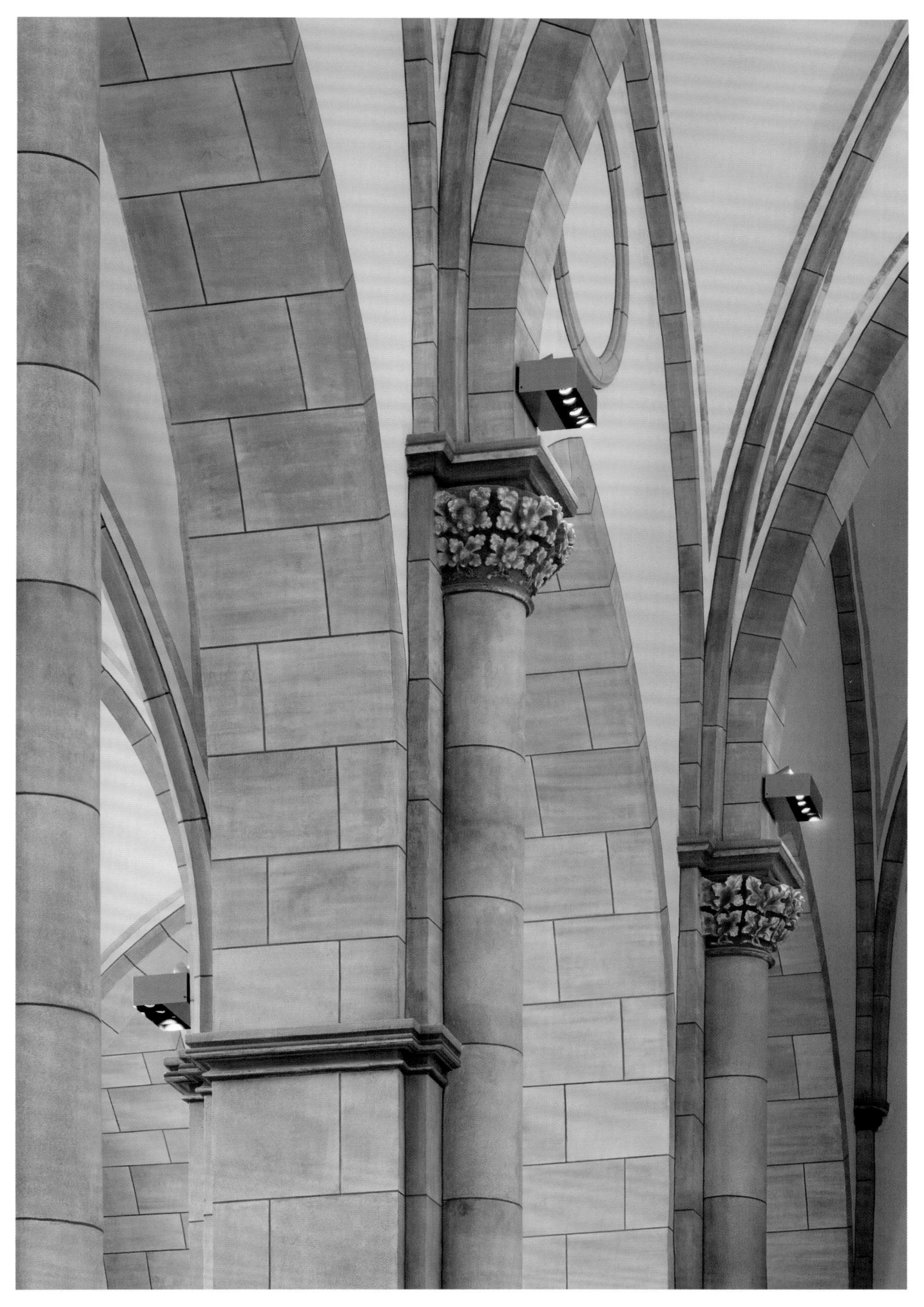

灯具采用紧凑型块状外壳，可容纳6盏可调式投光灯，其中4盏用于地面照明，另外2盏用来强调天花拱顶。这些照明组件基本都隐藏在灯体内。

The luminaires are designed with a compact block-shaped housing, accommodating six adjustable projectors, four for the illumination of the floor surface and two for the emphasis of the ceiling vault. The lighting components are largely concealed in this housing.

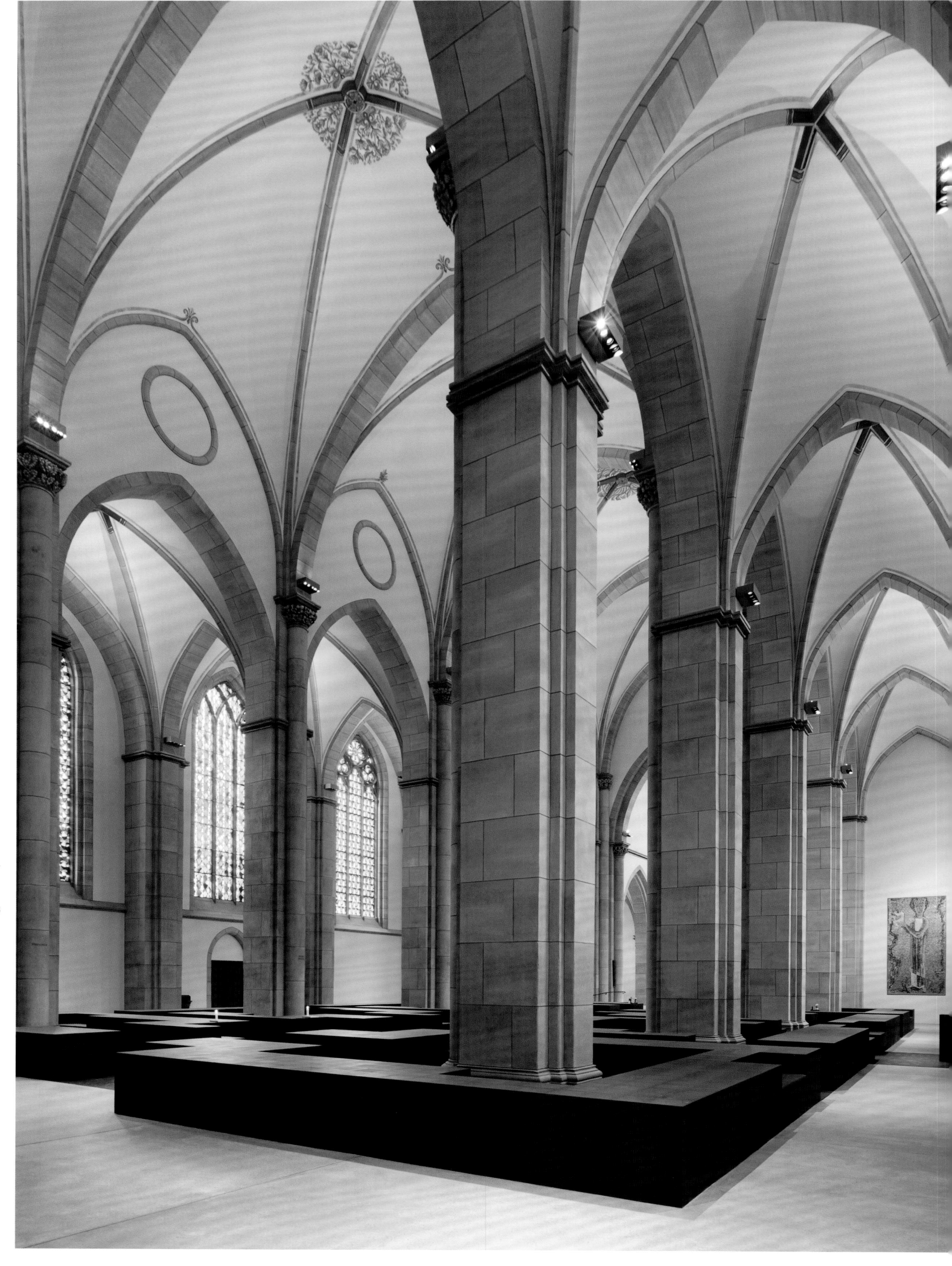

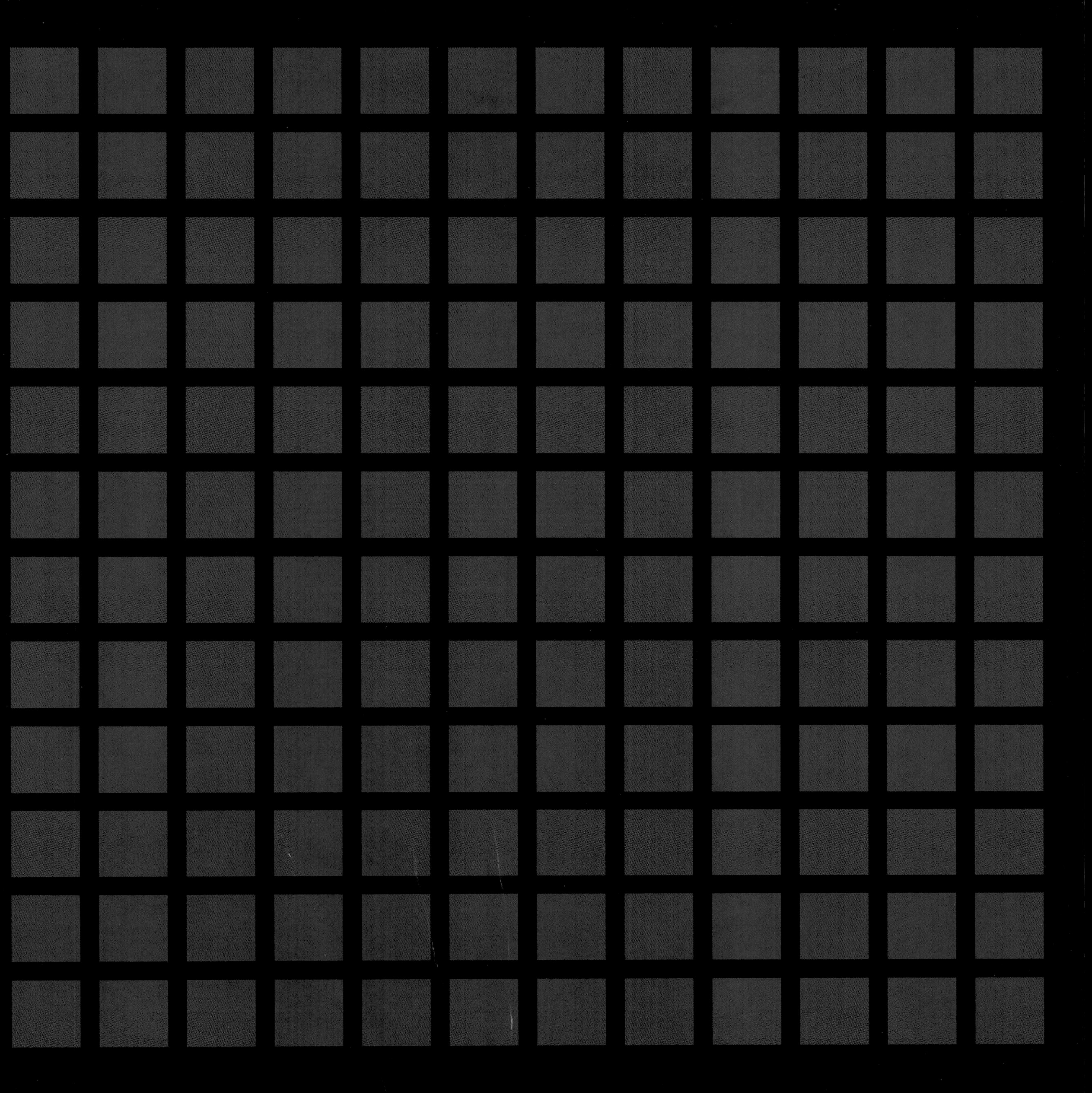

LICHT KUNST LICHT 4

威斯滕德校区的高校建筑群

法兰克福

UNIVERSITY BUILDINGS ON CAMPUS WESTEND

FRANKFURT AM MAIN

欢迎来到知识之城——位于威斯滕德校区的高校建筑群

于法兰克福的歌德大学威斯滕德校区是德国规模大的教育类建筑群。随着2008年法律与经济学大楼的竣工，以及2013年心理、教育和社会科学院大楼与行政管理大楼的相继竣工，歌德大学大模扩建项目中的三座新建筑均已经竣工验收。三建筑中包含演讲厅、大堂、研讨室和图书馆的多能空间分布形式为其照明设计方案提出了多种不的任务与要求。

律与经济学院大楼于2008年建成，是该项目的期开发工程。带有一个计算机中心的心理、教育社会科学学院大楼和服务于大学管理部门的行政理大楼均属于二期工程，于2013年完工。与校里的其他建筑一样，这三座由托马斯·缪勒·伊·赖曼建筑设计事务所设计的大楼与历史上赫赫名的IG法本公司大楼形成了强烈的呼应。IG法本司大楼由汉斯·普尔希（1869—1936）在1928至1931年间建造，是该校区的标志性建筑。据了，新建筑群需要在设计的各个方面与校区整体建环境保持和谐统一。

律与经济学院大楼

了这两个学院下设的众多院系办公与教学场所，法律与经济学院大楼还包括一个拥有近千个座的图书馆、一个自助餐厅和众多研讨室。在建筑层的公共区域中，室内房间都拥有良好的通透，借助大面积的落地玻璃，打造迷人的视觉联。谢绝进入、仅供观赏的庭院不仅优化了室内的光条件，而且还使室内与户外景观形成互动，具美化环境、提高使用者舒适度的作用。

门厅之中的宏伟入口

对融入校区整体面貌的校园建筑，其自身形象的达也是照明设计的主题。人工光照明有时是显露，有时是隐蔽的，但始终与建筑形式和谐统一，满足空间所有的功能需求。在这一总体设计构思，照明设计师在两层楼挑高的门厅中，使用经典环形凹槽灯带对墨绿色的天花完成面进行重点照，以突出门厅宽敞、优雅的空间特征。中央天花夹层天花开口的边缘为柔和的圆形倒角，与下方实的木质栏杆的圆形拐角相似。不显眼的微型天嵌入式下照灯为中央区域提供了适当的功能性和氛性照明。走廊和地面层也采用了下照灯和筒式墙灯。照明设计师特意在这些区域采取了内敛的明方式，以免抢了中央大厅的风采。与占据校区端的前IG法本公司大楼的门厅相呼应，这个门厅整体视觉效果可以看作是其现代化的参照，彰显它低调的豪华气质。

在雕塑般的楼梯中穿行

通过两个与门厅相邻的宽敞的楼梯井，访客可以通往建筑上层的系部教室和研讨室。主楼梯的台阶采用独立的钢结构，周边设置围护结构。楼梯背面和木质镶板之间有一条开槽，凹槽灯带内嵌其中，均匀地铺设在整个楼梯的背面，赋予楼梯雕塑般的形式感。

与建筑相呼应的主题线形光

图书馆的照明设计由三种灯光元素组成：天花中的半嵌入式下照灯、线形灯带和工作阅读灯。线形灯带贯穿整个图书馆，通过图形线条来辅助和凸显建筑形式。依照这一照明构思，设计师使用三组发光边框分别勾勒了三个天花开口。灵活变化的折叠天窗结构位于阅读区上方，围绕其框架边缘的线形灯带将它们作为空间设计元素进行凸显。

此外，设计师在图书馆的中间区域增设了一组环绕整个空间的线形发光边框，在凸显建筑造型的同时，也强调了图书馆温润的木饰面质感。线形灯带的形式语言也出现在吊灯设计中。与空间中其他线形灯带元素一样，即使书架的照明有时由与其平行的灯具提供，有时由与其正交的灯具提供，但所有房间的吊灯只沿一个方向排列。

在图书馆的其他区域，由于结构限制，下照灯采用半嵌入式的形式安装在混凝土天花中，为空间提供辅助照明。设计师交替使用线形光和点状光元素，一方面确保了各区域的基本照明，另一方面可以防止使用者产生完全功能性的过度单调的光线感受。为了创造高品质的光环境氛围，并确保阅读区和工位区所需的照度水平，设计师在书桌上设置了简约时尚的阅读灯。阅读灯固定在桌面上，使用者可任意打开、旋转和倾斜，并且在需要的时候为桌面提供工作、阅读所需的500勒克斯的照度。

心理、教育和社会科学学院大楼

除了院系办公与教学场所外，心理、教育和社会科学学院大楼亦设有挑高两层的图书馆、演讲厅、研讨室、学生服务中心及自助餐厅等公用设施。该学院大楼和邻近的行政管理大楼构成了一个大型建筑群。

大楼中的两个庭院在其建筑整体设计中起着关键性作用。通过不同的观看视角，室内和室外空间得以关联和互补。比如，在室内庭院中，访客可以看到开放、透明的图书馆，感受到温暖的饰面材质和友好的光线气氛。从图书馆向外观望时，这种内外环境的协同效应更加明显。只有在这个观看视角中，访客才能切身体会到建筑立面上所采用的令人难以置信的大尺度落地玻璃。落地窗几乎贯穿了两层楼的高度，让访客有种仿佛置身于室外的感受。

现代性和开放性：心理、教育和社会科学学院大楼的门厅

在门厅内，灵活变化的折叠天窗结构在天花上形成了光影浮雕，呈现出迷人的明暗渐变图案，而直射的阳光则通过有趣的光影组合，将门厅衬托得光彩夺目。在夜间，天窗框架正下方内嵌的环绕边框的灯带勾勒出其形态，使建筑天花仍是整个空间设计的核心。环绕边框的灯带不仅突出了天窗浮雕般的结构，还极大辅助了空间的整体照明。

在首层，强有力的擦墙光围绕门厅四周带纹理的混凝土墙壁进行照明，营造了空间的光环境。安装在天花外围的灯具以活泼、生动的方式呈现出墙身本身的质感。擦墙光的一部分光线投在了与墙壁交接的地面边缘，勾勒出地面的轮廓，在引导空间方向的同时，突显了门厅的形态。

轮廓光是四周门廊中唯一的光源，营造出适度的戏剧性视觉效果。

在门廊上方的空间中，照明设计采用较为克制的表达方式。设计师使用下照灯提供照明，确保光线不会妨碍访客欣赏浮雕般的天窗。从大堂移步到走廊，照明设计方案随之改变。在这里，大型柔光灯具被内嵌在木质天花镶板中，这种设计语汇同样运用在了其他走廊空间。

业主：

黑森州，由黑森州建筑和施工管理部代表执行

使用人：

法兰克福歌德大学

建筑设计：

托马斯·缪勒·伊万·赖曼建筑设计事务所，柏林

LKL项目经理：

埃德温·斯米达

竣工时间：

2008年（法律与经济学院大楼）

2013年（心理、教育和社会科学学院大楼）

项目规模：

30 000 平方米（法律与经济学院大楼）

41 700平方米（心理、教育和社会科学学院大楼）

项目总预算：

5300万欧元（法律与经济学院大楼）

9600万欧元（心理、教育和社会科学学院大楼）

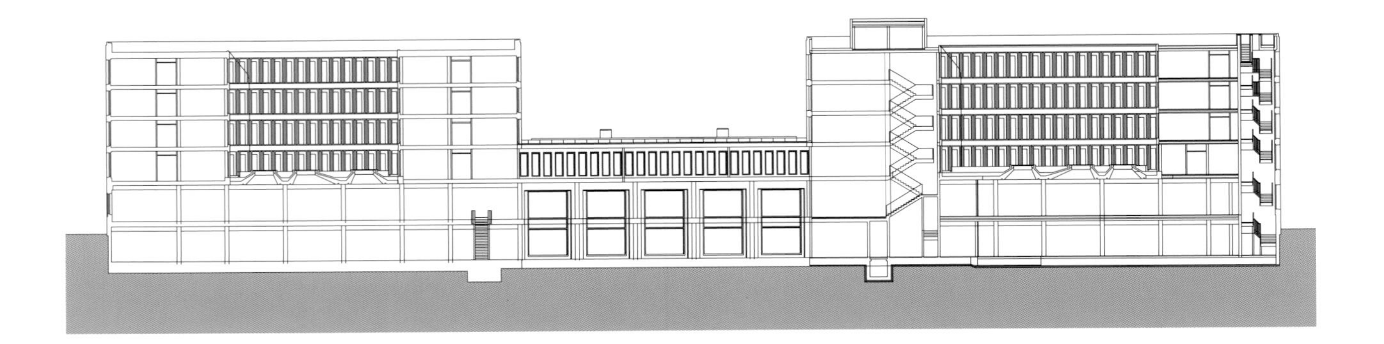

法律与经济学院大楼剖面图　Section RUW faculty building

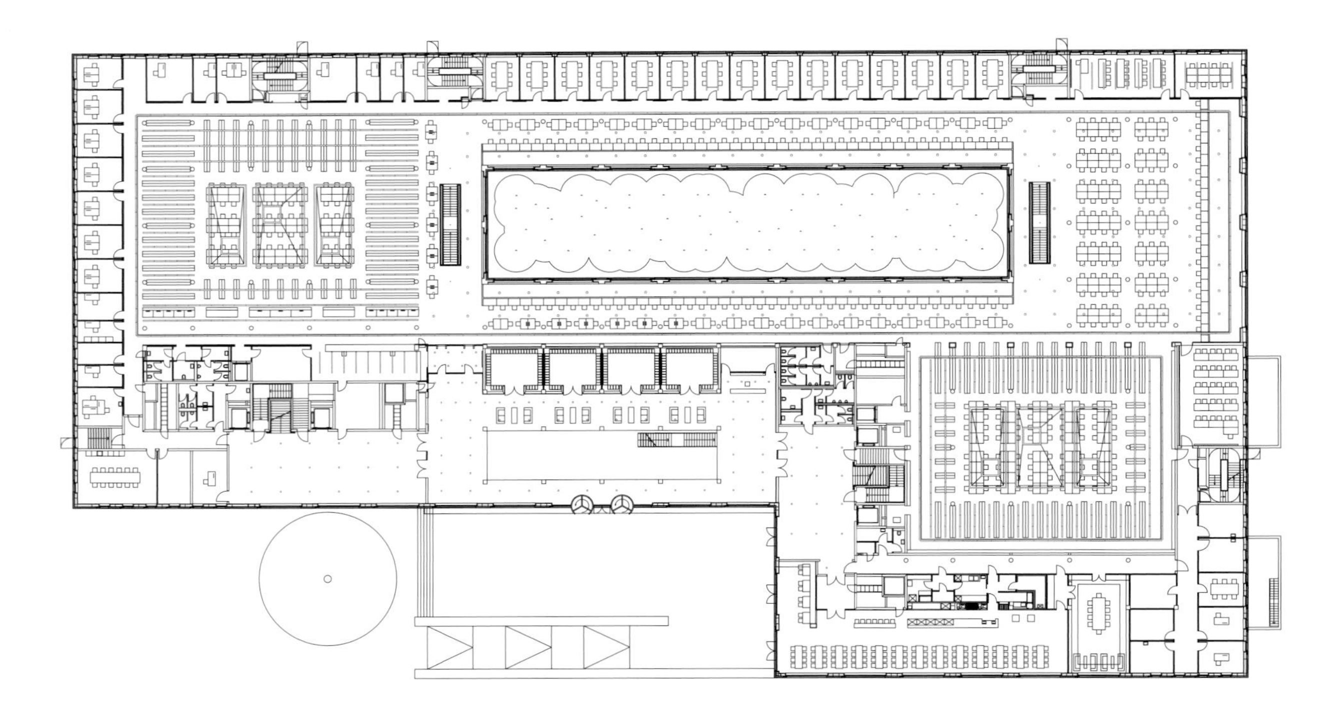

法律与经济学院大楼平面图　Floor plan RUW faculty building

e Westend Campus of the Goethe University in Frankfurt is nsidered the largest educational building site in Germany. th the faculty building for Law and Economics (RUW) finished 2008, and both the Department of Psychology, Education and cial Sciences (PEG) and Governing and Administration buildings mpleted in 2013, three further buildings of the university's large-ale development project have been handed over. The versatile ace allocation plan, consisting of lecture halls, lobbies, seminar oms, and libraries, presented the lighting design with a variety tasks.

e RUW faculty building was realized during the first develop-ent phase, in 2008. The PEG faculty building and the Governing d Administration building, which includes a computer centre d serves the university's management, were both finished in 13 during a second construction stage. Like all other buildings campus, these three buildings designed by Müller Reimann Ar-itekten, reflect heavily on the campus' landmarked and histori-lly-charged IG Farben Building, built by Hans Poelzig (1869 – 1936) tween 1928 and 1931. It was understood that the new buildings ould harmonize in all aspects with the overall campus ensemble.

CULTY BUILDING FOR LAW AND ECONOMICS (RUW)

addition to the various departments included in the two fac-ies, the construction of the Law and Economics facilities was include a department library with nearly 1,000 work stations, a feteria, and a large number of seminar rooms. In the public are-, located at the building's base, the rooms are marked by a high gree of transparency and allow fascinating visual relationships rough extensive glazing. Non-accessible courtyards optimize e daylight conditions and offer a relationship to the exterior, all ile providing amenity values.

RANDIOSE ENTRANCE THROUGH THE FOYER

e sensitivity for the buildings' integration into the entire cam-s ensemble was also the theme for the electric lighting design ncept. Sometimes visible, sometimes hidden, the lighting al-ays stays appropriate to the architecture and offers full func-nality to the rooms. Thus, the lighting concept supports the acious and elegant impact of the two-story foyers by accenting e billiard-green ceiling with a classic circular cove motif. The ge between the ceiling and the ceiling aperture is softly round-, similar to the rounded corners of the solid wooden balustrade low. Tiny, flush recessed downlights pleasantly provide the cen-al zone with an appropriate level of useable and atmospheric ht. The surrounding galleries and the ground floor are also illu-inated with downlights and downlight wall washers.These areas e intentionally restrained in order not to steal the show from e central interior foyers. The overall effect of the foyer is to be nderstood as a modern reference – of luxury, but not exorbitance o the foyer of the former IG Farben Building, which forms the uthern end of the campus.

ROGRESSION THROUGH SCULPTURAL STAIRCASES

e upper floors' department facilities and seminar rooms are ac-ssible via two spacious stairwells adjacent to the foyer. The flight stairs in the main staircase is a free-standing steel structure centred in its envelope. A slot exists between the underside of the stairs and the wood paneling, where a cove light uniformly stretches around all staircase soffits, adding to the sculptural presence of the stairs.

A THEME OF LIGHT LINES SUPPORTING THE ARCHITECTURE

The concept in the library is comprised of three lighting elements: semi-recessed downlights in the ceiling, light ribbons, and task lights. The light lines are integrated throughout, occurring in all areas where the architectural forms want to be supported and ac-cented by a graphic line. As such, three light frames delineate the perimeter of three dynamically angled ceiling pop-ups above the reading area, underscoring them as design elements.

The entire room is also framed by an additional light line running around the whole middle section of the library, simultaneously accentuating the warm, wood finishes encircling the room. The light lines appear again as pendant luminaires, and are, as always, oriented in each room to run in only one direction, even when the stacks are arranged both parallel and perpendicular to the linear illumination .

Not entirely recessed in the concrete ceilings due to structural rea-sons, downlights are located in the remaining areas to complement the space. The alternation of light lines and light points offers the general lighting in all areas, and at the same time prevents exces-sive monotony and functionality. For ambient quality and to ensure the required light levels for reading and work stations are met, sleek reading lights are fixed to the desks, where they can be freely switched on, rotated, and tilted to bring the required 500 lux to any desired target needed on the table tops.

FACULTY BUILDING FOR THE DEPARTMENT OF PSYCHOLOGY, EDUCATION AND SOCIAL SCIENCES (PEG)

In addition to the faculty departments, the Psychology, Education and Social Sciences building also houses communal facilities such as the two-storey library, lecture halls, seminar rooms, a stu-dent service center, and the cafeteria. Together with the neigh-bouring Governing and Administration building, a large complex of buildings is formed.

The two courtyards of the PEG building play a prominent role in the building's overall design, as their different perspectives dis-play how much the indoor and outdoor space benefit from each other. One example is the view from the interior courtyard into an open, transparent, and active library with warm finishes and inviting light. The synergy is even more strongly supported when looking from the library to the exterior, because only here are the incredibly large dimensions of the selected windows first revealed. Spanning nearly the entire two-storey height, the glazing effec-tively creates the feeling of sitting outdoors.

MODERN AND OPEN: THE FOYER OF THE PEG BUILDING

In the foyer, dynamically folded skylights form a relief on the ceiling with their impressive daylight gradients, while direct sunlight cre-ates brilliance and life through exciting play of light and shadow. At night, the ceiling remains the center of the design, as the lighting concept incorporates a circular band of light just underneath the skylight profiles. It not only accentuates the structure of the reliefs, but also provides no small contribution to the general lighting.

On the ground floor, the scene is framed by an intense graze light illumination on the textured concrete walls surrounding the foyer. A perimeter ceiling light profile displays the walls in a lively and viv-id manner. Part of its luminous flux draws a peripheral contour on the ground near the wall, thus making the large foyer tangible as it provides orientation.

The light profile is the only light source under the surrounding gal-lery, resulting in an appropriately dramatic impression.

Above the gallery however, the lighting concept is rather restrained, using downlights to ensure that the relief-like design of the sky-lights experiences no competition. Moving from the lobby towards the corridors, the concept changes. Here, large-scale diffuse lumi-naires are recessed into the wood-panelled ceilings, which can be found repeated throughout all corridors.

CLIENT:
Federal State of Hesse, represented by
Hessisches Baumanagement (Hessian Building
and Construction Management)
OCCUPANT:
Goethe-University Frankfurt am Main
ARCHITECTS:
Thomas Müller Ivan Reimann Gesellschaft
von Architekten mbH, Berlin
TEAM LEADER LICHT KUNST LICHT:
Edwin Smida
COMPLETION:
2008 (RUW)
2013 (PEG)
PROJECT SIZE:
30,000 m^2 (RUW)
41,700 m^2 (PEG)
OVERALL BUILDING BUDGET:
53 million euros (RUW)
96 million euros (PEG)

类似公共广场的入口平台位于广阔的校园和学院楼之间。低矮的护墙呈现出欢迎的姿态，仿佛是在邀请人们在此休憩和停留。

An entrance terrace that resembles a public square mediates between the vast site and the faculty building. The low height of the parapet walls offer an invitation to sit and linger.

夜晚，柔和的灯光从护墙石材砌面的缝隙和雨篷的接口
空间散出，未使用高杆灯对此区域进行照明，以免影响
观感。

At night, the reveals in the stone cladding of the parapet wall and canopy unfold as luminous lines, thus avoiding the use of disruptive light poles.

INFORMATION

和门厅一样，圆角设计也出现在了楼梯区域。通过巧妙的设计，设计师以尽可能短的荧光灯管总长度，将凹槽灯沿着楼梯台阶的曲线均匀排布。

As in the foyer, the design motif of rounded corners is repeated in the stair flights. The cove follows the curves by means of a clever choice and layout of the shortest possible fluorescent lamps.

通过使用与建筑一体化的凹槽灯，设计师从视觉上突出了庞大而富有魅力的独立式楼梯，提升了楼梯的雕塑感。

By use of its integrated light cove, the massive and charismatic freestanding stair is visually underlined and thus emerges as a stair sculpture.

线形发光边框与下照灯交替使用，在提供功能性照明的同时，避免了设计的单调性。

Framing light lines alternating with downlights provide functional light and prevent design monotony.

PE
PE
Bitte leise sein
Jura
PE 482-
Jura
PE 695-
PD
PD
PD

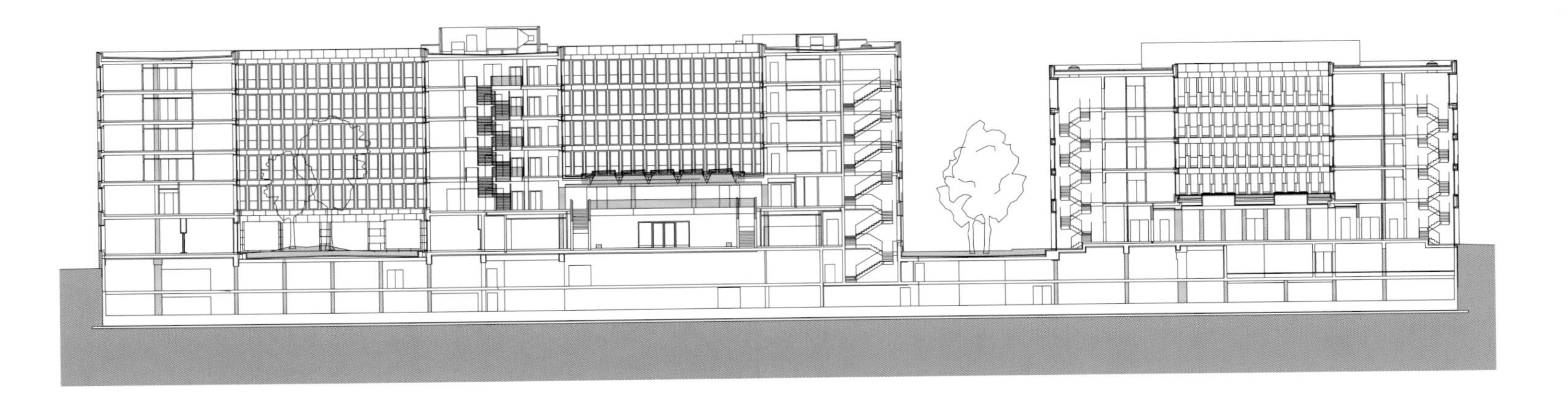

心理、教育和社会科学学院大楼和行政管理大楼剖面图

Sections PEG institute building / PA Governing and Administration building

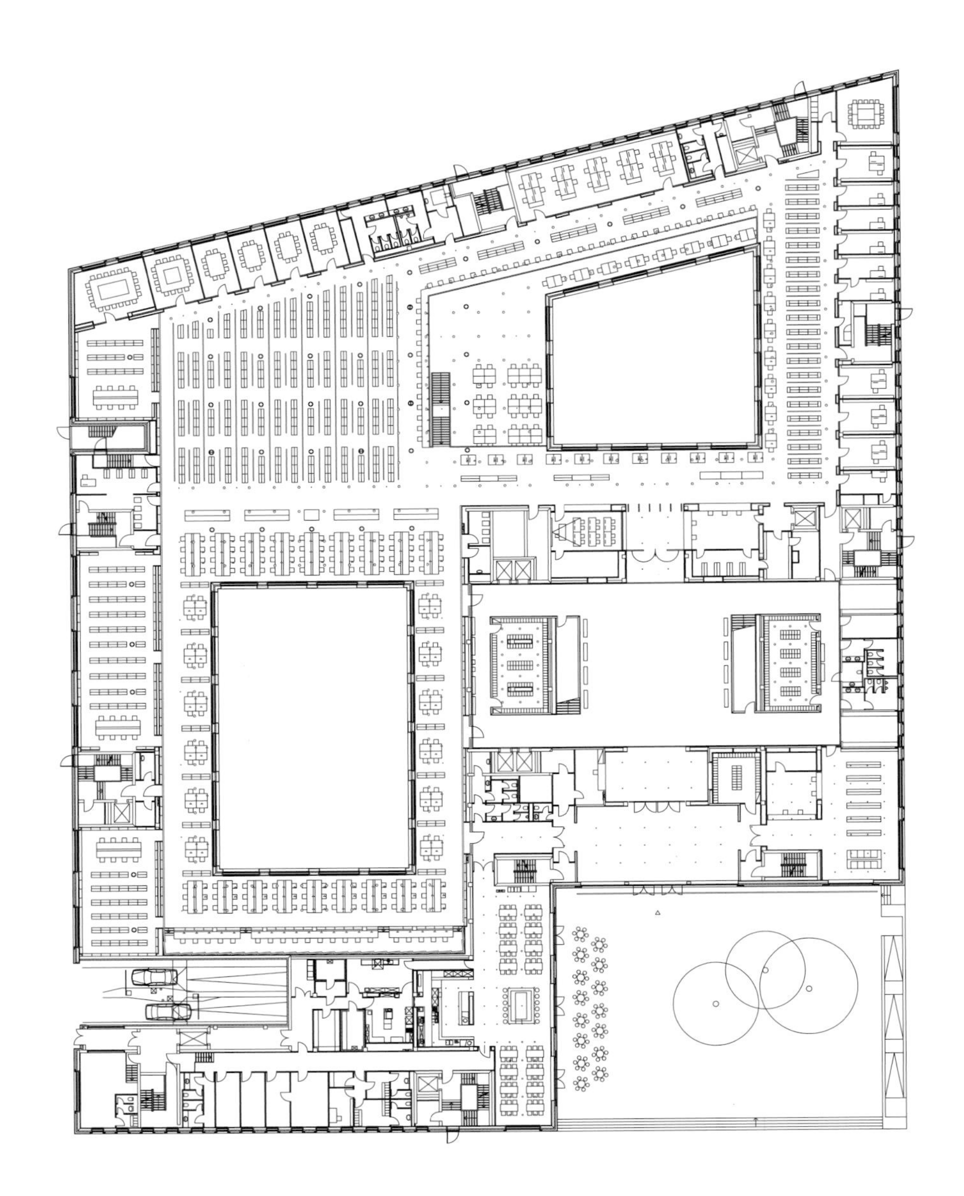

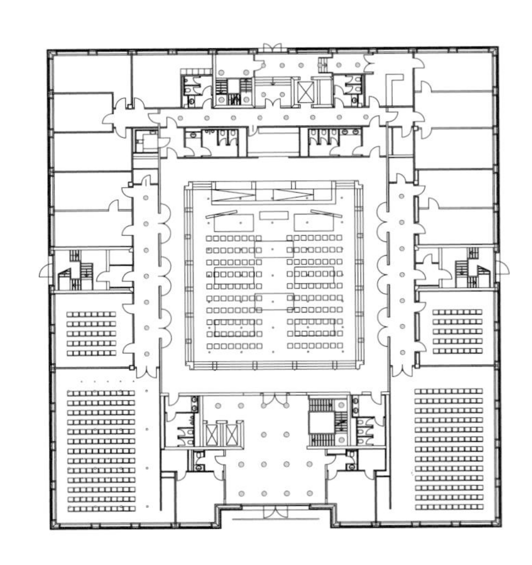

心理、教育和社会科学学院大楼和行政管理大楼平面图

Floor plans PEG institute building / PA Governing and Administration building

外围强有力的擦墙光彰显了首层裸露的荔枝面混凝土墙面，让人对空间的整体氛围印象深刻。

A powerful perimeter graze light marks the ground floor's exposed bush-hammered concrete walls and impressively enframes the ensemble.

立面上强有力的擦墙光与浅色地面、黄色天花板相辅相成，清晰地勾勒出门厅的轮廓，同时营造出温馨、友好的氛围。

The interplay of a powerful graze light, a light-coloured floor, and yellow ceiling panels, provides the foyer a striking surrounding with a warm and friendly feel.

谢绝进入的内部庭院为图书馆内的用户提供了绝佳的观赏景观，而图书馆内明亮的光线也向外展现出开放、好客的姿态。

The inaccessible inner courtyard benefits from the view of the library interior, that invitingly radiates from within.

书馆面向内部庭院的落地窗设计为室内引入了大量的然光，并与室外景观形成了重要的视觉联系。

The generous fenestration of the library towards the inner courtyards yields an unusually high daylight intake and an important visual relationship with the outdoor environment.

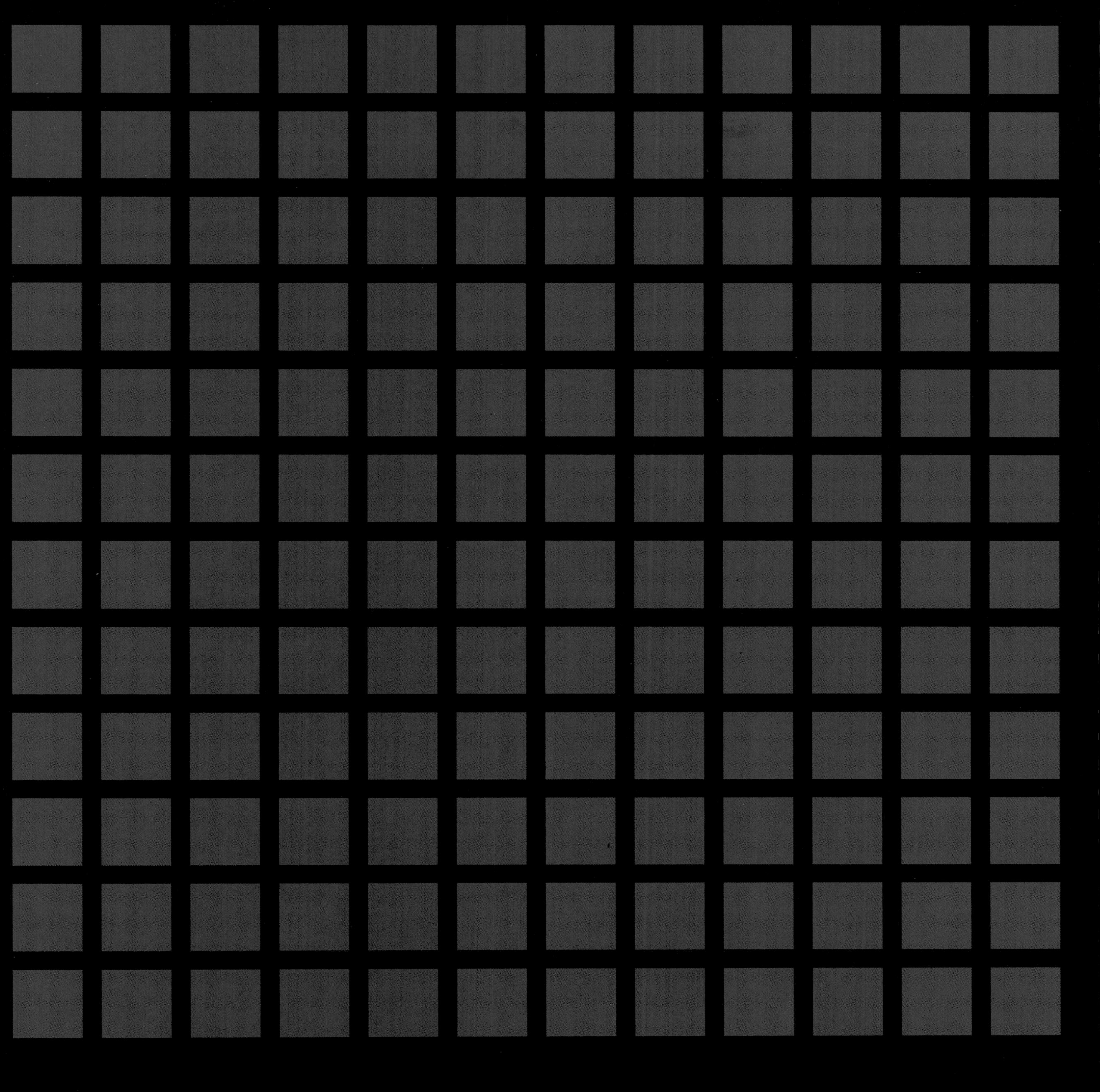

LICHT KUNST LICHT 4

会议中心 慕尼黑

CONFERENCE CENTRE MUNICH

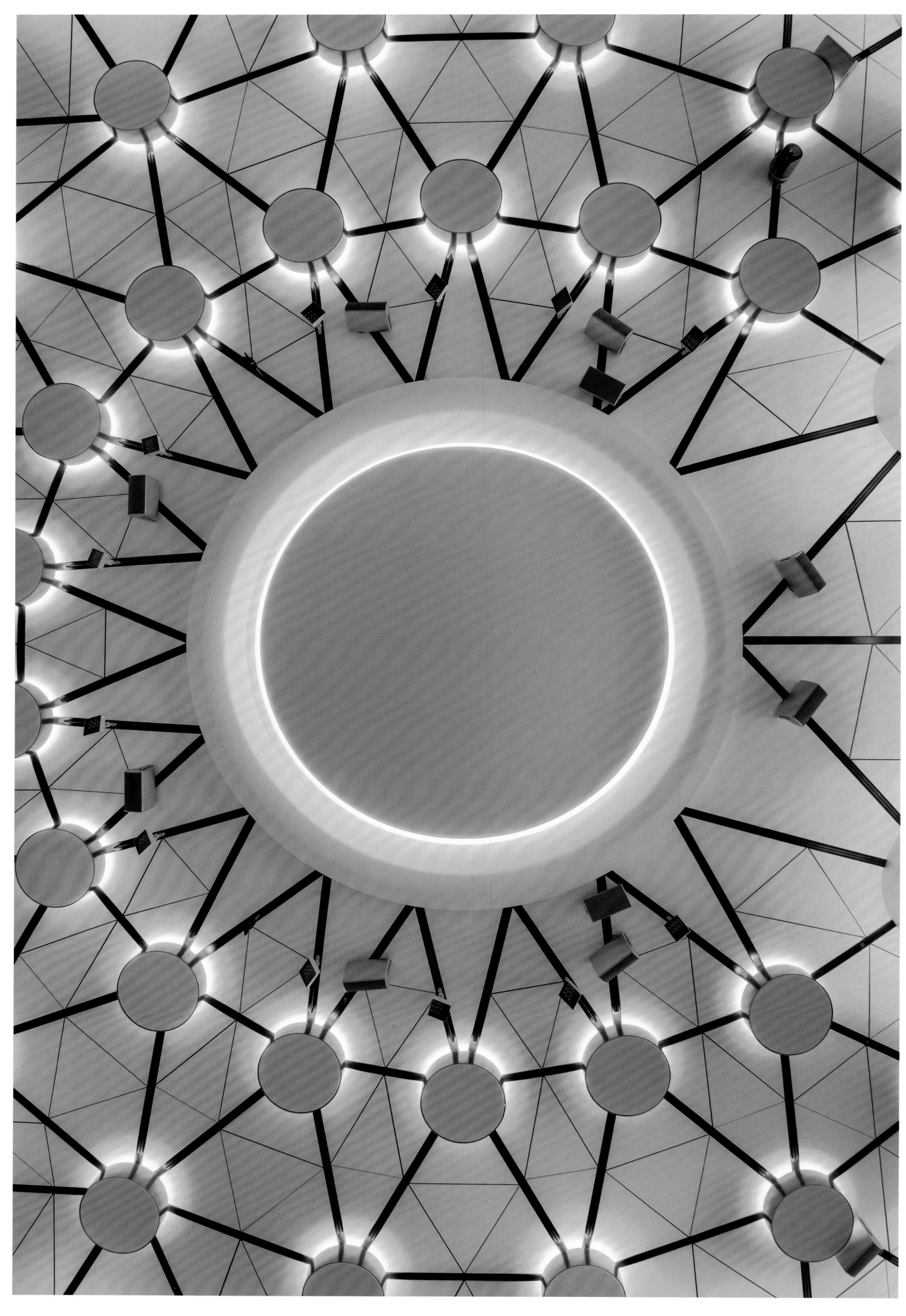

建筑诠释企业精神——拥有动态彩光和可调白光场景的慕尼黑会议中心

座坐落在慕尼黑黄金地段的庄严气派的建筑始建1955年，是一家大型服务公司的总部。根据业的要求，建筑设计师重建了部分楼体，增设了一带有大型活动空间的多功能会议中心。优雅时尚建筑造型，搭配由全方位LED技术带来的多样化照明场景，给人留下了非同寻常的空间印象。

自慕尼黑的WKP建筑设计事务所是该项目的规单位，他们在2009年为地下会议中心研发了建方案。在地方当局的全面配合下，项目于2012动工。2011年，新建筑的室内设计方案通过一设计竞赛选出，来自柏林的丹·珀尔曼品牌形建筑设计股份有限公司和赖希瓦尔德·舒尔茨筑设计事务所的联合设计方案最终脱颖而出，获大奖。

竞赛的主要设计任务是让人一进入室内就可以感到公平透明、合作互助的企业精神。对于大部分于地下且只能从旁侧建筑入口进入的项目来说，个任务充满挑战。设计师欣然接受了挑战，并为内设计和企业精神的传达研发了一套整体性的设方案。根据这一构思，他们在保持建筑本身优雅尚造型的同时，加入了鲜明的设计元素。

庸置疑的是，项目的设计核心是设有300个座位礼堂。礼堂墙壁以经过打磨的可丽耐面板作为饰材料，设计细节旨在创造一种轻盈宽敞的视觉印，进而弱化访客身处地下空间的感受。通过升降台阶，“多用空间”可以单独进行重组，并根据同需求转变为礼堂、会议厅、会客室、活动场、会堂或展示区。然而，营造礼堂中恢宏壮阔空体验的决定性设计要素，当属拥有丰富变化场景照明系统，该系统通过照明编程进行控制。

色光营造空间氛围，自然白光迎接访客

扁平圆柱体照明元素组成的系列灯具为礼堂提供体照明，该灯具以礼堂的穹顶为视觉中心，呈同式环绕分布。这套灯具包含三种尺寸，其中安装演讲区附近的灯具为吸顶式下照灯，另外两种是吊高度不一的造型吊灯。

具统一采用LED光源，可允许其中的直接照明组进行明暗调光和色温的调节控制。灯具的底面是光面，光线透过一层半透明的柔光膜为座椅和通区域提供柔和的照明。与此同时，系列灯具以高质的外观造型为缺乏日光的地下空间增添了一种机勃勃的设计元素。

了直接照明组件，灯具还设有内置隐蔽式RGB-ED灯环，为周围的天花提供彩光照明。空间穹顶于中央演讲区的正上方，设计师对此区域采用了小型圆柱体灯具相同的LED照明技术，以便提供不同亮度和色温的白光照明。

凭借这一照明系统，空间可以实现全方位的功能转变。值得一提的是，灯具所具备的RGB间接照明元素能够提供高饱和度的彩光照明，可营造出多种光环境氛围。令人惊喜的是，灯具的直接照明组件确保了在任何空间氛围中，访客都可以沐浴在高显色度的白光照明之下，以满足他们视觉上的功能性需求。

以可调白光作为照明设计的主基调

不仅在礼堂空间，设计师将可调白光照明作为贯穿所有空间的关键性设计主题，在挑高两层的中庭内，3000开尔文、4000开尔文色温的细长形下照灯被集成在通风凹槽中，向地面投射强有力的暖白光和冷白光。仿佛雕塑般伫立在中庭外露式立柱结构之间的黑色楼梯背面设有发光薄膜，同样通过可调色温的白光LED灯具进行照明。

地下层的房间分为开放式办公区和休息区，可同时为身处室内和室外首层露台空间的访客提供用餐场所。在此区域，整体照明由方形LED下照灯提供。

与之相似的是首层的整体照明方式，由与天花完成面齐平的内嵌式方形下照灯提供。下照灯配备了能根据一天中日光光色的变化进行调节的LED照明技术。尤其在休息室和等待区，通过LED照明技术，人工光可以在白天提供纯净的中性白光照明，而在晚间活动时提供暖白光的环境照明。

访客从主入口进入首层空间，当他们朝后墙走去时，可以欣赏到建筑后方的特色花园。一盏吊灯成为接待区的视觉焦点，确保后墙前方的接待台被氛围光线重点突出。

慕尼黑这个项目证明了搭配智能控制系统的LED照明技术不仅可以节约建筑的运营成本，还能帮助照明设计师拓展设计思路。无论是礼堂中引人入胜的灯光，还是在仅有人工光照明的地下空间引入辅助人体昼夜节律需求的动态可调白光照明，都体现了这一特征。

建筑设计：
丹·珀尔曼品牌形象建筑设计股份有限公司，柏林
赖希瓦尔德·舒尔茨建筑设计事务所，柏林
LKL项目经理：
迈克·沙尼亚克
竣工时间：
2014年
项目规模：
1170 平方米

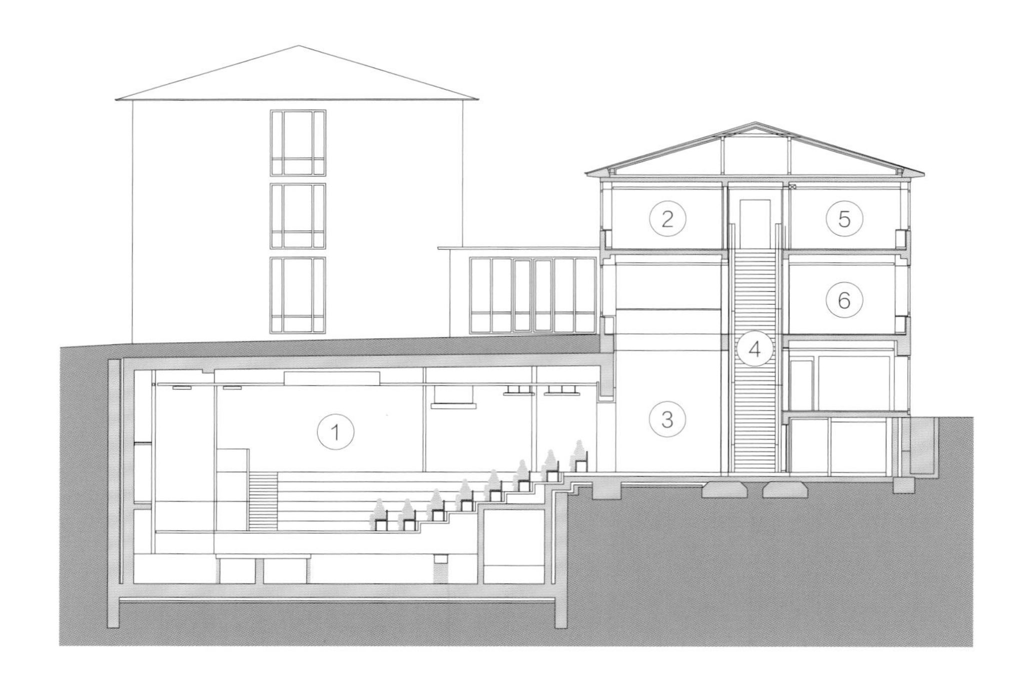

会议中心剖面图

Cross section of conference centre

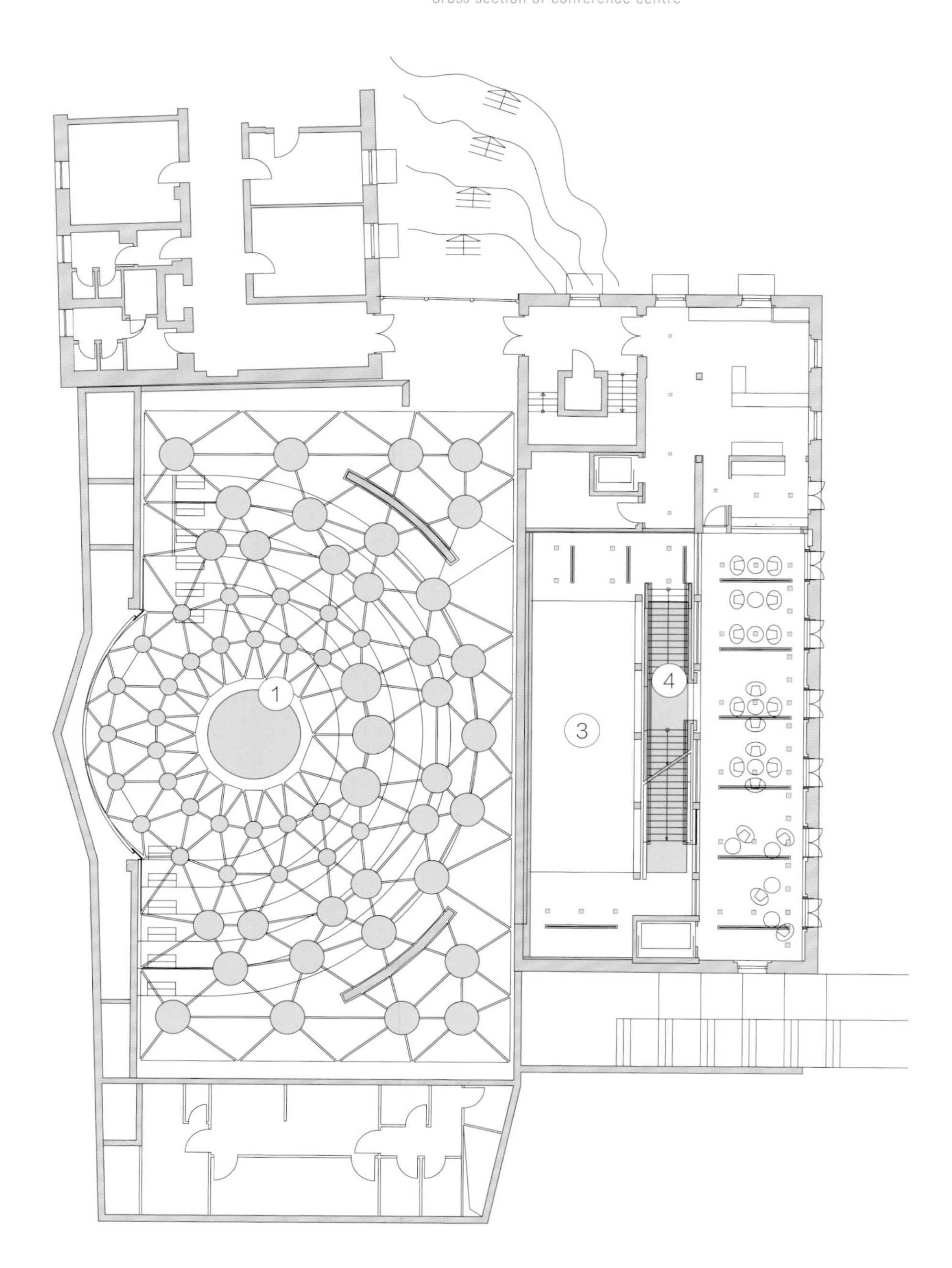

包含照明布局的地下室平面图

Basement floor plan with lighting layout

1.礼堂　　4.楼梯间
2.会议室　5.开放空间
3.中庭　　6.接待处和休息室

1. Auditorium
2. Meeting room
3. Atrium
4. Stairs
5. Open space
6. Reception and lounge

e dignified headquarters of a large service company was built one of Munich's prime locations in 1955. According to the cli- t's wishes, a partial building was to be redeveloped and a mul- unctional conference centre with a big event space was to be signed. The result is impressive with its sleek architecture and traordinarily variable illumination, exclusively using LED tech- logy.

e building concept for the underground conference centre was veloped in 2009 by the Munich based architectural firm WKP eickenmeier, Kunz + Partner as the architect of record. Following thorough coordination with local authorities, construction work gan in 2012. An international competition for the interior design as won in 2011 by the Berlin based office for Markenarchitektur n pearlman in cooperation with Reichwald Schultz Architekten.

ne of the competition brief's requirements was to make corpo- te values, such as transparency and cooperation, perceivable on entering the building – a challenge in a building that is large- underground and accessed through a neighbouring building. e designers accepted the challenge and developed a holistic ncept for the interior design and mediation of information. In ing so, they have maintained the sleek building architecture hile complementing it with distinct design elements.

e heart of the project is undoubtedly the 300-seat auditorium ntre. Its walls are lined with milled Corian panels, while all de- gn details aim at conveying an impression of lightness and spa- ousness, thus evading the feeling of being underground. By vir- e of lowerable tiers, the 'convertible room' can be transformed dividually and used as an auditorium, conference hall, meeting om, event space, assembly hall, or performance location, de- ending on the function. However, the decisive contribution to e overwhelming spatial experience in the auditorium occurs rough the highly variable illumination system, which is run via a ogrammable lighting control system.

OLOUR FOR THE SPACE, NATURAL WHITE FOR THE VISITOR

ne general lighting in the auditorium originates from a lumi- aire family consisting of flat, cylindrical lighting objects that are ranged concentrically around an optical centre of the room, the culus. The luminaire family consists of three different housing siz- s. The light fittings mounted near the speaker's area are surface ounted versions whereas the remaining two types have been de- gned as pendant luminaires with two varying suspension heights.

l fixtures are exclusively lamped with LEDs, allowing direct lighting omponents to be dimmable and controllable between warm and ool white. The light emission occurs at the luminaire's bottom ace through a semi-transparent membrane, thus creating a gen- e light atmosphere in the seating and access areas while creating n invigorating element in this daylight-deprived space through its igh-quality appearance.

eyond this direct lighting component, all luminaires have an tegral hidden ring of RGB-LEDs, thereby offering the ability to uminate the immediately surrounding ceiling with coloured light. ne oculus is located directly above the central speaker's area and fitted with the same LED technology as the small cylindrical lu- minaires in order to create direct white light at varying brightness values and colour temperatures.

With this lighting configuration, the space can be completely transformed. In particular, the indirect RGB illumination offers the ability to customize a variety of different ambiences through the implementation of highly saturated colours. At the same time, the visitor is always bathed in pure white light with very high colour rendering from the direct light component– a surprising effect.

TUNABLE WHITE LIGHT AS LIGHTING DESIGN LEITMOTIF

Tunable white lighting continues as a pivotal design theme throughout the entire project: In the two-storey atrium, slender linear downlights in both 3,000 K and 4,000 K have been integrated in the ventilation slots and create a resonant warm white and cool white light on the floor. The black staircase, which forms a sculpture amidst the atrium's open pillar structure, is fitted with a recessed luminous membrane ceiling on its underside–again equipped with colour temperature controllable white LEDs.

In the basement, the rooms are organized as open work and relaxation zones that offer gastronomical venues both indoors and at the adjacent exterior ground floor garden terrace. Here, the general illumination is provided by square LED downlights.

Similarly, the general illumination of the ground floor is provided by square downlights installed flush with the ceiling. The downlights are equipped with an LED technology that permits to respond adequately to the shift of daylight colours over the course of the day. Particularly at the lounge / waiting area, a crisp neutral white light colour is suitable during the day while a warm white light ambience is applied for evening events.

Through the main entrance, the visitors arrive on the ground floor and approach the rear wall with a view of the featured garden behind. A pendant light marks the reception area and ensures that the counter in front of the rear wall is atmospherically emphasized.

The Munich project demonstrates how LED technology not only saves energy but also expands the lighting designers' creative reach when combined with intelligent control systems. This is seen in the compelling light orchestrations of the auditorium as well as in the implementation of dynamic colour temperature lighting to support the circadian rhythm in an artificial-only illuminated space.

ARCHITECTS:
dan pearlman Markenarchitektur GmbH, Berlin
Reichwald Schultz Architekten, Berlin
TEAM LEADER LICHT KUNST LICHT:
Maik Czarniak
COMPLETION:
2014
PROJECT SIZE:
1,170 m^2

由扁平圆柱体照明灯具组成，三种不同尺寸的系列灯具为礼堂提供整体照明。通过灯具内集成的RGB- LED间接照明组件，系列灯具可以投射出白光或带有色彩倾向的光晕。因此，在这个缺乏自然光的空间中，灯具成为一种生机勃勃的设计元素。

The general lighting in the auditorium emanates from a luminaire family consisting of flat, cylindrical lighting objects in three different sizes. Through an RGB-LED indirect light component they create a corona of white or coloured light. Thus, the luminaires become an invigorating element in this daylight-deprived space.

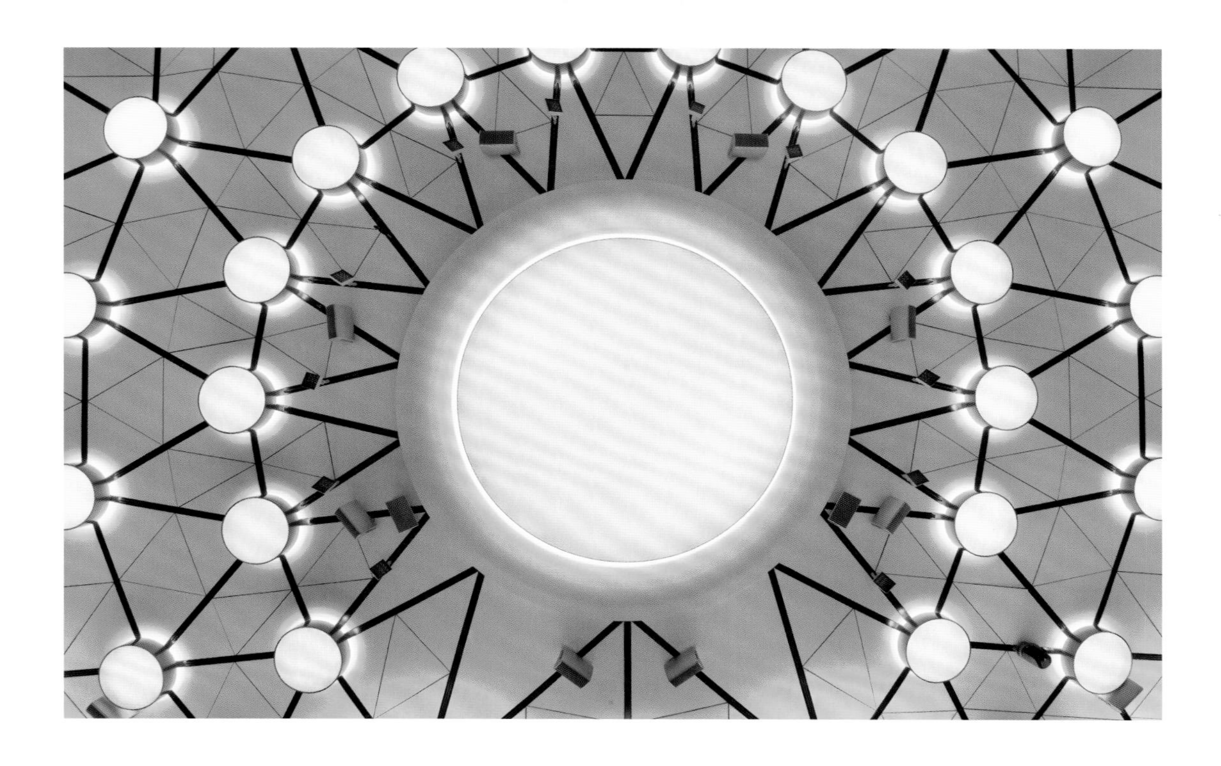

圆柱体灯具和礼堂穹顶呈同心式环绕分布，形成了一个发光的网状结构。通过照明控制系统，可以对位于LED灯具半透明薄膜内的直接照明组件进行明暗和冷暖色温的调节。

Arranged concentrically around the oculus, the cylindrical luminaires create a luminous web. Fitted with a semi-transparent membrane, the direct components of the LED light fixtures are both dimmable and warm and cool white controllable.

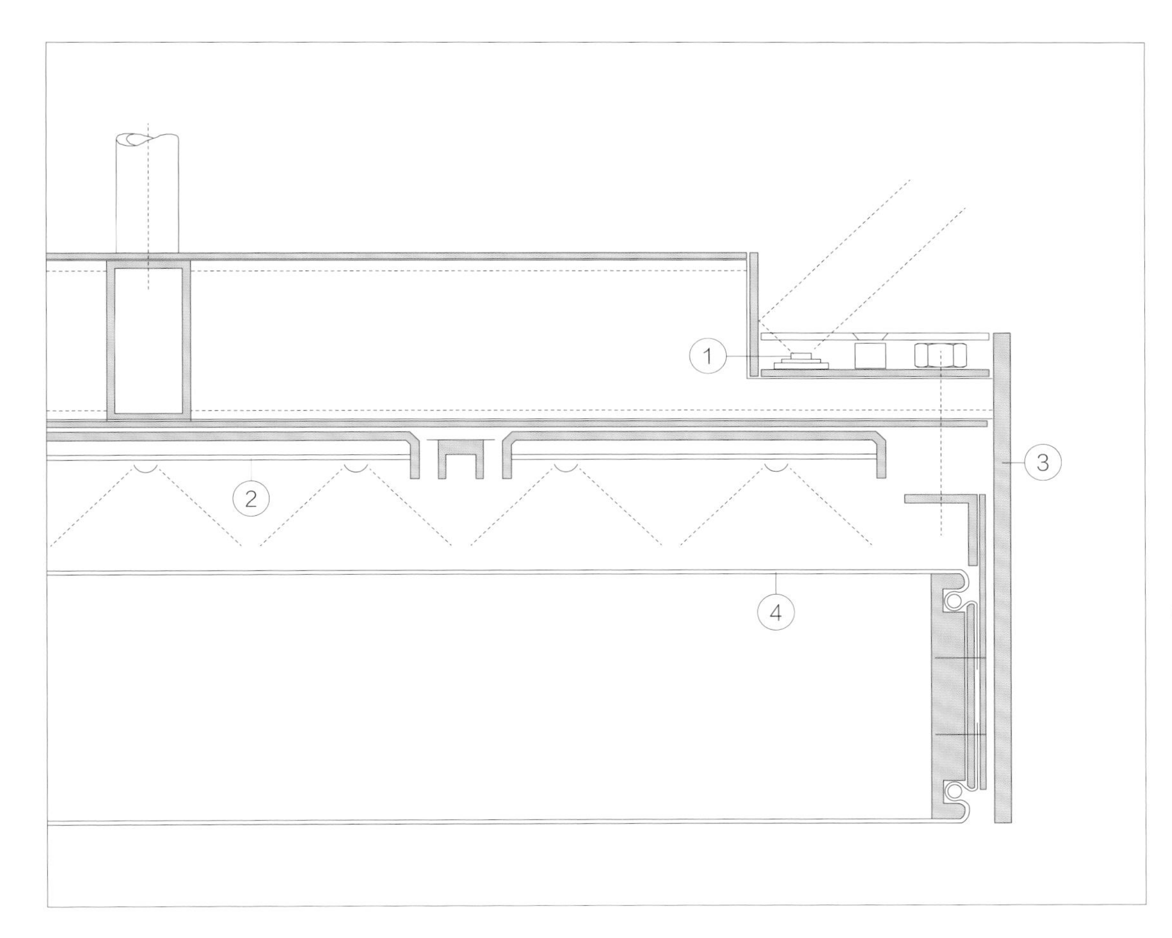

吊灯的细节剖面图
1.间接照明的RGB- LED上照灯
2.可调白光的 LED模块（白光色温范围 2700–5000开尔文）
3.铝制灯壳
4.80%透光率薄膜

Detail section of pendant luminaire
1. Indirect RGB-LED uplight
2. Tunable white LED modules 2,700 – 5,000 K
3. Aluminium housing
4. 80 % light transmission membrane

AWARD
WINNER

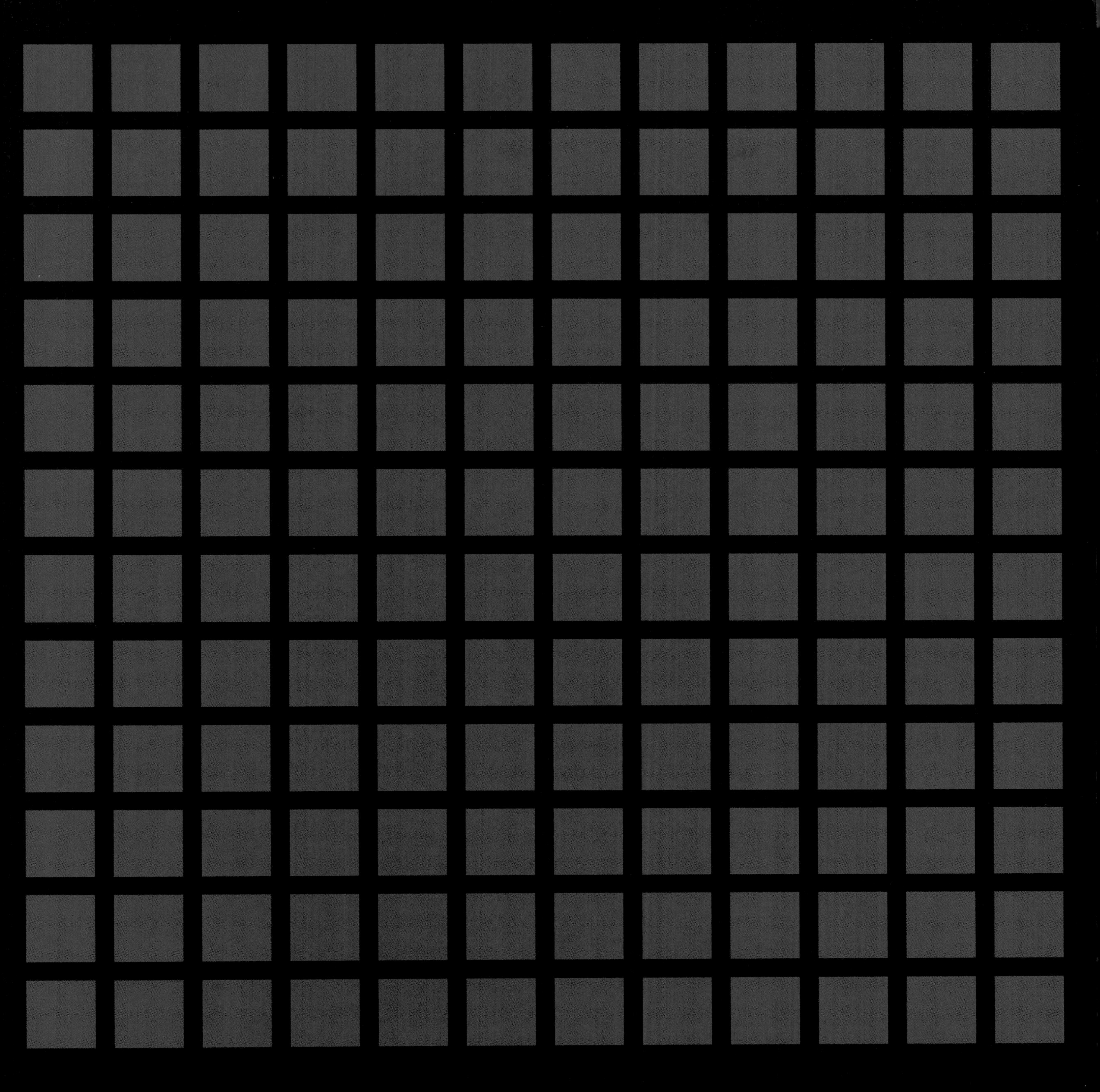

奖项

AWARDS

LKL
德国，波恩/柏林

LICHT KUNST LICHT AG
Bonn / Berlin, Germany

2015年度最佳照明设计师
德国灯光设计奖

LIGHTING DESIGNER OF THE YEAR 2015
Der Deutsche Lichtdesign-Preis

Abadia Retuerta LeDomaine
酒店和水疗中心
西班牙，萨尔东德杜埃罗

HOTEL & SPA ABADIA RETUERTA LEDOMAINE
Sardón de Duero, Spain

2015年度优胜奖
GE爱迪生奖

AWARD OF MERIT 2015
GE Edison Awards

赛恩钢铁铸件馆
德国，本多夫

SAYN IRON WORKS FOUNDRY
Bendorf, Germany

2015年度卓越奖
GE爱迪生奖

AWARD OF EXCELLENCE 2015
GE Edison Awards

阿伦斯霍普艺术博物馆
德国，阿伦斯霍普

AHRENSHOOP MUSEUM OF ART
Ahrenshoop, Germany

2016年度最佳室内建筑奖
德国设计大奖
德国设计协会

2015年度最佳灯光装置奖
AZ 奖
《蔚蓝》杂志，加拿大

2015年度评审团日光奖
德国灯光设计奖

2015年度优胜奖
IALD国际照明设计奖
国际照明设计师协会

2014年度优胜奖
GE爱迪生奖

WINNER INTERIOR ARCHITECTURE 2016
German Design Award
German Design Council

WINNER BEST LIGHTING INSTALLATIONS 2015
AZ Awards
Azure Magazine, Canada

WINNER JURY AWARD DAYLIGHT 2015
Der Deutsche Lichtdesign-Preis

AWARD OF MERIT 2015
IALD International Lighting Design Awards
International Association of Lighting Designers

AWARD OF MERIT 2014
GE Edison Awards

龙岩山高原餐厅
德国，克尼格斯温特尔

DRACHENFELSPLATEAU
Königswinter, Germany

2015年度最佳酒店与餐饮类设计奖
德国灯光设计奖

AWARD WINNER HOTEL AND GASTRONOMY 2015
Der Deutsche Lichtdesign-Preis

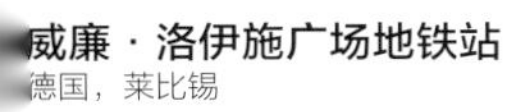

威廉·洛伊施广场地铁站
德国，莱比锡

SUBWAY STATION, 'WILHELM-LEUSCHNER-PLATZ'
Leipzig, Germany

2015年度最佳交通与运输类设计奖
德国灯光设计奖

AWARD WINNER TRAFFIC AND TRANSPORTATION 2015
Der Deutsche Lichtdesign-Preis

LWL艺术与文化博物馆
德国，明斯特

LWL-MUSEUM FÜR KUNST UND KULTUR
Münster, Germany

2015年度优胜奖和最受欢迎奖
AZ奖
《蔚蓝》杂志，加拿大

2014年度卓越奖
GE爱迪生奖

AWARD OF MERIT AND PEOPLE'S CHOICE WINNER 2015
AZ Awards
Azure Magazine, Canada

AWARD OF EXCELLENCE 2014
GE Edison Awards

巴登–符腾堡州内政部新大楼
德国，斯图加特

BADEN-WÜRTTEMBERG MINISTRY OF THE INTERIOR
Stuttgart, Germany

2014年度优胜奖
GE爱迪生奖

AWARD OF MERIT 2014
GE Edison Awards

理查德–瓦格纳广场
德国，莱比锡

RICHARD-WAGNER-PLATZ
Leipzig, Germany

2014年度灯光设计奖一等奖
“城市 · 居民 · 灯光” 奖

1ST PRICE LIGHTING DESIGN 2014
city.people.light Awards

施泰德博物馆
德国，法兰克福

STÄDELSCHES KUNSTINSTITUT UND STÄDTISCHE GALERIE
Frankfurt am Main, Germany

2014年度最佳照明项目奖
WIN英国世界室内新闻奖

2013年度优胜奖
IES照明奖
北美照明工程协会

2013年度最佳照明设计奖和最受欢迎奖
AZ 奖
《蔚蓝》杂志, 加拿大

2013年度评审团日光奖
德国灯光设计奖

2012年度银奖（文化类）
无限流明奖

WINNER LIGHTING PROJECT 2014
World Interiors News Annual Award

AWARD OF MERIT 2013
IES Illumination Awards
Illuminating Engineering Society of North America

WINNER LIGHTING DESIGN AND PEOPLE'S CHOICE AWARD 2013
AZ Awards
Azure Magazine, Canada

WINNER JURY AWARD DAYLIGHT 2013
Der Deutsche Lichtdesign-Preis

WINNER SILVER (CULTURAL) 2012
illumni infinity Awards

新展览馆
德国，卡塞尔

NEW GALLERY
Kassel, Germany

2013年度最佳博物馆类设计奖
德国灯光设计奖

AWARD WINNER MUSEUMS 2013
Der Deutsche Lichtdesign-Preis

巴伐利亚国王博物馆
德国，霍恩施万高

MUSEUM OF THE BAVARIAN KINGS
Hohenschwangau, Germany

2012年度铜奖（文化类）
无限流明奖

2011年度卓越奖
GE爱迪生奖

WINNER BRONZE (CULTURAL) 2012
illumni infinity Awards

AWARD OF EXCELLENCE 2011
GE Edison Awards

利布弗劳恩哥伦比亚圣玛利亚教堂骨灰安置所
德国，多特蒙德

COLUMBARIUM LIEBFRAUENKIRCHE
Dortmund, Germany

2012年度最佳国际项目（室内设计）奖
LDA照明设计奖

WINNER INTERNATIONAL PROJECT (INTERIORS) 2012
Lighting Design Awards

蒂森克虏伯新总部园区
德国，埃森

THYSSENKRUPP QUARTER
ssen, Germany

2012年度最佳城市照明——公共区域照明奖
德国建筑师协会

2011年度卓越成就奖
AL设计大奖

2011年度最佳室内照明奖
LAMP照明解决方案奖

2011年度推荐国际项目奖
LDA照明设计奖

2010年度GE爱迪生奖
GE爱迪生奖

2010年度卓越奖
GE爱迪生奖

AWARD WINNER PUBLIC SPACES - URBAN LIGHTING 2012
Der Deutsche Lichtdesign-Preis

OUTSTANDING ACHIEVEMENT AWARD 2011
AL Light & Architecture Design Awards

WINNER INTERIOR LIGHTING 2011
LAMP Lighting Solutions Awards

COMMENDED INTERNATIONAL PROJECT 2011
Lighting Design Awards

GE EDISON AWARD WINNER 2010
GE Edison Awards

AWARD OF EXCELLENCE 2010
GE Edison Awards

LKL的安德烈亚斯 · 舒尔茨先生
德国，波恩/柏林

ANDREAS SCHULZ
LICHT KUNST LICHT AG
onn / Berlin, Germany

2011年度灯光设计师
德国灯光设计奖

LIGHTING DESIGNER OF THE YEAR 2011
Der Deutsche Lichtdesign-Preis

TELEKOM大桥
德国，波恩

TELEKOM BRIDGE
onn, Germany

2011年度光辉奖
IALD国际照明设计奖
国际照明设计师协会

2011年度卓越奖
IALD国际照明设计奖
国际照明设计师协会

2011年度最佳国际项目奖
LDA照明设计奖

2010年度杰出成就奖
AL设计大奖

RADIANCE AWARD 2011
IALD International Lighting Design Awards
International Association of Lighting Designers

AWARD OF EXCELLENCE 2011
IALD International Lighting Design Awards
International Association of Lighting Designers

WINNER INTERNATIONAL PROJECT 2011
Lighting Design Awards

OUTSTANDING ACHIEVEMENT AWARD 2010
AL Light & Architecture Design Awards

ENBW CITY办公大楼
德国，斯特加特

ENBW CITY
tuttgart, Germany

2009年GE爱迪生奖
GE爱迪生奖

2009年度卓越奖
GE爱迪生奖

GE EDISON AWARD WINNER 2009
GE Edison Awards

AWARD OF EXCELLENCE 2009
GE Edison Awards

诺华制药园区MAKI办公楼
瑞士，巴塞尔

NOVARTIS CAMPUS
MAKI OFFICE BUILDING
asel, Switzerland

2011年度最佳办公与行政大楼类设计奖
德国灯光设计奖

2010年度优胜奖
IALD国际照明设计奖
国际照明设计师协会

2009年度卓越奖
GE爱迪生奖

AWARD WINNER OFFICE AND
ADMINISTRATION BUILDINGS 2011
Der Deutsche Lichtdesign-Preis

AWARD OF MERIT 2010
IALD International Lighting Design Awards
International Association of Lighting Designers

AWARD OF EXCELLENCE 2009
GE Edison Awards

北威州K20艺术收藏馆
德国，杜塞尔多夫

K20 KUNSTSAMMLUNG NRW
Düsseldorf, Germany

2010年度优胜奖
GE爱迪生奖

AWARD OF MERIT 2010
GE Edison Awards

埃米尔·舒马赫博物馆
德国，哈根

EMIL SCHUMACHER MUSEUM
Hagen, Germany

2009年度卓越奖
GE爱迪生奖

AWARD OF EXCELLENCE 2009
GE Edison Awards

瓦杜兹联邦议会厅
列支敦士登，瓦杜兹

THE FEDERAL STATE PARLIAMENT VADUZ
Vaduz, Liechtenstein

2009年度力荐国际项目
LDA照明设计奖

2009年度优胜奖
IALD国际照明设计奖
国际照明设计师协会

2008年度卓越奖
GE爱迪生奖

HIGHLY COMMENDED INTERNATIONAL PROJECT 2009
Lighting Design Awards

AWARD OF MERIT 2009
IALD International Lighting Design Awards
International Association of Lighting Designers

AWARD OF EXCELLENCE 2008
GE Edison Awards

诺华制药园区KRISCHANITZ实验室大楼
瑞士，巴塞尔

NOVARTIS CAMPUS KRISCHANITZ LABORATORY BUILDING
Basel, Switzerland

2009年度优胜奖
IALD国际照明设计奖
国际照明设计师协会

2008年度优胜奖
GE爱迪生奖

AWARD OF MERIT 2009
IALD International Lighting Design Awards
International Association of Lighting Designers

AWARD OF MERIT 2008
GE Edison Awards

诺华制药园区SERRA接待大楼和停车场
瑞士，巴塞尔

NOVARTIS CAMPUS SERRA RECEPTION BUILDING AND PARKING
Basel, Switzerland

2009年度卓越奖
IES照明奖
北美照明工程协会

2009年度优胜奖
IALD国际照明设计奖
国际照明设计师协会

AWARD OF EXCELLENCE 2009
IES Illumination Awards
Illuminating Engineering Society of North America

AWARD OF MERIT 2009
IALD International Lighting Design Awards
International Association of Lighting Designers

尔韦林公园
国，埃森佐尔韦林煤矿区

LLVEREIN PARK
lverein Colliery, Essen, Germany

2008年度优胜奖
GE爱迪生奖

AWARD OF MERIT 2008
GE Edison Awards

兰克福纪念大厅
国，法兰克福

STIVAL HALL FRANKFURT
nkfurt am Main, Germany

2008年度优胜奖
IIDA国际照明设计奖
北美照明工程协会

AWARD OF MERIT 2008
IIDA International Illumination Design Awards
Illuminating Engineering Society of North America

尼卡大厦
地利，维也纳

IQA TOWER
nna, Austria

2007年度优胜奖
IALD国际照明设计奖
国际照明设计师协会

2007年度优胜奖
SEGD全球设计大奖
美国环境平面设计协会

2007年度卓越奖
IIDA国际照明设计奖
北美照明工程协会

AWARD OF MERIT 2007
IALD International Lighting Design Awards
International Association of Lighting Designers

AWARD OF MERIT 2007
SEGD Global Design Awards
Society for Environmental Graphic Design

AWARD OF EXCELLENCE 2007
IIDA International Illumination Design Awards
Illuminating Engineering Society of North America

尔韦林煤矿区洗煤厂
国，埃森

AL WASHING PLANT,
LLVEREIN COLLIERY
sen, Gerrmany

2007年度优胜奖
IIDA国际照明设计奖
北美照明工程协会

2006年度卓越奖
GE爱迪生奖

AWARD OF MERIT 2007
IIDA International Illumination Design Awards
Illuminating Engineering Society of North America

AWARD OF EXCELLENCE 2006
GE Edison Awards

华制药园区
纳与迪纳建筑设计事务所办公
楼
士，巴塞尔

VARTIS CAMPUS
ENER & DIENER OFFICE BUILDING
sel, Switzerland

2006年度灯光设计奖三等奖
SLG瑞士照明协会

2005年度优胜奖
GE爱迪生奖

3RD RANK，PRIX LUMIÈRE' 2006
SLG Schweizer Licht Gesellschaft

AWARD OF MERIT 2005
GE Edison Awards

玛丽·伊丽莎白·吕德斯大楼
德国，柏林

MARIE ELISABETH LÜDERS BUILDING
Berlin, Germany

2005年度优胜奖
IIDA国际照明设计奖
北美照明工程协会

2004年度卓越奖
GE爱迪生奖

AWARD OF MERIT 2005
IIDA International Illumination Design Awards
Illuminating Engineering Society of North America

AWARD OF EXCELLENCE 2004
GE Edison Awards

菩提树下大街旧司令部，由贝塔斯曼股份公司重建
德国，柏林

KOMMANDANTUR UNTER DEN LINDEN REPRESENTATIONAL BUILDING OF BERTELSMANN AG
Berlin, Germany

2005年度优胜奖
IIDA国际照明设计奖
北美照明工程协会

AWARD OF MERIT 2005
IIDA International Illumination Design Awards
Illuminating Engineering Society of North America

万豪酒店
德国，柏林

MARRIOTT HOTEL
Berlin, Germany

2005年度特别提名奖
IIDA国际照明设计奖
北美照明工程协会

2004年度FX设计奖
英国《FX》杂志

SPECIAL CITATION 2005
IIDA International Illumination Design Awards
Illuminating Engineering Society of North America

FX DESIGN AWARD 2004
FX Magazine, UK

马克思·恩斯特博物馆
德国，布吕尔

MAX ERNST MUSEUM
Brühl, Germany

2005年度建筑照明卓越奖
德国建筑博物馆

DISTINCTION IN LIGHT-ARCHITECTURE AWARDS 2005
Deutsches Architekturmuseum DAM

“城市之光”大楼
德国，柏林

CITY LIGHT HOUSE
Berlin, Germany

2004年度优胜奖
IIDA国际照明设计奖
北美照明工程协会

AWARD OF MERIT 2004
IIDA International Illumination Design Awards
Illuminating Engineering Society of North America

波鸿世纪大厅
德国，波鸿

JAHRHUNDERTHALLE 'CENTURY HALL' BOCHUM
Bochum, Germany

2004年度优胜奖
IIDA国际照明设计奖
北美照明工程协会

AWARD OF MERIT 2004
IIDA International Illumination Design Awards
Illuminating Engineering Society of North America

乔治・谢弗博物馆
德国，施韦因富特

GEORG SCHÄFER MUSEUM
Schweinfurt, Germany

2001年度优秀实践纲要奖
IALD国际照明设计奖
国际照明设计师协会

COMPENDIUM OF GOOD PRACTICE 2001
IALD International Lighting Design Awards
International Association of Lighting Designers

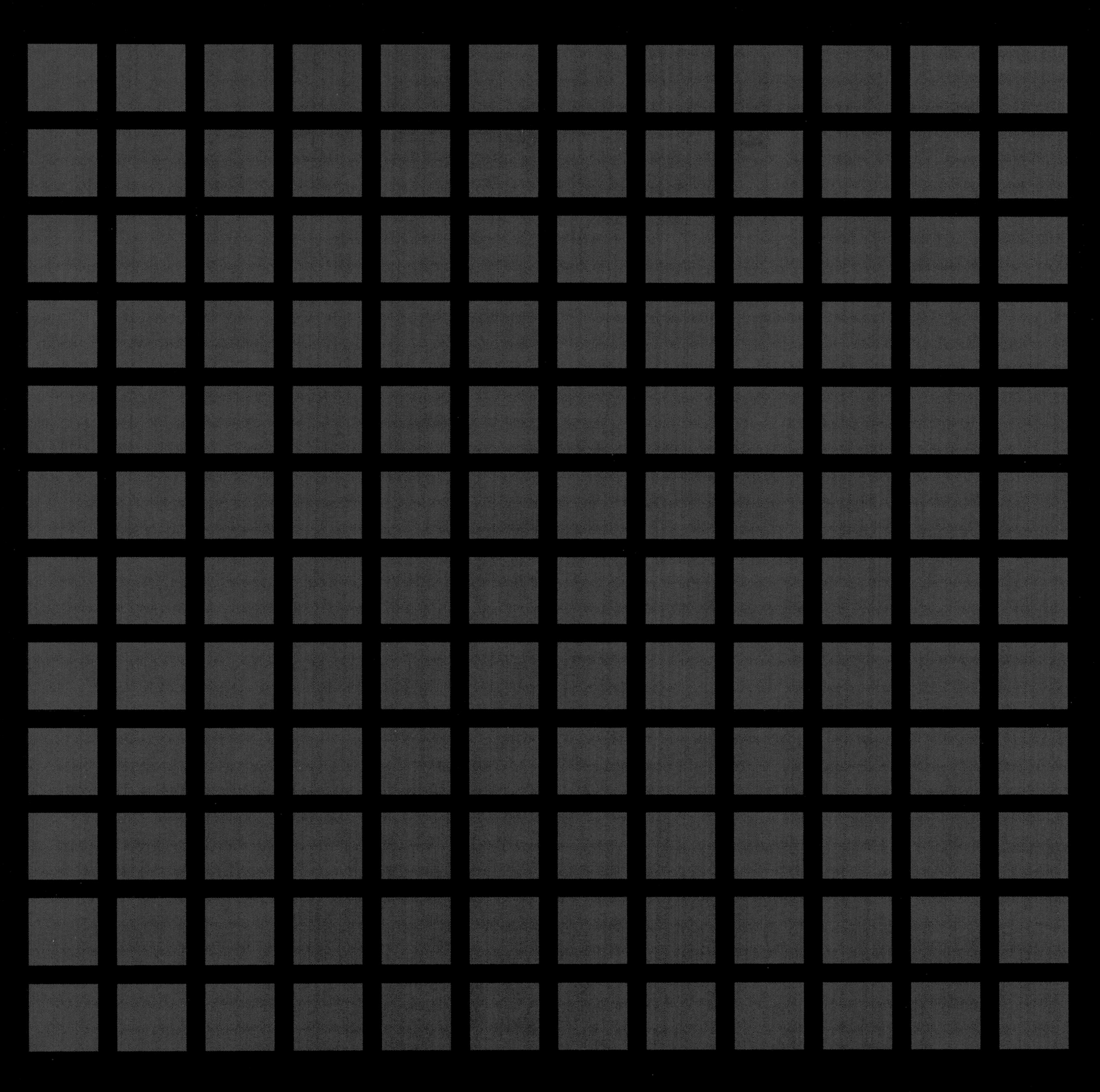

过往项目列表

LIST OF REFERENCE PROJECTS

博物馆

MUSEUMS

艺术博物馆，杜塞尔多夫
艺术博物馆基金会，杜塞尔多夫
2016年竣工

Museum Kunstpalast, Düsseldorf
Museum Kunstpalast Foundation, Düsseldorf
completed in 2016

犹太博物馆，法兰克福
施塔布建筑设计事务所，柏林
2020年竣工

Jewish Museum, Frankfurt am Main
Staab Architekten GmbH, Berlin
completed in 2020

托马斯·舒特雕塑馆，洪堡
RKW建筑城市规划事务所，杜塞尔多夫
2016年竣工

Skulpturenhalle Thomas Schütte, Hombroich
RKW Architektur + Städtebau, Düsseldorf
completed in 2016

巴贝里尼宫，波茨坦
希尔默与萨特勒和阿尔布雷特公司，柏林
2017年竣工

Palais Barberini, Potsdam
Hilmer & Sattler und Albrecht GmbH, Berlin
completed in 2017

巴伐利亚历史博物馆，雷根斯堡
沃纳·特拉克勒斯·里希特建筑设计事务所，法兰克福
HG梅尔茨博物馆建筑设计事务所，斯图加特
2019年竣工

Museum of Bavarian History, Regensburg
wörner traxler richter, Frankfurt am Main
HG Merz Architekten Museumsgestalter, Stuttgart
completed in 2019

金德尔啤酒厂，柏林
格里萨尔建筑设计事务所，苏黎世
2020年竣工

Brewhouse of the Kindl Brewery, Berlin
Grisard Architektur AG, Zurich
completed in 2020

沙克画廊，慕尼黑
艾希纳·卡泽建筑设计事务所，慕尼黑
2020年竣工

Schack Gallery, Munich
aichner kazzer architekten, Munich
completed in 2020

卡塔尔国家博物馆，多哈
让·努维尔建筑设计事务所，巴黎
2019年竣工

Qatar National Museum, Doha
Ateliers Jean Nouvel, Paris
completed in 2019

犹太博物馆，科隆
WHL旺德尔·洛奇建筑设计事务所
事务所，萨尔布吕肯
正在进行中

Archeologica zone / Jewish Museum, Cologne
Wandel Lorch WHL GmbH,
Architekten und Stadtplaner BDA, Saarbrücken
work in progress

Altes博物馆翻新项目，柏林
希尔默与萨特勒和阿尔布雷特有限公司，柏林
正在进行中

General Refurbishment of the 'Altes Museum', Berlin
Hilmer & Sattler und Albrecht GmbH, Berlin
work in progress

新斯潘道城堡博物馆，柏林
施塔布建筑设计事务所，柏林
2020年竣工

New museum citadel Spandau, Berlin
Staab Architekten GmbH, Berlin
completed in 2020

理查德·瓦格纳博物馆重建项目，拜罗伊特
HG梅尔茨博物馆建筑设计事务所，斯图加特
2015年竣工

Redevelopment of the Richard-Wagner-Museum, Bayreuth
HG Merz Architekten Museumsgestalter, Stuttgart
completed in 2015

医学史博物馆，英戈尔施塔特
施塔布建筑设计事务所，柏林
2015年竣工

Medical History Museum, Ingolstadt
Staab Architekten GmbH, Berlin
completed in 2015

斯普伦格尔博物馆，汉诺威
梅里和皮特建筑设计事务所，苏黎世
2015年竣工

Sprengel Museum, Hannover
Meili, Peter Architekten AG, Zurich
completed in 2015

威廉大帝博物馆，克雷费尔德
温弗里德·布雷纳建筑设计事务所，柏林
2015年竣工

Kaiser Wilhelm Museum, Krefeld
Winfried Brenne Architekten, Berlin
completed in 2015

施洛斯利市政博物馆，阿劳
迪纳与迪纳建筑设计事务所，巴塞尔
2014年竣工

Schlössli Municipal Museum, Aarau
Diener & Diener Architekten, Basel
completed in 2014

LWL艺术与文化博物馆，明斯特
施塔布建筑设计事务所，柏林
Space4有限公司，斯图加特
LDE贝尔兹纳·福尔摩斯，斯图加特
2014年竣工

LWL-Museum für Kunst und Kultur, Münster
Staab Architekten GmbH, Berlin
Space4 GmbH, Stuttgart
LDE Belzner Holmes, Stuttgart
completed in 2014

Ubiermonument展览馆，科隆
瓦伦丁建筑设计策划公司，科隆
阿德琳·里斯帕摄影工作室，巴黎
2014年竣工

Ubiermonument, Cologne
VALENTYNARCHITEKTEN
Planungsgesellschaft mbH, Cologne
Scenography studio adeline rispal, Paris
completed in 2014

阿尔特纳城堡冒险体验厅，阿尔特纳
克劳斯·霍伦贝克建筑设计事务所，科隆
2014年竣工

Lift Experience Burg Altena
Klaus Hollenbeck Architekten, Cologne
completed in 2014

洛温堡博物馆，卡塞尔
皮茨和霍尔建筑与景观设计事务所，柏林
2013年扩初设计

Löwenburg Museum, Kassel
Pitz & Hoh. Architektur und Denkmalpflege GmbH
Berlin
design development 2013

马赛文明博物馆场景设计，马赛
阿德琳·里斯帕摄影工作室，巴黎
鲁迪·里乔西蒂建筑设计事务所，邦多勒
2013年竣工

Musée des Civilisations de la Méditerranée, Marseille
Scenography
studio adeline rispal, Paris
Rudy Ricciotti architecte, Bandol
completed in 2013

阿伦斯霍普艺术博物馆，阿伦斯霍普
施塔布建筑设计事务所，柏林
2013年竣工

Ahrenshoop Museum of Art, Ahrenshoop
Staab Architekten GmbH, Berlin
completed in 2013

历史与纪念中心，德兰西
迪纳与迪纳建筑设计事务所，巴塞尔
2012年竣工

Center for History and Commemoration, Drancy
Diener & Diener Architekten, Basel
completed in 2012

施泰德艺术文化中心和施泰德画廊，法兰克福
施耐德+舒马赫建筑设计事务所，法兰克福
2012年竣工

Städelsches Kunstinstitut und Städtische Galerie, Frankfurt am Main
schneider + schumacher Planungsgesellschaft mbH
Frankfurt am Main
completed in 2012

瑞士国家博物馆，苏黎世
克里斯与甘滕贝恩建筑设计事务所，巴塞尔
2012年扩初设计

Swiss National Museum, Zurich
Christ & Gantenbein Architects, Basel
design development 2012

新加坡国家美术馆，新加坡
米洛建筑设计工作室，巴黎
2011年概念设计

National Art Gallery, Singapore
studioMilou architecture, Paris
design concept 2011

三星历史博物馆，龙仁市
布鲁克纳建筑设计工作室，斯图加特
2011年扩初设计

Samsung History Museum, Yongin-si City
Atelier Brückner GmbH, Stuttgart
design development 2011

达勒姆博物馆翻新项目，柏林
BHBVT建筑设计院，柏林
2011年竣工

Restoration of Dahlem Museum, Berlin
BHBVT Gesellschaft von Architekten mbH, Berlin
completed in 2011

巴伐利亚国王博物馆，霍恩施万高
施塔布建筑设计事务所，柏林
2011年竣工

Museum of the Bavarian Kings, Hohenschwangau
Staab Architekten GmbH, Berlin
completed in 2011

新美术馆，卡塞尔
施塔布建筑设计事务所，柏林
2011年竣工

New Gallery, Kassel
Staab Architekten GmbH, Berlin
completed in 2011

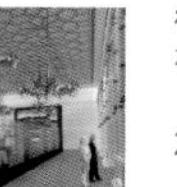

德累斯顿城堡北楼，德累斯顿
布鲁克纳建筑设计工作室，斯图加特
2010年扩初设计

Dresden Castle North Wing, Dresden
Atelier Brückner GmbH, Stuttgart
design development 2010

海洋研究所，摩纳哥
阿德琳·里斯帕工作室，巴黎

2010年竣工

Oceanographic Institute, Monaco
studio adeline rispal, Paris

completed in 2010

宝玑博物馆，苏黎世
阿德琳·里斯帕工作室，巴黎

2010年竣工

Breguet Museum, Zurich
studio adeline rispal, Paris

completed in 2010

北威州K20格拉布广场艺术博物馆现存建筑扩建与翻新项目，杜塞尔多夫
Dissing+Weitling建筑设计事务所，哥本哈根

2010年竣工

K20 am Grabbeplatz Kunstsammlung NRW, Düsseldorf
Extension and refurbishment of existing building
Dissing+Weitling architecture, Copenhagen

completed in 2010

沃邦别墅，卢森堡
黛安·海伦德和菲利普·施密特建筑设计事务所，卢森堡

2010年竣工

Villa Vauban, Luxembourg
Diane Heirend & Philippe Schmit architects, Luxembourg

completed in 2010

达尔海姆修道院基金会LWL修道院艺术文化博物馆，明斯特
菲佛·艾勒曼·普雷克尔建筑设计有限公司，明斯特

2010年竣工

Dalheim Abbey Foundation
LWL Museum of Abbey Culture
Pfeiffer Ellermann Preckel GmbH, Münster

completed in 2010

佐尔夫雷恩煤矿鲁尔博物馆，埃森
HG梅尔茨博物馆建筑设计事务所，斯图加特

2010年竣工

Ruhr Museum, Zollverein Colliery, Essen
HG Merz Architekten Museumsgestalter, Stuttgart

completed in 2010

沙克画廊，慕尼黑
慕尼黑公共建设局，慕尼黑

2009 年竣工

Schack Gallery, Munich
Public Construction Authority, Munich

completed in 2009

埃米尔·舒马赫博物馆，哈根
卡尔·恩斯特·奥斯特豪斯博物馆，哈根
林德曼建筑设计事务所，曼海姆

2009年竣工

Emil Schumacher Museum, Hagen
Karl Ernst Osthaus Museum, Hagen
Lindemann Architekten, Mannheim

completed in 2009

柏林艺术图书馆部分建筑，柏林
柏林国家博物馆，柏林

2009年竣工

Berlin Art Library, Berlin
Parts of the Building
Staatliche Museen zu Berlin, Berlin

completed in 2009

州立博物馆，什未林
M-V房地产和建筑服务机构，什未林

2009年竣工

State Museum, Schwerin
Real Estate and Construction Services M-V, Schwerin

completed in 2009

罗浮宫米洛画廊的维纳斯雕像，巴黎
斑点建筑设计事务所，巴黎

2008年概念设计

Musée du Louvre, Paris
Venus de Milo Gallery,
Repérages Architectures, Paris

design concept 2008

罗森加特博物馆，卢赛恩
迪纳与迪纳建筑设计事务所，巴塞尔

2008年竣工

Rosengart Museum, Lucerne
Diener & Diener Architekten, Basel

completed in 2008

伦巴赫美术馆，慕尼黑
福斯特建筑设计事务所，柏林

2008年扩初设计

Lenbachhouse, Munich
Foster + Partners, Berlin

design development 2008

冈曾豪瑟博物馆，开姆尼茨
施塔布建筑设计事务所，柏林

2008年竣工

Gunzenhauser Museum, Chemnitz
Staab Architekten GmbH, Berlin

completed in 2008

柏林洪堡大学自然历史博物馆，柏林
迪纳与迪纳建筑设计事务所，巴塞尔

2007年竣工

Museum of Natural History
Humboldt University Berlin, Berlin
Diener & Diener Architekten, Berlin

completed in 2007

老绘画陈列馆部分建筑，慕尼黑
施塔布建筑设计事务所，柏林

2005年竣工

Alte Pinakothek, Munich
Parts of the Building
Staab Architekten GmbH, Berlin

completed in 2005

罗兰塞克火车站，阿尔普博物馆，雷马根
理查德·迈耶建筑设计事务所，纽约
莱茵兰-普法尔茨州财政部

2004年竣工

Arp Museum, Bahnhof Rolandseck, Remagen
Richard Meier & Partners, New York
Financial Department Rheinland-Pfalz

completed in 2004

马克斯·恩斯特博物馆，布鲁尔
范登瓦伦丁建筑设计事务所，科隆
SMO建筑设计事务所，科隆

2004年竣工

Max Ernst Museum, Brühl
van den Valentyn Architektur, Cologne
SMO Architektur, Cologne

completed in 2004

科佩尼克宫殿装饰艺术博物馆，柏林
BASD建筑设计事务所，柏林

2004年竣工

Museum of Decorative Arts, Köpenick, Palace, Berlin
BASD Westphal+Schlotter, Berlin

completed in 2004

赫尔穆特·牛顿基金会，柏林
卡尔费尔特建筑设计有限公司，柏林

2004年竣工

Helmut Newton Foundation, Berlin
Kahlfeldt Architekten GmbH, Berlin

completed in 2004

阿尔贝廷宫翻新工程，德累斯顿
范登瓦伦丁建筑设计事务所，科隆

2004年竣工

Refurbishment of the Albertinum, Dresden
van den Valentyn Architektur, Cologne

completed in 2004

布洛翰博物馆，柏林
照明设备翻新、可行性研究
柏林参议院

2004年竣工

Bröhan Museum, Berlin
Lighting Refurbishment/Feasibility Study
Senate of Berlin

completed in 2004

新勃兰登堡州博物馆
贾斯特拉姆和巴特勒建筑设计事务所，罗斯托克

2002年概念设计

Neubrandenburg Regional Museum
Jastram+Buttler Architekten, Rostock

design concept 2002

北威州K21艺术收藏博物馆，杜塞尔多夫
基斯勒建筑设计事务所，慕尼黑

2002年竣工

K21 Kunstsammlung NRW 'Ständehaus', Düsseldorf
Kiessler+Partner Architekten GmbH, Munich

completed in 2002

旧国家画廊，柏林
HG梅尔茨教授，柏林

2001年竣工

Old National Gallery 'Alte Nationalgalerie', Berlin
Prof. HG Merz, Berlin

completed in 2001

艺术宫博物馆，杜塞尔多夫
O.M.安格斯教授工作室，科隆

2001年竣工

museum kunst palast, Düsseldorf
Prof. O.M. Ungers Architekt, Cologne

completed in 2001

格奥尔格·谢弗博物馆，施韦因富特
施塔布建筑设计事务所，柏林

2000年竣工

Georg Schäfer Museum, Schweinfurt
Staab Architekten GmbH, Berlin

completed in 2000

威廉港城堡博物馆，卡塞尔
斯蒂芬·布朗费尔斯建筑设计事务所，柏林

2000年竣工

Schloss Wilhelmshöhe Museum, Kassel
Stephan Braunfels Architekten, Berlin

completed in 2000

斯塔克别墅博物馆，慕尼黑
基斯勒建筑设计事务所，慕尼黑
伯迈斯特教授与瓦纳弗建筑设计事务所，慕尼黑

2000年竣工

Villa Stuck Museum, Munich
Kiessler + Partner Architekten GmbH, Munich
Prof. Burmeister, Wallnöfer und Partner, Munich

completed in 2000

弗里德里希斯哈根自来水厂，柏林
舍夫机械制造厂C地块改建项目
IBE，柏林

2000年竣工

Friedrichshagen Waterworks, Berlin
Reconstruction of the 'Schöpfmaschinenhauses' C
IBE, Berlin

completed in 2000

传媒博物馆，柏林
瓦赫金建筑设计事务所，布伦瑞克

1999年竣工

Museum for Communication, Berlin
Vahjen+Partner Architekten, Braunschweig

completed in 1999

马丁·格罗皮乌斯大厦，柏林
希尔默与萨特勒和阿尔布雷希特建筑设计有限公司，柏林

1999年竣工

Martin Gropius Building, Berlin
Hilmer & Sattler und Albrecht GmbH, Berlin

completed in 1999

伯格鲁恩博物馆，柏林
希尔默与萨特勒和阿尔布雷希特有限公司，柏林

1998年竣工

Berggruen Museum, Berlin
Hilmer & Sattler und Albrecht GmbH, Berlin

completed in 1998

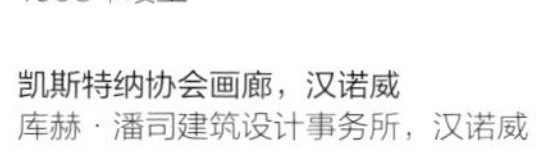
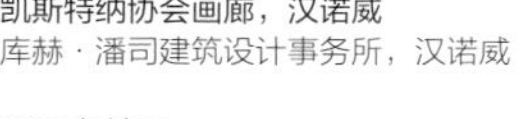

凯斯特纳协会画廊，汉诺威
库赫·潘司建筑设计事务所，汉诺威

1998年竣工

Kestnergesellschaft, Hannover
Koch Panse Architekten, Hannover

completed in 1998

什未林州立博物馆图像展览馆
联邦国家计划局和什未林图像展览馆建筑管理办公室

1998年竣工

State Museum Schwerin, Picture Gallery
Federal State Planning Department and Building Control Office of Schwerin Gemäldegalerie

completed in 1998

伊姆霍夫–斯托尔韦克博物馆泛光照明，科隆
埃勒·迈耶·沃尔特建筑设计事务所，杜塞尔多夫

1998年竣工

Imhoff-Stollwerck Museum, Cologne Exterior Illumination
Eller Maier Walter + Partner, Düsseldorf

completed in 1998

德国大教堂改建项目，柏林
于尔根·普莱瑟建筑设计事务所，柏林

1997年竣工

Renovation of the 'Deutscher Dom', Berlin
Jürgen Pleuser Architekten, Berlin

completed in 1997

不来梅联邦州政府办公室，柏林
莱昂·沃尔哈格维尔尼克建筑设计事务所，柏林

1999年竣工

Bremen Federal State Office, Berlin
Léon Wohlhage Wernik Architekten, Berlin

completed in 1999

FAZ新闻大楼，柏林
希尔默与萨特勒和阿尔布雷希特有限公司，柏林

1999年竣工

FAZ Press Building, Berlin
Hilmer & Sattler und Albrecht GmbH, Berlin

completed in 1999

ARD工作室（国家广播集团），柏林
奥特纳和奥特纳建筑艺术设计事务所，柏林

1999年竣工

ARD Studios (National Broadcasting Company Berlin
Ortner & Ortner Baukunst, Berlin

completed in 1999

阿德勒斯霍夫光子学中心新楼I，柏林
绍尔布鲁·赫顿通用规划有限公司，柏林

1998年竣工

Adlershof Photonics Center, New Building I, Ber
sauerbruch hutton
Generalplanungsgesellschaft mbH, Berlin

completed in 1998

阿德勒斯霍夫光子学中心新楼II与新楼III，柏林
奥特纳和奥特纳建筑艺术设计事务所，柏林

1998年竣工

Adlershof Photonics Center, New Buildings II and III, Berlin
Ortner & Ortner Baukunst, Berlin

completed in 1998

城市代表性建筑

REPRESENTATIVE BUILDINGS

国家储蓄银行，卢森堡
吉姆·克莱姆斯建筑与设计工作室，卢森堡

2016年竣工

Banque et Caisse d'Epargne de l'Etat, Luxembourg
Jim Clemes Associates, Esch-sur-Alzette, Luxembourg

completed in 2016

大主教住所，科隆
科隆大主教管区总督，科隆

2014年竣工

Archiepiscopal Domicile, Cologne
Archdiocese Cologne General Vicariate, Cologne

completed in 2014

美国学院，柏林
卡尔费尔特建筑设计有限公司，柏林

2006年竣工

American Academy, Berlin
Kahlfeldt Architekten GmbH, Berlin

completed in 2006

林登总部贝塔斯曼集团代表处，柏林
范登瓦伦丁建筑设计事务所，科隆

2003年竣工

Kommandantur Unter den Linden, Berlin Representative Office of Bertelsmann AG
van den Valentyn Architektur, Cologne

completed in 2003

亨尼格斯多夫市政厅，亨尼格斯多夫
绍尔布鲁·赫顿通用规划有限公司，柏林

2003年竣工

Hennigsdorf Town Hall, Hennigsdorf
sauerbruch hutton
Generalplanungsgesellschaft mbH, Berlin

completed in 2003

德国公务员协会大厦，柏林
卡尔–海因茨·肖默建筑设计事务所，波恩

2002年竣工

House of the German Civil Service Association, Berlin
Architekturbüro Karl-Heinz Schommer, Bonn

completed in 2002

新哈登堡皇宫，新哈登堡
KLA建筑设计事务所，杜塞尔多夫

2002年竣工

Neuhardenberg Palace, Neuhardenberg
KLA-Architektur GmbH, Düsseldorf

completed in 2002

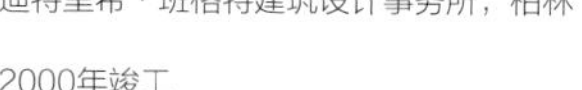

巴登–符腾堡州联邦办公大楼，柏林
迪特里希·班格特建筑设计事务所，柏林

2000年竣工

Baden-Württemberg Federal State Office, Berlin
Dietrich Bangert Architekturbüro, Berlin

completed in 2000

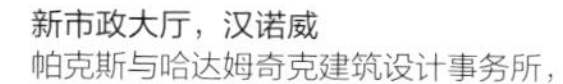

新市政大厅，汉诺威
帕克斯与哈达姆奇克建筑设计事务所，汉诺威

2000年竣工

New City Hall, Hannover
Pax + Hadamczik Architekten, Hannover

completed in 2000

哈登伯格之屋，汉诺威
ASP建筑设计事务所，汉诺威

2000年竣工

Hardenbergsches Haus, Hannover
ASP Architekten Schweger Partner, Hannover

completed in 2000

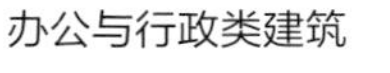

办公与行政类建筑

OFFICE AND ADMINISTRATION BUILDINGS

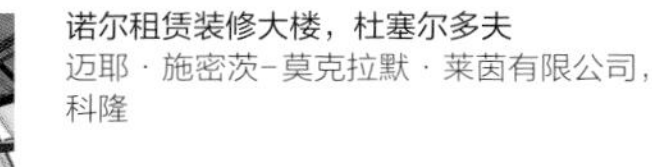

诺尔租赁装修大楼，杜塞尔多夫
迈耶·施密茨–莫克拉默·莱茵有限公司，科隆

2016年竣工

Tenant Fitout Noerr, Düsseldorf
msm meyer schmitz-morkramer rhein gmbh, Colog

completed in 2016

儿童医院，苏黎世
赫尔佐格与德梅隆建筑设计事务所，巴塞尔

正在进行中

Pediatric Hospital, Zurich
Herzog & de Meuron, Basel

work in progress

历史档案馆和影像资料收藏馆改建项目，科隆
韦希特与韦希特建筑设计事务所，达姆施塔特

2021年竣工

New Construction of the Historical Archive and the Rhenish Picture Library, Colog
Waechter + Waechter Architekten BDA, Darmsta

completed in 2021

新市政管理大厦，科隆
卡达维特菲尔德建筑设计有限公司，亚琛

2016年竣工

New Department, Cologne
kadawittfeldarchitektur GmbH, Aachen

completed in 2016

安永会计师事务所总部，卢森堡
绍尔布鲁·赫顿通用规划有限公司，柏林

2016年竣工

Ernst & Young Headquarter, Luxembourg
sauerbruch hutton
Generalplanungsgesellschaft mbH, Berlin

completed in 2016

大主教职业学院，科隆
3pass建筑设计事务所，科隆

2016年竣工

Erzbischöfliches Berufskolleg, Cologne
3pass Architekt/innen, Cologne

completed in 2016

瑞士保险公司新楼，苏黎世
迪纳与迪纳建筑设计事务所，巴塞尔

2016年竣工

New Construction of Swiss Re Next, Zurich
Diener & Diener Architekten, Basel

completed in 2016

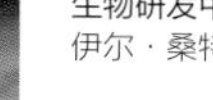

生物研发中心，巴塞尔
伊尔·桑特建筑设计事务所，苏黎世

2021年竣工

Biozentrum, Basel
IlgSanter Architekten, Zurich

completed in 2021

海格会议中心，奥贝奈
绍尔布鲁·赫顿通用规划有限公司，柏林

2015年竣工

Hager Forum, Obernai
sauerbruch hutton
Generalplanungsgesellschaft mbH, Berlin

completed in 2015

苏克总部大厦，曼海姆
RKW建筑城市规划事务所，杜塞尔多夫

2015年竣工

Südzucker Headquarter, Mannheim
RKW Architektur + Städtebau, Düsseldorf

completed in 2015

诺华制药实验室C5大楼，上海
非常建筑设计研究所，北京

2015年竣工

Novartis Laboratory Building C5, Shanghai
Atelier FCJZ, Beijing

completed in 2015

诺华制药实验室C11-2大楼，上海
迪纳与迪纳建筑设计事务所，巴塞尔

2015年竣工

Novartis Laboratory Building C11-2, Shanghai
Diener & Diener Architekten, Basel

completed 2015

诺华制药法国园区，巴黎
帕特里克·伯杰和雅克·安齐乌蒂建筑设计事务所，巴黎

2014年扩初设计

Novartis France, Paris
Patrick Berger and Jacques Anziutti architectes, Paris

design development 2014

EnBW行政楼，斯图加特
RKW建筑城市规划事务所，杜塞尔多夫

2014年竣工

EnBW Executive Floors, Stuttgart
RKW Architektur + Städtebau, Düsseldorf

completed in 2014

拜耳赌场，勒沃库森
RKW建筑城市规划事务所，杜塞尔多夫

2014年扩初设计

Bayer Casino, Leverkusen
RKW Architektur + Städtebau, Düsseldorf

design development 2014

Enervie集团总部，哈根
JSWD建筑设计事务所，科隆

2014年竣工

Enervie headquarter centralization, Hagen
JSWD Architekten, Cologne

completed in 2014

罗尔夫建筑设计事务所立方体办公楼，杜塞尔多夫
恩泽瑙尔建筑管理公司，杜塞尔多夫

2013年竣工

Rölfs & Partner Cubes, Düsseldorf
Enzenauer Architekturmanagement, Düsseldorf

completed in 2013

员工餐厅和会议大楼，西格斯多夫
格鲁兹卡和伊特曼建筑设计事务所，慕尼黑

2013年竣工

Canteen and Conference Building, Siegsdorf
Grudziecka & Itermann Architekten, Munich

completed in 2013

消防局控制中心，柏林
Ergoconcept有限公司，罗特克莱兹

2012年竣工

Fire department control centre, Berlin
Ergoconcept, Rotkreuz

completed in 2012

诺华制药园区安藤实验楼艺术照明，巴塞尔
安藤忠雄建筑与设计协会，大阪
布克哈特建筑设计有限公司，巴塞尔
马可·塞拉建筑设计事务所，巴塞尔
2012年竣工

Novartis Campus, Basel
Ando Laboratory Building, Art Illumination
Tadao Ando Architect & Associates, Osaka
Burckhardt + Partner AG, Basel
Marco Serra Architekt, Basel
completed in 2012

拜斯海姆中心部分建筑，柏林
希尔默与萨特勒和阿尔布雷希特建筑设计有限公司，柏林

2012年竣工

Beisheim Center, Berlin
Parts of the Building
Hilmer & Sattler und Albrecht GmbH, Berlin

completed in 2012

KfW德国复兴信贷银行特殊区域西拱廊，法兰克福
绍尔布鲁·赫顿通用规划有限公司，柏林

2010年竣工

KfW Bankengruppe
Special Areas Western Arcade, Frankfurt am Main
sauerbruch hutton
Generalplanungsgesellschaft mbH, Berlin

completed in 2010

蒂森克虏伯新总部园区，埃森
JSWD建筑设计事务所，科隆
柴克斯和莫雷尔建筑设计事务所，巴黎

2010年竣工

ThyssenKrupp Quarter, Essen
JSWD Architekten GmbH & Co. KG, Cologne
Atelier d'architecture Chaix & Morel et associés, Paris

completed in 2010

AachenMünchener保险公司大楼，亚琛
卡达维特菲尔德建筑有限公司，亚琛

2000年竣工

AachenMünchener Insurance, Aachen
kadawittfeldarchitektur GmbH, Aachen

completed in 2010

Nya Nordiska大楼，丹嫩贝格
施塔布建筑设计事务所，柏林

2010年竣工

Nya Nordiska, Dannenberg
Staab Architekten GmbH, Berlin

completed in 2010

Valiant银行大楼，巴塞尔
迪纳与迪纳建筑设计事务所，巴塞尔

2010年竣工

Valiant Bank, Basel
Diener & Diener Architekten, Basel

completed in 2010

城市大厅，巴塞尔
迪纳与迪纳建筑设计事务所，巴塞尔

2010年竣工

Citygate, Basel
Diener & Diener Architekten, Basel

completed in 2010

CGSH主楼，法兰克福
舒尔兹建筑设计事务所，达姆施塔特

2009年竣工

CGSH Main Tower, Frankfurt am Main
Schulze & Assoziierte, Darmstadt

completed in 2009

Interpipe办公楼，基辅
m2r建筑设计事务所，伦敦

2009年概念设计

Interpipe Office, Kiev
m2r-architecture, London

design concept 2009

诺华制药园区兰普尼亚尼办公楼，巴塞尔
di建筑设计工作室，英格·维托里奥教授、博士，马格戈·兰普尼亚尼，米兰

2009年竣工

Novartis Campus, Basel
Lampugnani Office Building
Studio di Architettura, Prof. Dr. Ing. Vittorio Magnago Lampugnani, Milan

completed in 2009

诺华制药园区麦奇办公楼，巴塞尔
麦奇建筑设计事务所，东京
茨威普菲尔建筑设计事务所，巴塞尔

2009年竣工

Novartis Campus, Basel
Maki Office Building
Maki and Associates, Tokyo
Zwimpfer Partner Architekten, Basel

completed in 2009

L银行施洛斯广场21号新建项目，卡尔斯鲁厄
威默勒建筑设计事务所，柏林

2009年竣工

New Contruction of Schloßplatz 21
L-Bank, Karlsruhe
Weinmiller Architekten, Berlin

completed in 2009

费尔德贝格大街办公楼，法兰克福
科利尼翁建筑与设计有限公司，柏林

2009年竣工

Office Building Feldbergstrasse, Frankfurt am Main
Collignon Architektur und Design GmbH, Berlin

completed in 2009

赫尔维蒂亚中心，米兰
美里和皮特建筑设计事务所，苏黎世

2009年竣工

Centro Helvetia, Milan
Meili, Peter Architekten AG, Zurich

completed in 2009

KfW德国复兴信贷银行E301会议室，法兰克福
泰斯建筑设计事务所，法兰克福

2009年竣工

KfW Bankengruppe
Conference Room E 301, Frankfurt am Main
Architekten Theiss, Frankfurt am Main

completed in 2009

KfW德国复兴信贷银行北拱廊入口大厅，法兰克福
泰斯建筑设计事务所，法兰克福

2009年竣工

KfW Bankengruppe
Entrance Lobby of Northern Arcade, Frankfurt am Main
Architekten Theiss, Frankfurt am Main

completed in 2009

KfW德国复兴信贷银行北拱廊员工餐厅，法兰克福
JSK建筑设计事务所，法兰克福

2009年竣工

KfW Bankengruppe
Staff Restaurant of Northern Arcade, Frankfurt am Main
JSK Dipl. Ing. Architekten, Frankfurt am Main

completed in 2009

AXA办公楼首层和5层特殊照明，科隆
范登瓦伦丁建筑设计事务所，科隆

2009年竣工

AXA-Office Building, Cologne
Special Illumination of Ground Floor & 5th Floor
van den Valentyn Architektur, Cologne

completed in 2009

EnBW City办公大楼，斯图加特
RKW建筑城市规划事务所，杜塞尔多夫

2009年竣工

EnBW City, Stuttgart
RKW Architektur + Städtebau, Düsseldorf

completed in 2009

国家储蓄银行大厦，奥尔登堡
RKW建筑城市规划事务所，杜塞尔多夫

2009年竣工

State Savings Bank, Oldenburg
RKW Architektur + Städtebau, Düsseldorf

completed in 2009

T-Home总部大厦，波恩
范登瓦伦丁建筑设计事务所，科隆

2009年竣工

T-Home Headquarters, Bonn
van den Valentyn Architektur, Cologne

completed in 2009

诺华制药园区克里斯沙尼茨实验大楼，巴塞尔
克利尚尼兹建筑设计有限公司，维也纳

2008年竣工

Novartis Campus, Basel
Krischanitz Laboratory Building
Architekt Krischanitz ZT GmbH, Vienna

completed in 2008

访客中心，特尼斯福斯特
恩格尔伯特·汉森建筑设计事务所，格尔德尔恩
2008年竣工

Visitors' Center, Tönisvorst
Architekt Engelbert Hanßen, Geldern
completed in 2008

新丸之内大楼，东京
霍普金斯建筑设计事务所，伦敦
三菱地产，东京
2007年竣工

Shin-Marunouchi Building, Tokyo
Hopkins Architects, London with
Mitsubishi Jisho Sekkei, Tokyo
completed in 2007

蒙绍尔大街办公楼，杜塞尔多夫
科利尼翁建筑与设计有限公司，柏林
2007年竣工

Office Building, Monschauer Straße, Düsseldorf
Collignon Architektur und Design GmbH, Berlin
completed in 2007

Vogelsanger Weg办公楼，杜塞尔多夫
科利尼翁建筑与设计有限公司，柏林
2007年竣工

Office Building, Vogelsanger Weg, Düsseldorf
Collignon Architektur und Design GmbH, Berlin
completed in 2007

贝斯广场Oval联合投资租赁式写字楼，法兰克福
MOW建筑设计事务所，法兰克福
2007年竣工

Tenant Interior Work, Union Investment Oval
Baseler Platz, Frankfurt am Main
MOW Architekten –
Olschok Westenberger + Partner, Frankfurt am Main
completed in 2007

KfW德国复兴信贷银行振兴总部，法兰克福
RKW建筑城市规划事务所，杜塞尔多夫
2006年竣工

KfW Bankengruppe
Revitalization of Headquarters, Frankfurt am Main
RKW Architektur + Städtebau, Düsseldorf
completed in 2006

诺华制药园区SANAA办公楼，巴塞尔
SANAA建筑设计事务所，东京
安德烈·普特曼经纪公司，巴黎
2006年竣工

Novartis Campus, Basel
SANAA Office Building
SANAA Architects, Tokyo
Agence Andrée Putman, Paris
completed in 2006

莱茵河创新园罗门勒，波恩
卡尔-海因茨·肖默建筑设计事务所，波恩
2006年竣工

Innovation Park on the Rhine
Rohmühle Bonn
Architekturbüro Karl-Heinz Schommer, Bonn
completed in 2006

赫蒂管理学院，柏林
克劳斯-彼得·德卢斯建筑设计工作室，柏林
2006年竣工

Hertie School of Governance, Berlin
Klaus-Peter Deluse, Berlin
completed in 2006

项目2号办公楼，布鲁塞尔
迪纳与迪纳建筑设计事务所，柏林
2006年竣工

Project 2 Headoffice, Brussels
Diener & Diener Architekten, Berlin
completed in 2006

莱比锡广场1—6号保险公司办公与住宅楼，柏林
穆勒·莱曼建筑设计有限公司，柏林
2006年竣工

Württembergische Versicherung
(Insurance Company) Office and Residential
Building Leipziger Platz 1–6, Berlin
Müller Reimann Architekten GmbH, Berlin
completed in 2006

诺华制药园区乔斯和马蒂斯入口大堂，巴塞尔
乔斯和马蒂斯建筑设计事务所，苏黎世
2005年竣工

Novartis Campus, Basel
Joos & Mathys Entrance Lobby
Joos & Mathys, Zurich
completed in 2005

诺华制药园区迪纳与迪纳建筑设计事务所办公楼，巴塞尔
迪纳与迪纳建筑设计事务所，巴塞尔
2005年竣工

Novartis Campus, Basel
Diener & Diener Office Building
Diener & Diener Architekten, Basel
completed in 2005

德国储蓄银行协会改建项目，波恩
卡尔-海因茨·肖默建筑设计事务所，波恩
2005年竣工

Conversion of the Association of Savings and
Transfer Banks Building, Bonn
Architekturbüro Karl-Heinz Schommer, Bonn
completed in 2005

城市活动中心1号塔楼，法兰克福
让·努维尔建筑设计事务所，巴黎
ABB建筑设计事务所，法兰克福
2004年概念设计

Urban Entertainment Center, Frankfurt am Main
Tower 1
Ateliers Jean Nouvel, Paris
ABB Architekten, Frankfurt am Main
design concept 2004

主机场中心，法兰克福
ABB建筑设计事务所，法兰克福
2004年竣工

Main Airport Center, Frankfurt am Main
ABB Architekten, Frankfurt am Main
completed in 2004

"波恩-视觉"莱茵河畔创新公园入口大厅，波恩
卡尔-海因茨·肖默建筑设计事务所，波恩
2004年竣工

Entrance Hall, 'Bonn – Visio' Innovation Park
on the Rhine, Bonn
Architekturbüro Karl-Heinz Schommer, Bonn
completed in 2004

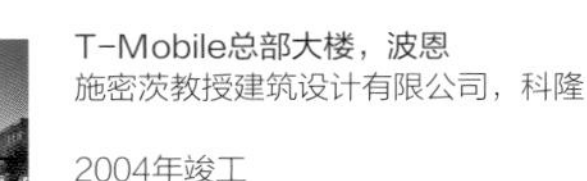
T-Mobile总部大楼，波恩
施密茨教授建筑设计有限公司，科隆
2004年竣工

Headquarters of T-Mobile, Bonn
Prof. Schmitz Architekten GmbH, Cologne
completed in 2004

E.ON总部大楼，杜塞尔多夫
昂格斯终身荣誉教授建筑设计有限公司，科隆
2004年竣工

E.ON Headquarters, Düsseldorf
Prof. O.M. Ungers GmbH, Cologne
completed in 2004

E-Werk旧电站，柏林
霍耶、申德勒和希尔斯缪勒建筑设计事务所，柏林
2003年概念设计

'E-Werk' Former Power Station, Berlin
Hoyer, Schindele, Hirschmüller + Partner, Berlin
design concept 2003

德国电信公司总部，波恩
佩津卡·品客建筑技术公司，杜塞尔多夫
2003年概念设计

Headquarters of German Telekom AG, Bonn
Petzinka Pink Technologische Architektur, Düsseldorf
design concept 2003

RWE总部大楼，多特蒙德
KSP尤尔根·恩格尔建筑设计事务所，科隆
2003年概念设计

RWE Main Building, Dortmund
KSP Jürgen Engel Architekten, Cologne
design concept 2003

办公与住宅大楼莱内德雷克A3栋，柏林
ASP建筑设计事务所，柏林
2003年竣工

Office and Residential Building
Lennédreieck A3, Berlin
ASP Architekten Schweger Partner, Berlin
completed in 2003

办公与住宅大楼莱内德雷克A4栋，柏林
科利尼翁建筑与设计有限公司，柏林
2003年竣工

Office and Residential Building
Lennédreieck A4, Berlin
Collignon Architektur und Design GmbH, Berlin
completed in 2003

办公与住宅大楼莱内德雷克A5栋，柏林
科利尼翁建筑与设计有限公司，柏林
2003年竣工

Office and Residential Building
Lennédreieck A5, Berlin
Collignon Architektur und Design GmbH, Berlin
completed in 2003

卡尔西奥多大街"伽利略"办公楼，杜塞尔多夫
HPP建筑设计事务所，杜塞尔多夫
2003年竣工

'Galileo' Office Building
Karl-Theodor-Straße, Düsseldorf
HPP Architekten, Düsseldorf
completed in 2003

格罗斯埃申海默大街"席勒故居"办公与住宅大楼，法兰克福
莱昂·沃尔哈格·维尔尼克建筑设计事务所，柏林
2003年竣工

'Schillerhaus' Office and Residential Building
Große Eschenheimerstr., Frankfurt am Main
Léon Wohlhage Wernik Architekten, Berlin
completed in 2003

德国社会协会总部Rolandufer6办公楼，柏林
莱昂·沃尔哈格·维尔尼克建筑设计事务所，柏林
2003年竣工

Rolandufer 6 Office Building, Berlin
Headquarters of the Sozialverband e.V., Germany
Léon Wohlhage Wernik Architekten, Berlin
completed in 2003

陶本大街商业大厦，柏林
卡尔费尔特建筑设计事务所，柏林
2003年竣工

Taubenstraße Business Building, Berlin
Kahlfeldt Architekten, Berlin
completed in 2003

特里尔应用科技大学，特里尔
LBB莱茵兰-普法尔茨建筑与地产公司
2003年竣工

Trier University of Applied Sciences
LBB Rheinland-Pfalz
completed in 2003

康拉德·阿登瑙尔·乌费尔街道办公与酒店大楼，科隆
范登瓦伦丁建筑设计事务所，科隆
2002年概念设计

Office Building and Hotel
on Konrad-Adenauer-Ufer, Cologne
van den Valentyn Architektur, Cologne
design concept 2002

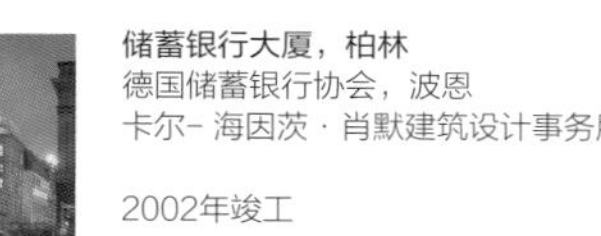
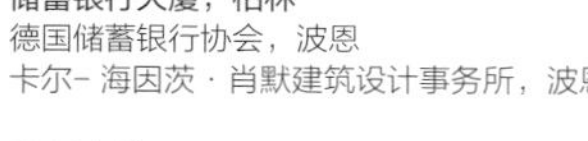
储蓄银行大厦，柏林
德国储蓄银行协会，波恩
卡尔-海因茨·肖默建筑设计事务所，波恩
2002年竣工

Sparkassen Building (Savings Bank), Berlin
Sparkassen- und Giroverband, Bonn
Architekturbüro Karl-Heinz Schommer, Bonn
completed in 2002

御林广场1号楼30区选定区域灯光设计，柏林
瓦斯科尼联合建筑设计事务所，巴黎
2001年竣工

Building 1, Quartier 30, Gendarmenmarkt, Berlin
Lighting Consultation for Selected Areas
Vasconi Associés Architectes, Paris
completed in 2001

茨温格大楼，德累斯顿
海因茨·泰萨尔建筑设计工作室，柏林
2000年竣工

'Haus am Zwinger', Dresden
Atelier Heinz Tesar, Berlin
completed in 2000

EWAG客户服务中心办公楼，纽伦堡
豪斯曼和穆勒建筑设计事务所，科隆
2000年竣工

Office Building,
EWAG Customer Service Center, Nürnberg
Hausmann + Müller Architekten, Cologne
completed in 2000

马克斯普朗克学会总部大楼，慕尼黑
波普·斯特雷布伯爵建筑设计事务所，慕尼黑
1999年竣工

Headquarters of the
Max Planck Society, Munich
Graf Popp Streib Architekten, Munich
completed in 1999

West LB大楼，杜塞尔多夫
KLA建筑设计事务所，杜塞尔多夫
1999年竣工

West LB, Düsseldorf
KLA-Architektur, Düsseldorf
completed in 1999

恩格尔哈特广场零售及行政大楼，柏林
卡尔费尔特建筑设计事务所，柏林
1998年竣工

Engelhardt Courts, Berlin
Retail and Administration Building
Kahlfeldt Architekten, Berlin
completed in 1998

108街区零售及行政大楼，柏林
范登瓦伦丁建筑设计事务所，科隆
1998年竣工

Quartier 108, Berlin
Retail and Administration Building
van den Valentyn Architektur, Cologne
completed in 1998

弗里德里希沙根水厂企业餐厅，柏林
维尔纳·吕肯建筑设计事务所，柏林
1997年竣工

Company Restaurant
of the Friedrichshagen Waterworks, Berlin
Werner Lüken Architekten, Berlin
completed in 1997

特里贡行政大楼中的执行楼层，柏林
巴森格，普汉-舒尔茨，海因里希，
施赖伯建筑设计事务所，柏林
1996年竣工

Executive Storeys of the Trigon
Administration Building, Berlin
Bassenge, Puhan-Schulz, Heinrich,
Schreiber Architekten, Berlin
completed in 1996

韦伯银行，柏林
玛吉特·弗莱茨室内设计，柏林
1996年竣工

Weberbank, Berlin
Margit Flaitz Innenarchitektin, Berlin
completed in 1996

行政大楼楼梯井的特殊照明，柏林
梅德巴赫和雷德莱特建筑设计事务所，
柏林
1995年竣工

Administration Building, Berlin
Special Illumination of the Stair Wells
Maedebach & Redeleit Architekten, Berlin
completed in 1995

市政项目

GOVERNMENT PROJECTS

图嫩研究所，不来梅港
施塔布建筑设计事务所，柏林
2017年竣工

Thünen-Institute, Bremerhaven
Staab Architekten GmbH, Berlin
completed in 2017

下萨克森州议会，汉诺威
布芬赫建筑设计事务所，斯图加特
2017年竣工

The State Parliament of Lower Saxony, Hannover
Blocher Blocher Partners, Stuttgart
completed in 2017

玛丽·伊丽莎白·吕德斯大楼扩建项目，
柏林
斯蒂芬·布朗费尔斯建筑设计事务所，柏林
正在进行中

Extension of the Marie Elisabeth Lüders Building,
Berlin
Stephan Braunfels Architekten, Berlin
work in progress

梅克伦堡-前波美拉尼亚州议会全会厅，
什未林
丹海默和乔斯建筑设计有限公司，慕尼黑
2017年竣工

Plenary Hall of the State Parliament of Mecklen-
burg-Western Pomerania, Schwerin
Dannheimer & Joos Architekten Gmbh, Munich
completed in 2017

巴登-符腾堡州议会，斯图加特
施塔布建筑设计事务所，柏林
2016年竣工

The State Parliament of Baden-Württemberg,
Stuttgart
Staab Architekten GmbH, Berlin
completed in 2016

德国大使馆整体装修，华盛顿特区
HPP建筑设计事务所，杜塞尔多夫
2015年竣工

General Renovation of German Embassy,
Washington D.C.
HPP Hentrich-Petschnigg & Partner GmbH + Co. KG,
Düsseldorf
completed in 2015

杜塞尔多夫区政府，杜塞尔多夫
北莱茵-威斯特法伦州建筑与房地产联盟，
杜塞尔多夫
2014年扩初设计

District Council Düsseldorf, Düsseldorf
Construction and Real Estate Union
North Rhine-Westphalia, Düsseldorf
design development 2014

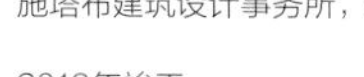

巴登-符腾堡州内政部，斯图加特
施塔布建筑设计事务所，柏林
2013年竣工

Baden-Württemberg Ministry of the Interior,
Stuttgart
Staab Architekten GmbH, Berlin
completed in 2013

巴伐利亚州议会扩建项目，慕尼黑
莱昂·沃尔哈格·维尔尼克建筑设计事务所，柏林
2012年竣工

Extension of the Bavarian State Parliament, Munich
Léon Wohlhage Wernik Architekten, Berlin
completed in 2012

德国联邦环境、自然保护与核安全部，柏林
于尔根·普莱瑟建筑设计事务所，柏林
2011年竣工

Federal Ministry for the Environment,
Nature Conservation and Nuclear Safety, Berlin
Jürgen Pleuser Architekten, Berlin
completed in 2011

瓦杜兹联邦州议会，列支敦士登
汉斯约格·格里茨建筑设计事务所，
汉诺威
弗里克建筑设计事务所，沙恩
2009年竣工

The Federal State Parliament, Vaduz, Liechtenstein
Hansjörg Göritz Architektur, Hannover
Frick Architektur, Schaan
completed in 2009

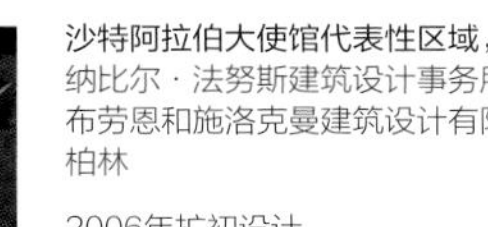

沙特阿拉伯大使馆代表性区域，柏林
纳比尔·法努斯建筑设计事务所，伦敦
布劳恩和施洛克曼建筑设计有限公司，
柏林
2006年扩初设计

Saudi Arabian Embassy, Berlin
Representative Areas
Nabil Fanous Architects, London
Braun & Schlockermann und Partner GmbH, Berlin
design development 2006

巴伐利亚州议会和马克西米利安纪念馆，
慕尼黑
施塔布建筑设计事务所，柏林
2005年竣工

The Bavarian State Parliament,
Maximilianeum, Munich
Staab Architekten GmbH, Berlin
completed in 2005

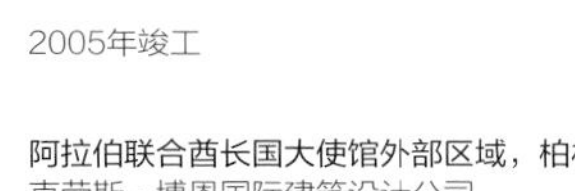

阿拉伯联合酋长国大使馆外部区域，柏林
克劳斯·博恩国际建筑设计公司，
埃施韦勒
2004年竣工

Exterior Areas of the Embassy of the
United Arab Emirates, Berlin
Krause Bohne Architects Planners International,
Eschweiler
completed in 2004

玛丽·伊丽莎白·吕德斯大厦德国联邦议院的新议会大楼，柏林
斯蒂芬·布朗费尔斯建筑设计事务所，
柏林
2004年竣工

New Parliamentary Buildings for the
'German Bundestag':
Marie Elisabeth Lüders Building, Berlin
Stephan Braunfels Architekten, Berlin
completed in 2004

保罗·洛贝大厦德国联邦议院的新议会大楼，柏林
斯蒂芬·布朗费尔斯建筑设计事务所，
柏林
2001年竣工

New Parliamentary Buildings for the
'German Bundestag':
Paul Löbe Building, Berlin
Stephan Braunfels Architekten, Berlin
completed in 2001

联邦总理府，柏林
阿克塞尔·舒尔茨建筑设计事务所，柏林
2001年竣工

Federal Chancellery, Berlin
Axel Schultes Architekten, Berlin
completed in 2001

联邦交通运输部，柏林
马克斯·杜德勒建筑设计事务所，柏林
2000年竣工

Federal Ministry of Traffic and Transport, Berlin
Max Dudler Architekt, Berlin
completed in 2000

国际海洋法法庭，汉堡
伊曼纽拉·弗莱因·冯·布兰卡和亚历山大·弗莱赫尔·冯·布兰卡，慕尼黑
2000年竣工

International Tribunal for the Law of the Sea,
Hamburg
Eamnuela Freiin von Branca and
Alexander Freiherr von Branca, Munich
completed in 2000

联邦外交部，柏林
穆勒·莱曼建筑设计有限公司，柏林
1999年竣工

The Federal Foreign Office, Berlin
Müller Reimann Architekten GmbH, Berlin
completed in 1999

柏林“国务院大楼”联邦总理临时办公室，柏林
吕特尼克建筑工程公司，柏林

1998年竣工

'Staatsratsgebäude' Berlin
Interim Use as Offices for the Federal Chancellor, Berlin
Rüthnick Architekten Ingenieure, Berlin

completed in 1998

展览陈设空间

EXHIBITIONS

实验展览，海尔布隆
绍尔布鲁·赫顿通用规划有限公司，柏林

2018年竣工

Experimenta Exhibition, Heilbronn
sauerbruch hutton
Generalplanungsgesellschaft mbH, Berlin

completed in 2018

联邦总理阿登纳故居常设展览馆，巴特洪内夫
布鲁克纳建筑设计工作室，斯图加特

2017年竣工

Permanent Exhibition
Chancellor-Adenauer-House, Bad Honnef
Atelier Brückner GmbH, Stuttgart

completed in 2017

格哈德·里希特收藏馆ERGO保险照明，杜塞尔多夫
HG梅尔茨博物馆建筑设计事务所，斯图加特

2015年竣工

ERGO Insurance Illumination of
Gerhard Richter Collection, Düsseldorf
HG Merz Architekten Museumsgestalter, Stuttgart

completed in 2015

施韦比施展览厅住房建设信贷社陈列室，施韦比施
丹·珀尔曼品牌形象建筑设计公司，柏林

2013年竣工

Bausparkasse Schwäbisch Hall
Showroom, Schwäbisch Hall
dan pearlman Markenarchitektur GmbH, Berlin

completed in 2013

达尔文动物园，罗斯托克
布鲁克纳建筑设计工作室，斯图加特

2012年竣工

Darwineum zoological garden, Rostock
Atelier Brückner GmbH, Stuttgart

completed in 2012

俄罗斯联邦储蓄银行，莫斯科
布鲁克纳建筑设计工作室，斯图加特

2011年竣工

Sberbank, Moscow
Atelier Brückner GmbH, Stuttgart

completed in 2011

欧罗巴利亚Le grand工作室，布鲁塞尔
Repérages建筑设计事务所，巴黎

2007年竣工

Europalia 'Le grand Atelier', Brussels
Repérages Architectures, Paris

completed in 2007

北威州杉本博司K20艺术收藏展，杜塞尔多夫

2007年概念设计

Hiroshi Sugimoto
K20 Art Collection NRW, Düsseldorf

design concept 2007

威尼斯双年展第十届国际建筑展德国馆
格兰特·恩斯特建筑设计事务所，柏林

2006年竣工

The Venice Biennale
The 10th International Architecture Exhibition German Pavilion
Grüntuch Ernst Architekten, Berlin

completed in 2006

德意志银行长廊艺术馆，科隆
卡里姆·拉希德公司，纽约
Uniplan国际公司，克彭

2006年竣工

Deutsche Bank Lounge, Art Cologne, Cologne
Karim Rashid Inc., New York City
Uniplan International, Kerpen

completed in 2006

马丁·格罗皮乌斯大厦的伯恩哈德·海利格展览，柏林
卡尔费尔特建筑设计有限公司，柏林

2005年竣工

Bernhard Heiliger Exhibition at the Martin Gropius Building, Berlin
Kahlfeldt Architekten GmbH, Berlin

completed in 2005

汉堡火车站当代艺术博物馆纪念约瑟夫·保罗·克莱休斯70岁诞辰，柏林
JPK70 项目组，柏林

2003年竣工

JPK 70 Marking the 70th Birthday of
Josef Paul Kleihues, Hamburger Bahnhof –
Museum for Contemporary Art, Berlin
Project Bureau JPK 70, Berlin

completed in 2003

威尼斯双年展第八届国际建筑展德国馆
莱昂·沃尔哈格·维尔尼克建筑设计事务所，柏林

2002年竣工

The Venice Biennale
The 8th International Architecture Exhibition German Pavilion
Léon Wohlhage Wernik Architekten, Berlin

completed in 2002

柏林旧博物馆“洪水之后——德累斯顿美术馆收藏品”，柏林
卡尔费尔特建筑设计有限公司，柏林

2002年竣工

After the Flood – Masterpieces of the Dresden Picture Gallery, Berlin
Old Museum ('Altes Museum'), Berlin
Kahlfeldt Architekten GmbH, Berlin

completed in 2002

柏林旧博物馆“柏林时代——密斯·凡·德罗1907—1938”，柏林
卡尔费尔特建筑设计有限公司，柏林

2001年竣工

The Berlin Years – Mies van der Rohe
1907–1938 Old Museum ('Altes Museum'), Berlin
Kahlfeldt Architekten GmbH, Berlin with P. Dvo

completed in 2001

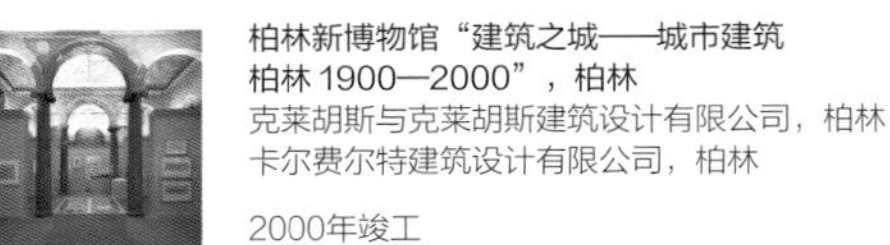

柏林新博物馆“建筑之城——城市建筑柏林 1900—2000”，柏林
克莱胡斯与克莱胡斯建筑设计有限公司，柏林
卡尔费尔特建筑设计有限公司，柏林

2000年竣工

City of Architecture – Architecture of the City, Berlin 1900 – 2000, New Museum ('Neues Museum'), Berlin
Kleihues + Kleihues GmbH, Berlin
Kahlfeldt Architekten GmbH, Berlin
completed in 2000

沃尔特·德·玛丽亚2000年的雕塑，汉堡火车站当代艺术博物馆，柏林

2000年竣工

The 2000 Sculpture, Walter de Maria
Hamburger Bahnhof –
Museum for Contemporary Art, Berlin

completed in 2000

马丁·格罗皮乌斯大厦德国图像特别展，柏林
卡尔费尔特建筑设计有限公司，柏林

1999年竣工

Pictures of Germany, Special Exhibition,
Martin Gropius Building, Berlin
Kahlfeldt Architekten GmbH, Berlin

completed in 1999

媒体立面和灯光艺术

MEDIA FACADES AND LIGHT ART

水塔，卢森堡
吉姆·克莱姆斯建筑与设计工作室，卢森堡

2018年竣工

Water Tower, Luxembourg
Atelier d'Architecture et de Design
Jim Clemes s.a., Luxembourg

completed in 2018

鹏瑞集团深圳湾壹号T1—T6和T8媒体立面，深圳
福克斯建筑设计事务所，纽约

2018年竣工

Parkland Center, One Shenzhen Bay, Shenzhen
Media facades Towers T1 – T6 and T8
Kohn Pedersen Fox Associates, New York

completed in 2018

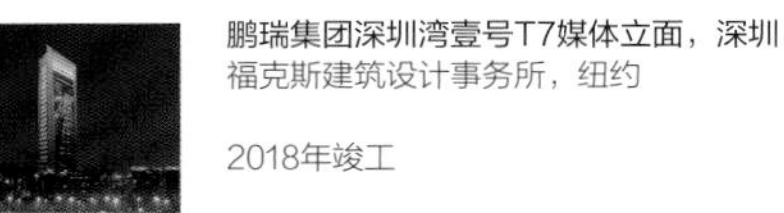

鹏瑞集团深圳湾壹号T7媒体立面，深圳
福克斯建筑设计事务所，纽约

2018年竣工

Parkland Center, One Shenzhen Bay, Shenzhen
Media facade Tower T7
Kohn Pedersen Fox Associates, New York

completed in 2018

德国储蓄银行灯光艺术装置ICOON，汉诺威
舒尔茨建筑设计事务所，汉诺威

2015年竣工

Sparkasse, Hannover
Light art 'icoon'
schulze & partner. architektur., Hannover

completed in 2015

“太阳之光”声光艺术装置展览厅，法兰克福

2012年竣工

'suns shine' Light and Sound Installation
Showroom, Frankfurt am Main

completed in 2012

OLED灯光艺术装置
国际照明与建筑展览会，法兰克福

2010年竣工

OLED Installation
Light + Building, Frankfurt am Main

completed in 2010

Telekom步行桥，波恩
布劳恩·施洛克曼建筑设计事务所，斯图加特

2010年竣工

Telekom Bridge, Bonn
Schlaich Bergermann und Partner, Stuttgart

completed in 2010

阿斯特电影休息室，柏林
马斯克和苏伦建筑与设计事务所，柏林

2009年竣工

Astor Film Lounge, Berlin
Maske + Suhren Architekten und Designer, Ber

completed in 2009

斯巴达银行，法兰克福
bgf建筑设计事务所，威斯巴登

2009年竣工

Sparda Bank, Frankfurt am Main
bgf + architekten, Wiesbaden

completed in 2009

“四球”声光艺术装置
过渡“移动中的光”

2007年竣工

'Four Spheres' Light and Sound Installation
Transition 'Light on the move'

completed in 2007

尤尼卡大厦，维也纳
诺伊曼建筑设计事务所，维也纳

2006年竣工

Uniqa Tower, Vienna
Neumann + Partners, Vienna

completed in 2006

柏林万豪酒店部分室内空间灯光设计
伯恩·阿尔伯斯教授，柏林

2004年竣工

**Berlin Marriott Hotel
Parts of the Building**
Prof. Bernd Albers, Berlin

completed in 2004

"城市之光"大楼，柏林
科利尼翁建筑与设计有限公司，柏林

2003年竣工

City Light House, Berlin
Collignon Architektur und Design GmbH, Berlin

completed in 2003

文化类建筑

CULTURAL BUILDINGS

救世主教区中心教堂，巴特戈德斯贝格
DEEN建筑设计工作室，明斯特

2020年竣工

**Church of the Redeemer Parish Center,
Bad Godesberg**
DEEN architects, Münster

completed in 2020

圣海德薇大教堂，柏林
西肖和沃尔特建筑设计有限公司，富尔达

正在进行中

St. Hedwigs-Kathedrale, Berlin
Sichau & Walter Architekten GmbH, Fulda

work in progress

慕尼黑设计应用技术大学，慕尼黑
施塔布建筑设计事务所，柏林

2015年竣工

University of Design, Munich
Staab Architekten GmbH, Berlin

completed in 2015

校园景观照明，科隆
格诺·舒尔茨建筑设计有限公司，科隆

正在进行中

Educational Landscape, Cologne
gernot schulz : architektur GmbH, Cologne

work in progress

汉莎中学，科隆
IAA建筑设计有限公司，恩斯赫德

正在进行中

Hansa Secondary School, Cologne
IAA Architecten GmbH, Enschede

work in progress

阿尔滕贝格青年教育中心，奥登塔尔阿尔滕贝格
格诺·舒尔茨建筑设计有限公司，科隆

2018年竣工

**Youth Education Center House Altenberg,
Odenthal Altenberg**
gernot schulz : architektur GmbH, Cologne

completed in 2018

迈克尔斯贝格天主教社会研究所，西格堡
MSM建筑设计有限公司，科隆

2017年竣工

**Catholic-Social Institute Michaelsberg,
Siegburg**
MSM Meyer Schmitz-Morkramer Rhein GmbH,
Cologne

completed in 2017

SOS教育与会议中心，柏林
路德洛夫与路德洛夫建筑设计事务所，
柏林

2017年竣工

SOS Education and Meeting Center, Berlin
ludloff + ludloff Architekten, Berlin

completed in 2017

格尼格尔教育校区，萨尔茨堡
斯托奇·埃勒斯建筑设计民法公司，汉诺威

2017年竣工

Educational Campus, Salzburg
Storch Ehlers Partner Architekten GbR, Hannover

completed in 2017

市民和媒体中心，斯图加特
亨宁·拉森建筑设计事务所，慕尼黑

2017年竣工

Citizen and Media Center, Stuttgart
Henning Larsen Architects, Munich

completed in 2017

北莱茵-威斯特法伦州健康校园，波鸿
莱昂·沃尔哈格·维尔尼克建筑设计事务所，柏林

2017年竣工

Health Campus North Rhine Westphalia, Bochum
Léon Wohlhage Wernik Architekten, Berlin

completed in 2017

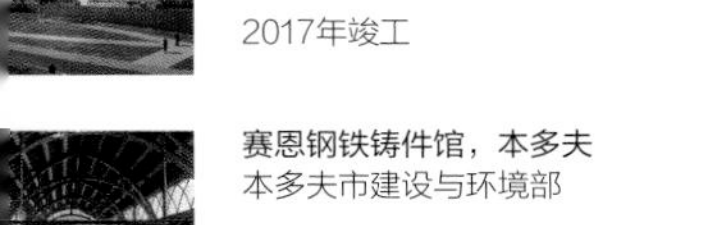

赛恩钢铁铸件馆，本多夫
本多夫市建设与环境部

2015年竣工

Sayn Iron Works Foundry, Bendorf
Department of Construction and Environment,
Municipality of Bendorf

completed in 2015

圣雷诺迪教堂，多特蒙德
多特蒙德联合教堂协会

2014年竣工

St. Reinoldi Church, Dortmund
Vereinigte Kirchenkreise Dortmund

completed in 2014

世界遗产地弗尔克林根小屋的钢铁厂熔炉台，弗尔克林根
于佩尔与于佩尔建筑规划有限公司，萨尔布吕肯

2014年竣工

**Furnace Platform of the Sintering Plant at the
Ironworks of World Heritage Site
Völklinger Hütte, Völklingen**
Huppert & Huppert Bauplanungs GmbH,
Saarbrücken
completed in 2014

福音复活教堂门厅扩建项目，西格堡
马提尼建筑设计事务所，波恩

2012年扩初设计

**Extension Foyer of the Ev.
Auferstehungskirche, Siegburg**
Martini Architekten, Bonn

design development 2012

西校区社会与教育科学学院和校长楼新楼扩建项目，法兰克福
托马斯·穆勒和伊万·赖曼建筑设计有限责任公司

2012年竣工

**New Development for the Faculty of Social and
Educational Sciences and Presidency,
Campus Westend, Frankfurt am Main**
Thomas Müller Ivan Reimann Gesellschaft
von Architekten mbH
completed in 2012

克拉森大街礼堂中心，亚琛
施密特·哈默·拉森建筑设计事务所，
奥胡斯
施莱尔建筑工程事务所，亚琛

2011年扩初设计

Auditorium Center Claßenstrasse, Aachen
schmidt hammer lassen architects, Aarhus
Höhler + Partner Architekten und Ingenieure,
Aachen

design development 2011

考夫霍夫画廊办公区展厅和连接通道，
科隆
海因里希与斯坦哈特建筑设计有限公司，
本多夫

2011年竣工

**Galeria Kaufhof executive area, Cologne
Auditorium and connecting passage**
Heinrich + Steinhardt GmbH, Bendorf

completed in 2011

萨尔茨堡大学文化与社会科学学院
斯托奇·埃勒斯建筑设计民法公司，汉诺威

2011年竣工

**Faculty of Cultural and Social Sciences,
University of Salzburg**
Storch Ehlers Partner Architekten GbR, Hannover

completed in 2011

约瑟夫·贝克技术高中扩建项目，格雷文马赫
北极星建筑设计事务所，卢森堡

2010年扩初设计

**Extension of the 'Lycée technique Joseph Bech',
Grevenmacher**
Polaris Architects, Luxembourg

design development 2010

比撒列学院，耶路撒冷
Studyo建筑设计事务所，科隆

2010年扩初设计

Bezalel Academy, Jerusalem
Studyo Architects, Cologne

design development 2010

科隆歌剧院和剧场翻新项目，科隆
JSWD建筑设计事务所，科隆
柴克斯和莫雷尔建筑设计事务所，巴黎

2010年扩初设计

Refurbishment of Opera House and Theatre, Cologne
JSWD Architekten, Cologne
Atelier d'architecture Chaix & Morel et
associés, Paris

design development 2010

利布弗劳恩圣玛利亚教堂骨灰安置所，
多特蒙德
施塔布建筑设计事务所，柏林

2011年竣工

Columbarium Liebfrauenkirche, Dortmund
Staab Architekten GmbH, Berlin

completed in 2011

高等音乐学校，艾因西德伦
迪纳与迪纳建筑设计事务所，巴塞尔

2010年竣工

Donation Music School, Einsiedeln
Diener & Diener Architekten, Basel

completed in 2010

煤矿厂洗煤区，埃森
雷姆·库哈斯大都会建筑办公室，鹿特丹
海因里希·博尔建筑设计事务所，埃森

2009年竣工

Coal Washing Plant, Zeche Colliery, Essen
Rem Koolhaas, The Office for Metropolitan
Architecture, Rotterdam
Heinrich Böll Architekten, Essen

completed in 2009

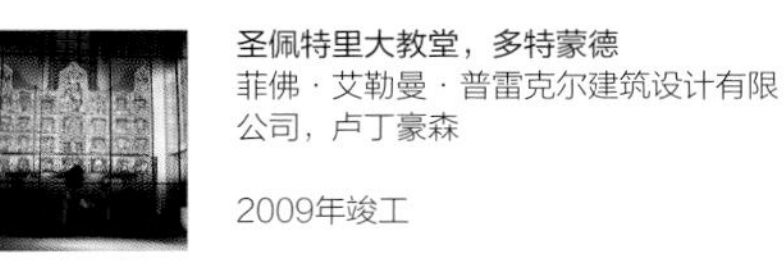

圣佩特里大教堂，多特蒙德
菲佛·艾勒曼·普雷克尔建筑设计有限公司，卢丁豪森

2009年竣工

St. Petri, Dortmund
Pfeiffer Ellermann Preckel GmbH, Lüdinghausen

completed in 2009

法律与经济学院西校区，法兰克福
托马斯·穆勒和伊万·赖曼建筑设计有限责任公司

2008年竣工

**Department of Law and Economics,
Campus Westend, Frankfurt am Main**
Thomas Müller Ivan Reimann Gesellschaft
von Architekten mbH

completed in 2008

包卡德米国际建筑学院，柏林
国际建筑学院协会，柏林

2007年概念设计

The International 'Bauakademie', Berlin
Association of the International Building Academy,
Berlin

design concept 2007

皇家罗马浴场入口，特里尔
昂格斯终身荣誉教授建筑设计有限公司，
科隆

2007年竣工

Entrance to Imperial Roman Baths, Trier
Prof. O.M. Ungers GmbH, Cologne

completed in 2007

齐默大街90、91号，daad柏林艺术家画廊计划，柏林
JWB建筑设计事务所，布伦瑞克

2005年竣工

daad Gallery of the Berlin Artists' Program
Zimmerstr. 90/91, Berlin
Architekten JWB, Braunschweig

completed in 2005

安娜·阿玛利亚公爵夫人图书馆扩建项目，魏玛
巴尔兹、里特曼斯佩格与施密茨建筑设计事务所，魏玛

2005年竣工

Extension of the Duchess Anna Amalia Library, Weimar
Barz, Rittmannsperger, Schmitz Architekten, Weimar

completed in 2005

歌剧院部分建筑，波恩
卡尔-海因茨·肖默建筑设计事务所，波恩

2004年概念设计

Opera House, Bonn
Parts of the Building
Architekturbüro Karl-Heinz Schommer, Bonn

design concept 2004

斯泰勒传教士协会，圣奥古斯丁
马提尼建筑设计事务所，波恩

2004年竣工

Steyler Missionaries Association, St. Augustin
Martini Architekten, Bonn

completed in 2004

洪堡大学演讲厅，柏林
米勒·赖曼建筑设计有限公司，柏林

2003年竣工

Lecture Theatre of Humboldt University, Berlin
Müller Reimann Architekten GmbH, Berlin

completed in 2003

圣约翰新教修道院，柏林
罗伯特·凯特勒建筑设计事务所，柏林

2003年竣工

Protestant Monastery of St. John, Berlin
Robert Ketterer Architekten, Berlin

completed in 2003

基督教堂，科隆-德尔布吕克
马提尼建筑设计事务所，波恩

2003年竣工

Christchurch, Cologne-Dellbrück
Martini Architekten, Bonn

completed in 2003

柏林勃兰登堡科学与人文学院，柏林
安德哈尔滕建筑设计事务所，柏林

2001年竣工

Berlin-Brandenburg Academy of Sciences and Humanities, Berlin
Anderhalten Architekten, Berlin

completed in 2001

圣约翰教堂，特罗斯多夫
马提尼建筑设计事务所，波恩

2001年竣工

St. John's Church, Troisdorf
Martini Architekten, Bonn

completed in 2001

魏玛包豪斯大学
范登瓦伦丁建筑设计事务所，科隆

2000年竣工

Bauhaus University Weimar (Henry van de Velde)
van den Valentyn Architektur, Cologne

completed in 2000

海伦豪森花园餐厅，汉诺威
ASP建筑设计事务所，汉诺威

2000年竣工

Restaurant, Herrenhäuser Gardens, Hannover
ASP Architekten Schweger Partner, Hannover

completed in 2000

世博馆，沃尔夫斯堡
莱昂·沃尔哈格·维尔尼克建筑设计事务所，柏林

1999年竣工

Expo Pavilion, Wolfsburg
Léon Wohlhage Wernik Architekten, Berlin

completed in 1999

菩提树大街国家图书馆，柏林
BBD，柏林
吕特尼克建筑工程公司，柏林

1997年竣工

State Library, Unter den Linden, Berlin
BBD Berlin
Rüthnick Architekten Ingenieure, Berlin

completed in 1997

音乐高中学院景观楼层，魏玛
范登瓦伦丁建筑设计事务所，科隆

1997年竣工

Music High School, Schloss Belvedere, Weimar
van den Valentyn Architektur, Cologne

completed in 1997

新警所，柏林
德意志联邦共和国中央纪念碑
希尔默与萨特勒和阿尔布雷希特有限公司，柏林

1996年竣工

Neue Wache, Berlin
Central Memorial of the Federal Republic of Germany
Hilmer & Sattler und Albrecht GmbH, Berlin

completed in 1996

购物中心

MALLS / SHOPPING CENTRES

Bikini购物中心，柏林
SAQ工作室，布鲁塞尔

2014年竣工

Bikini Berlin, Berlin
SAQ Studio Arne Quinze, Brussels

completed in 2014

斯图基购物中心，巴塞尔
RKW建筑城市规划事务所，杜塞尔多夫

2013年竣工

Stücki Shopping Centre, Basel
RKW Architektur + Städtebau, Düsseldorf

completed in 2013

Boulevard购物中心，柏林
奥特纳和奥特纳建筑艺术设计事务所，柏林

2012年竣工

Boulevard Berlin, Berlin
Ortner & Ortner Baukunst, Berlin

completed in 2012

Höfe am Brühl购物中心，莱比锡
格兰特·恩斯特建筑设计事务所，柏林

2011年扩初设计

Höfe am Brühl, Leipzig
Grüntuch Ernst Architekten, Berlin

design development 2011

杜斯曼文化百货，柏林
罗伯特纽恩建筑设计事务所，柏林

2010年竣工

KulturKaufhaus Dussmann, Berlin
Robertneun, Berlin

completed in 2010

阿尼肯画廊，希尔德斯海姆
格尔德·林德曼建筑设计事务所，布伦瑞克

2010年概念设计

Arneken Gallery, Hildesheim
Gerd Lindemann + Partner, Braunschweig

design concept 2010

教堂花园商业大厦，巴塞尔
迪纳与迪纳建筑设计事务所，巴塞尔

2010年竣工

Kirschgarten Commercial Building, Basel
Diener & Diener Architekten, Basel

completed in 2010

Galeria Kaufhof旗舰店
布雷古拉建筑设计事务所，格里斯海姆

2010年竣工

Galeria Kaufhof Model Branches
Architekturbüro Bregulla, Griesheim

completed in 2010

Galeria Kaufhof玛丽亚广场店部分建筑，慕尼黑
斯塔瓦尔德·马加尔建筑设计事务所，杜塞尔多夫

2010年竣工

Galeria Kaufhof Marienplatz, Munich
Parts of the Building
Statwald Magar Architekten, Düsseldorf

completed in 2010

光影拱廊，杜塞尔多夫
恩泽瑙建筑管理公司，杜塞尔多夫

2009年竣工

Shadow Arcades, Düsseldorf
Enzenauer Architekturmanagement, Düsseldorf

completed in 2009

第一装饰店，吉达
博纳特·莱纳特建筑设计事务所，柏林

2009年扩初设计

1st for Decoration Shopdesign, Jeddah
Bernardy°Lehnert Architekten, Berlin

design development 2009

Maishaa购物中心，印度
博纳特·莱纳特建筑设计事务所，柏林

2009年竣工

Maishaa Shops, India
Bernardy°Lehnert Architekten, Berlin

completed in 2009

Galeria Kaufhof购物中心，杜塞尔多夫
萨特瓦尔德·马加尔建筑设计事务所，杜塞尔多夫

2009年扩初设计

Galeria Kaufhof, Düsseldorf
Statwald Magar Architekten, Düsseldorf

design development 2009

Galeria Kaufhof购物中心，卡塞尔
布雷古拉建筑设计事务所，格里斯海姆

2009年竣工

Galeria Kaufhof, Kassel
Architekturbüro Bregulla, Griesheim

completed in 2009

Galeria Kaufhof蒙克贝格大街店，汉堡
海涅策划公司，汉堡

2009年竣工

Galeria Kaufhof Mönckebergstrasse, Hamburg
Heine Planungsgesellschaft, Hamburg

completed in 2009

法兰克福查尔大街Galeria Kaufhof购物中心
斯塔尼采克建筑设计事务所，波恩
2009年竣工

Galeria Kaufhof, Frankfurt-Zeil
Architekt Stanitzek, Bonn
completed in 2009

慕尼黑卡尔广场Galeria Kaufhof购物中心美食中心
米歇尔集团，乌尔姆
2009年竣工

Galeria Kaufhof, Munich-Stachus
Food Department
Michelgroup, Ulm
completed in 2009

Galeria Kaufhof购物中心美食中心，曼海姆
斯塔尼采克建筑设计事务所，波恩
2009年竣工

Galeria Kaufhof, Mannheim
Food Department
Architekt Stanitzek, Bonn
completed in 2009

Galeria Kaufhof购物中心部分建筑，奥伯豪森
海涅策划公司，汉堡
2009年竣工

Galeria Kaufhof, Oberhausen
Parts of the Building
Heine Planungsgesellschaft, Hamburg
completed in 2009

海滨城，不来梅
RKW建筑城市规划事务所，杜塞尔多夫
2009年竣工

Waterfront, Bremen
RKW Architektur + Städtebau, Düsseldorf
completed in 2009

REAL未来城
麦德龙集团资产管理有限责任公司，杜塞尔多夫
伍尔夫建筑设计事务所，斯图加特
2009年竣工

Future Store REAL
METRO Group Asset Management GmbH & Co. KG, Düsseldorf
Wulf & Partner, Stuttgart
completed in 2009

Galeria Kaufhof购物中心美食中心，科隆魏登
斯塔尼采克建筑设计事务所，波恩
2008年竣工

Galeria Kaufhof, Cologne-Weiden
Food Department
Architekt Stanitzek, Bonn
completed in 2008

切拉兹零售公园，波兰
RKW建筑城市规划事务所，杜塞尔多夫
2008年竣工

Retail Park in Czeladz, Poland
RKW Architektur + Städtebau, Düsseldorf
completed in 2008

卡罗莱纳购物中心，俄斯特拉发
大都会建筑事务所，鹿特丹
海因里希·博尔建筑设计事务所，埃森
2008年概念设计

Karolina Shopping Centre, Ostrava
Rem Koolhaas, The Office for Metropolitan Architecture, Rotterdam
Heinrich Böll Architekt, Essen
design concept 2008

安妮·方丹水疗中心，巴黎
安德烈·普特曼经纪公司，巴黎
2008年竣工

Spa Anne Fontaine, Paris
Agence Andrée Putman, Paris
completed in 2008

Galeria Kaufhof购物中心，汉诺威
布雷古拉建筑设计事务所，格里斯海姆
2007年竣工

Galeria Kaufhof, Hannover
Architekturbüro Bregulla, Griesheim
completed in 2007

Galeria Kaufhof购物中心部分建筑，美因茨
斯塔尼采克建筑设计事务所，波恩
2007年竣工

Galeria Kaufhof, Mainz
Parts of the Building
Architekt Stanitzek, Bonn
completed in 2007

Galeria Kaufhof购物中心，波恩
萨特瓦尔德工作室，杜塞尔多夫
2007年竣工

Galeria Kaufhof, Bonn
Office Statwald, Düsseldorf
completed in 2007

Galeria Kaufhof购物中心，维尔茨堡
斯塔尼采克建筑设计事务所，波恩
2007年竣工

Galeria Kaufhof, Würzburg
Architekt Stanitzek, Bonn
completed in 2007

Galeria Kaufhof购物中心亚历山大广场店，柏林
克莱休斯与克莱休斯建筑设计有限公司，柏林
2006年竣工

Galeria Kaufhof Alexanderplatz, Berlin
Kleihues + Kleihues Architekten GmbH, Berlin
completed in 2006

Galeria Kaufhof购物中心，法兰克福
JSK建筑设计事务所，法兰克福
2004年竣工

Galeria Kaufhof, Frankfurt am Main
JSK Architekten, Frankfurt am Main
completed in 2004

Galeria Kaufhof购物中心，慕逊加柏
斯塔尼采克建筑设计事务所，波恩
2004年竣工

Galeria Kaufhof, Mönchengladbach
Architekt Stanitzek, Bonn
completed in 2004

Galeria　Kaufhof购物中心整体照明规划
凯撒教授
Kaufhof有限公司建筑规划部，科隆
2004年竣工

General Shop Lighting Concept
Galeria Kaufhof
Prof. Cesarz,
Building Department Kaufhof AG, Cologne
completed in 2004

Apropos Cöln体验店，科隆
恩泽瑙尔建筑管理公司，杜塞尔多夫
2004年竣工

Apropos Cöln, Cologne
Enzenauer Architekturmanagement, Düsseldorf
completed in 2004

Peek & Cloppenburg购物中心整体照明概念规划
3号建筑
2003年竣工

General Shop Lighting Concept
Peek & Cloppenburg
Building Department III
completed in 2003

德国邮政中心整体照明概念规划
德国邮政零售服务有限公司
2002年竣工

General Shop Lighting Concept
Post-Shop / McPaper
Deutsche Post Retail Service GmbH
completed in 2002

布尔托工作室，波恩
迪特沙伊德与莫泽西伊建筑设计事务所，波恩
2002年竣工

Bulthaup Studio, Bonn
Ditscheid & Modelsee Architekten, Bonn
completed in 2002

Kaiserpassage购物中心，波恩
卡尔－海因茨·肖默建筑设计事务所，波恩
2001年竣工

Kaiserpassage Shopping Mall, Bonn
Architekturbüro Karl-Heinz Schommer, Bonn
completed in 2001

康托尔豪斯施丁伯克店照明概念设计，不来梅
舒默·舒尔曼建筑设计事务所，不来梅
2001年竣工

Kontorhaus Stintbrücke, Bremen
Lighting Concept
Architekten schomers.schürmann, Bremen
completed in 2001

交通和运输设施

TRAFFIC AND TRANSPORT FACILITIES

欧洲桥，柏林
ACME建筑设计事务所，柏林
2021年竣工

Bridge Europacity, Berlin
acme, Berlin
completed in 2021

法兰克福机场T3航站楼，法兰克福
克里斯托夫·麦克勒建筑设计事务所，法兰克福
正在进行中

Frankfurt Airport, Frankfurt am Main
Terminal 3
Christoph Mäckler Architekten, Frankfurt am Main
work in progress

诺华制药园区广场地下自行车车库及展馆，巴塞尔
马可·塞拉建筑设计事务所，巴塞尔
2017年竣工

Novartis Campus Square, Basel
Underground Bike Garage and Pavilion
Marco Serra Architekt, Basel
completed in 2017

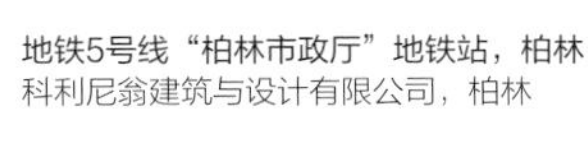

地铁5号线“柏林市政厅”地铁站，柏林
科利尼翁建筑与设计有限公司，柏林
2020年竣工

'Berliner Rathaus' Subway Station, Line 5 Berlin
Collignon Architektur und Design GmbH, Berlin
completed in 2020

南北城市地铁线“波恩墙”地铁站，科隆
BHBFH建筑设计协会有限公司，科隆
2014年竣工

North-South Cityline, Cologne
Subway Station 'Bonner Wall'
BHBFH Gesellschaft von Architekten mbH, Cologne
completed in 2014

“威廉·洛伊施纳广场”地铁站城市隧道，莱比锡
马克斯·杜德勒建筑设计事务所，柏林
2014年竣工

City Tunnel, 'Wilhelm Leuschner Platz' Subway Station, Leipzig
Max Dudler Architekt, Berlin
completed in 2014

城市火车站大楼，巴特洪堡
米兰德工程咨询有限公司，卡尔斯鲁厄
2013年竣工

Railway station building, Bad Homburg
Mailänder Ingenieur Consult GmbH, Karlsruhe
completed in 2013

中央汽车站，古默斯巴赫
Pool 2建筑设计事务所，卡塞尔
2013年扩初设计

Central Bus Station, Gummersbach
Pool 2 Architekten, Kassel
design development 2013

弗里德里希斯桥，柏林
格拉斯尔工程有限公司，柏林
2011年竣工

Friedrichsbrücke, Berlin
Ingenieurbüro Grassl GmbH, Berlin
completed in 2011

南北城市地铁线布雷斯劳尔广场地铁站，科隆
比德尔与河泽尔建筑设计事务所，科隆
2011年竣工

North-South Cityline, Cologne
Breslauer Platz Subway Station
Büder + Menzel Architekten BDA, Cologne
completed in 2011

“动物园/植物园”地铁站，科隆
鲁布萨门建筑设计事务所，波鸿
2011年竣工

Station 'Zoo/Flora', Cologne
Rübsamen + Partner Architekten, Bochum
completed in 2011

贸易展览会入口大门，法兰克福
英戈·施拉德建筑设计事务所，柏林
2010年竣工

Trade Fair Entrance Gates, Frankfurt am Main
Ingo Schrader Architekt, Berlin
completed in 2010

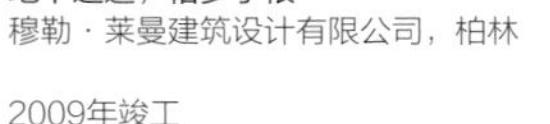

地下通道，格罗宁根
穆勒·莱曼建筑设计有限公司，柏林
2009年竣工

Underpass, Groningen
Müller Reimann Architekten GmbH, Berlin
completed in 2009

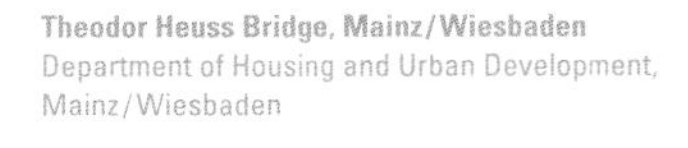

特奥多尔·豪斯桥，美因茨/威斯巴登
住房和城市发展部，美因茨/威斯巴登
2008年竣工

Theodor Heuss Bridge, Mainz/Wiesbaden
Department of Housing and Urban Development, Mainz/Wiesbaden
completed in 2008

施泰因米勒站点环形交叉路口，古默斯巴赫
greenbox景观建筑设计事务所，杜塞尔多夫
2008年竣工

Traffic Circle, Steinmüller Site, Gummersbach
greenbox Landschaftsarchitekten, Düsseldorf
completed in 2008

诺华制药园区塞拉接待大楼和停车场，巴塞尔
马可·塞拉建筑设计事务所，巴塞尔
恩斯特·巴塞尔建筑设计事务所，苏黎世
2007年竣工

Novartis Campus, Basel
Serra Reception Building and Car Park
Marco Serra Architekt, Basel
Ernst Baseler & Partner AG, Zurich
completed in 2007

地铁5号线“国会大楼”地铁站，柏林
阿克塞尔·舒尔茨建筑设计事务所，柏林
2006年竣工

'Reichstag' Subway Station, Line 5, Berlin
Axel Schultes Architekten, Berlin
completed in 2006

“波茨坦广场”火车站，柏林
莫德松和弗赖斯莱本建筑设计事务所，柏林
希尔默与萨特勒和阿尔布雷特建筑设计有限公司，柏林
2006年竣工

'Potsdamer Platz' Railway Station, Berlin
Modersohn & Freiesleben Architekten, Berlin
Hilmer & Sattler und Albrecht GmbH, Berlin
completed in 2006

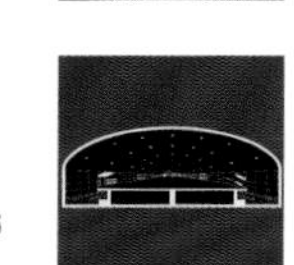

地铁5号线“施普雷岛”地铁站，柏林
马克斯·杜德勒建筑设计事务所，柏林
2005年扩初设计

'Spreeinsel' Subway Station, Line 5, Berlin
Max Dudler Architekt, Berlin
design development 2005

地铁5号线“林登大道”地铁站，柏林
亨舍尔-奥地利建筑设计事务所，柏林
2005年扩初设计

'Unter den Linden' Subway Station, Line 5, Berlin
Hentschel-Oestreich Architekten, Berlin
design development 2005

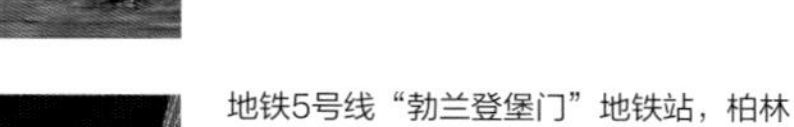

地铁5号线“勃兰登堡门”地铁站，柏林
马克斯·杜德勒建筑设计事务所，柏林
2005年扩初设计

'Brandenburger Tor' Subway Station, Line 5, Berlin
Max Dudler Architekt, Berlin
design development 2005

斯图加特机场T3航站楼
GMP建筑设计事务所，汉堡
2004年竣工

Stuttgart Airport, Terminal 3
von Gerkan, Marg + Partner, Hamburg
completed in 2004

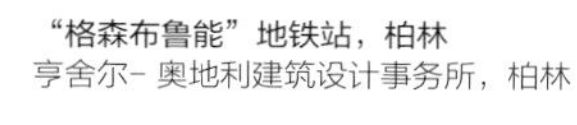

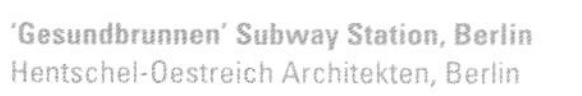

“格森布鲁能”地铁站，柏林
亨舍尔-奥地利建筑设计事务所，柏林
2003年竣工

'Gesundbrunnen' Subway Station, Berlin
Hentschel-Oestreich Architekten, Berlin
completed in 2003

巴伐利亚火车站城市隧道，莱比锡
彼得库尔卡建筑设计事务所，德累斯顿
2002年概念设计

City Tunnel, 'Bayerischer Bahnhof' Railway Station, Leipzig
Peter Kulka Architektur, Dresden
design concept 2002

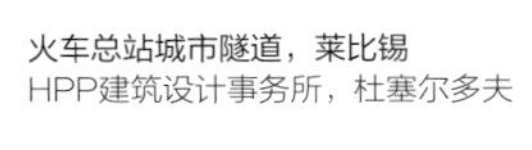

火车总站城市隧道，莱比锡
HPP建筑设计事务所，杜塞尔多夫
2002年概念设计

City Tunnel, Main Railway Station, Leipzig
HPP Architekten, Düsseldorf
design concept 2002

城市集市火车站城市隧道，莱比锡
ksw建筑设计与城市规划有限公司，汉诺威
2002年概念设计

City Tunnel, 'Markt' Railway Station, Leipzig
kellner schleich wunderling architekten + urban design gmbh, Hannover
design concept 2002

2000年世博线海因霍尔茨火车站，汉诺威
汉斯约格·戈利茨建筑设计事务所，汉诺威
1998年竣工

'Hainholz' Railway Station, Expo 2000 Line, Hannover
Hansjörg Göritz Architektur, Hannover
completed in 1998

光环境设计总体规划

MASTERPLANS

Rail Boxes国际机场，多哈
大都会建筑事务所，鹿特丹
2014年竣工

International Airport Rail Boxes, Doha
Rem Koolhaas, The Office for Metropolitan Architecture, Rotterdam
completed in 2014

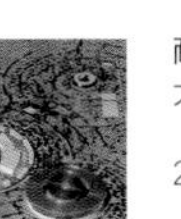

耐力社区村，卡塔尔
大都会建筑事务所，鹿特丹
2014年竣工

Endurance Community Village, Qatar
Rem Koolhaas, The Office for Metropolitan Architecture, Rotterdam
completed in 2014

公路沿线艺术景观光环境整体规划，卡塔尔
大都会建筑事务所，鹿特丹
2015年竣工

Orbital Highway Artscape, Qatar
Rem Koolhaas, The Office for Metropolitan Architecture, Rotterdam
completed in 2015

西夫鲁赛尔滨水项目，多哈
古斯塔夫森·波特建筑设计事务所，伦敦
正在进行中

Seef Lusail Waterfront, Doha
Gustafson Porter, London
work in progress

HIA国际机场社区，多哈
大都会建筑事务所，鹿特丹
2014年竣工

HIA International Airport City, Doha
Rem Koolhaas, The Office for Metropolitan Architecture, Rotterdam
completed in 2014

诺华制药园区整体规划，巴塞尔
英格·维托里奥教授、博士，马格戈·兰普尼亚尼，di建筑设计工作室，米兰
正在进行中

Novartis Campus, Basel
Masterplan
Studio di Architettura Prof. Dr. Ing. Vittorio Magnago Lampugnani, Milan
work in progress

中央火车站周边光环境整体规划，柏林
参议院城市发展部，柏林
2015年竣工

Lighting the Area Surrounding the Central Station, Berlin
Senate Department for Urban Development, Berlin
completed in 2015

电视塔前广场，柏林
莱文·蒙西尼景观建筑设计事务所，柏林
2014年竣工

Forecourt, Television Tower, Berlin
Levin Monsigny Landschaftsarchitekten, Berlin
completed in 2014

关税同盟煤矿工业区关税同盟公园，埃森
奥伯豪森规划设计有限公司，奥伯豪森
2014年竣工

Zollverein Park, Zollverein Colliery, Essen
Planergruppe Oberhausen GmbH, Oberhausen
completed in 2014

卡塔尔文化与运动中心，多哈
大都会建筑事务所，鹿特丹
2014年竣工

Qatar Cultural and Sports Hub, Doha
Rem Koolhaas, The Office for Metropolitan Architecture, Rotterdam
completed in 2014

卡塔尔大学，多哈
大都会建筑事务所，鹿特丹
2013年扩初设计

Qatar University, Doha
Rem Koolhaas, The Office for Metropolitan Architecture, Rotterdam
design development 2013

鲁赛尔奥林匹克街区，多哈
大都会建筑事务所，鹿特丹

2013年扩初设计

Lusail Olympic Precinct, Doha
Rem Koolhaas, The Office for Metropolitan Architecture, Rotterdam

design development 2013

赫斯佩朗日中心，卢森堡
布鲁克与韦克尔建筑设计事务所，卢森堡

2013年竣工

Center for Hesperange, Luxembourg
Bruck+Weckerle Architekten, Luxembourg

completed in 2013

苏迪克东城光环境整体规划，开罗
大都会建筑事务所，鹿特丹

2010年竣工

Sodic Eastown, Cairo Masterplan
Rem Koolhaas, The Office for Metropolitan Architecture, Rotterdam

completed in 2010

明辛格公园，柏林
TOPOS景观建筑设计事务所，柏林

2010年竣工

Münsinger Park, Berlin
TOPOS Landschaftsplanung, Berlin

completed in 2010

的黎波里绿党区，利比亚
莱昂·沃尔哈格·维尔尼克建筑设计事务所，柏林

2008年扩初设计

Tripoli Greens, Libya
Léon Wohlhage Wernik Architekten, Berlin

design development 2008

诺华制药园区4号园区，巴塞尔
沃格特景观建筑设计事务所，苏黎世

2008年竣工

Novartis Campus, Basel Park Area 4
Vogt Landschaftsarchitekten, Zurich

completed in 2008

慕尼黑工业大学加兴校区中央校园
莱昂·沃尔哈格·维尔尼克建筑设计事务所，柏林

2005年概念设计

Zentrale Mitte, Garching University Campus of the Technical University of Munich
Léon Wohlhage Wernik Architekten, Berlin

design concept 2005

帕绍城市光环境总体规划“夜间的帕绍”灯光设计
帕绍住房和城市发展部，帕绍

2001年概念设计

Lighting of the 'Passau at Night' Urban Masterplan of Passau
Department of Housing and Urban Development, Passau

design concept 2001

科研项目

RESEARCH PROJECTS

Tegel照明可行性研究，柏林
Tegel Projekt有限责任公司

2015年竣工

Tegel Feasibility Study, Berlin
Tegel Projekt GmbH

completed in 2015

试点项目“参数化（梦想）空间设计”，柏林
夏洛特CFM设备管理有限责任公司
GRAFT建筑设计有限责任公司
ART + COMAG，由AFI运营推广，柏林

2013年扩初设计

Pilot Project 'parametric (dream)space design', Berlin
Charité CFM Facility Management GmbH
GRAFT-Gesellschaft von Architekten mbH
ART+COM AG; promoted by AFI, Berlin
design development 2013

活动场所

EVENT LOCATIONS

贸易中心9T入口，法兰克福，
英戈·施拉德建筑设计事务所，柏林

2017年竣工

Trade Fair, Frankfurt am Main, Entrance 9T
Ingo Schrader Architekt, Berlin

completed in 2017

2015年世博会法国国家馆，米兰
XTU建筑设计事务所，巴黎
阿德琳·里斯帕尔设计工作室，巴黎

2015年竣工

Pavilion France Expo 2015, Milan
XTU architects, Paris
studio adeline rispal, Paris

completed in 2015

会议中心，慕尼黑
丹·珀尔曼品牌形象建筑设计股份有限公司，柏林
赖希瓦尔德·舒尔茨建筑设计事务所，柏林

2014年竣工

Conference centre, Munich
dan pearlman Markenarchitektur GmbH, Berlin
Reichwald Schultz Architekten, Berlin

completed in 2014

“水房”，波鸿
波鸿城市公共事业有限责任公司

2011年竣工

'Wassersaal', Bochum
Stadtwerke Bochum GmbH

completed in 2011

埃尔策霍夫活动中心，美因茨
ATELIER 30 建筑设计有限责任公司，卡塞尔
BDA费舍尔－克罗伊茨格建筑设计公司，卡塞尔

2011年扩初设计

Eltzer Hof, Mainz
ATELIER 30 Architekten GmbH, Kassel
Fischer – Creutzig BDA, Kassel

design development 2011

“斯坦德豪斯”翻新项目，卡塞尔
ATELIER 30建筑设计有限责任公司，卡塞尔

2011年竣工

Refurbishment of the 'Ständehaus', Kassel
ATELIER 30 Architekten GmbH, Kassel

completed in 2011

苏黎世体育场
梅丽，彼得建筑设计股份有限公司，苏黎世

2010年扩初设计

Zurich Stadium
Meili, Peter Architekten AG, Zurich

design development 2010

科隆大教堂文化交流中心，科隆
马提尼建筑设计事务所，波恩

2010年竣工

Domforum, Cologne
Martini Architekten, Bonn

completed in 2010

莫斯科咖啡馆，柏林
HSH建筑设计事务所，柏林

2009年竣工

Café Moskau, Berlin
HSH Hoyer Schindele Hirschmüller Architekten, Berlin

completed in 2009

法兰克福节日中心外立面照明和部分室内照明
ASP建筑设计有限责任公司，法兰克福

2007年竣工

Frankfurt am Main Festival Hall Facade Illumination and Partial Interior Illumination
Albert Speer und Partner GmbH, Frankfurt am Main

completed in 2007

“世纪大厦”波鸿音乐厅
佩津卡·品客建筑技术公司，杜塞尔多夫

2003年竣工

'Jahrhunderthalle' Contenary Hall Bochum
Petzinka Pink, Technologische Architektur, Düsseldorf

completed in 2003

汉诺威会议中心HCC
ASP建筑设计事务所，汉诺威

1999年竣工

HCC, Hannover Congress Center
ASP Architekten Schweger Partner, Hannover

completed in 1999

体育与文化多功能厅，迈宁根
彼得·库尔卡建筑设计事务所，德累斯顿

1997年竣工

Multifunctional Hall for Sports and Culture, Meiningen
Peter Kulka Architektur, Dresden

completed in 1997

酒店和餐饮

HOTELS AND GASTRONOMY

布尔根施托克温泉度假村，欧贝尔根
帕特里克、迪尔克斯、诺伯特和萨克斯建筑设计事务所，柏林

2017年竣工

Bürgenstock Resort Spa, Obbürgen
Patrik Dierks Norbert Sachs Architekten BDA, Berlin

completed in 2017

The Ash美式牛排馆，特罗斯多夫
尤瑟里默建筑设计事务所，波恩

2016年竣工

The Ash – American Steakhouse, Troisdorf
Oezen-Reimer+Partner, Bonn

completed in 2016

汇丰银行员工餐厅，杜塞尔多夫
ttsp hwp seidel规划有限责任公司，法兰克福

2017年竣工

Staff restaurant, Düsseldorf
ttsp hwp seidel Planungsgesellschaft mbH, Frankfurt am Main

completed in 2017

BER公共休息室，柏林
克莱默·诺伊曼建筑设计事务所，柏林

2018年竣工

Common Use Lounge BER, Berlin
Cramer Neumann Architekten, Berlin

completed in 2018

恩贡贝安缦度假村，加蓬
丹尼斯顿国际建筑师事务所，吉隆坡

2015年竣工

Amanresort, Ngombé, Gabun
Denniston International Architects, Kuala Lumpur

completed in 2015

Abadia Retuerta LeDomaine酒店和水疗中心，西班牙
迪纳与迪纳建筑设计事务所，巴塞尔

2015年竣工

Spa Abadia Retuerta LeDomaine, Sardón de Duero, Spain
Diener & Diener Architekten, Basel

completed in 2015

阿尔卑斯玫瑰酒店，霍恩施万高
施塔布建筑设计事务所，柏林
2013年竣工

Hotel 'Alpenrose', Hohenschwangau
Staab Architekten GmbH, Berlin
completed in 2013

龙岩山高原餐厅，克尼格斯温特尔
佩普与佩普建筑设计事务所，卡塞尔
2013年竣工

Drachenfelsplateau, Königswinter
pape + pape architekten bda, Kassel
completed in 2013

舒伯特林香格里拉大酒店，维也纳
ARGE鲍姆施拉格-艾兴格建筑设计事务所，维也纳
弗兰克建筑设计有限责任公司，维也纳
2011年竣工

Shangri-La Hotel on Schubertring, Vienna
ARGE Baumschlager-Eichinger, Vienna
Architekten Frank + Partner GmbH, Vienna
completed in 2011

Abadia Retuerta LeDomaine酒店和水疗中心，西班牙
马可·塞拉建筑设计事务所，巴塞尔
2012年竣工

Hotel Abadia Retuerta LeDomaine, Sardón de Duero, Spain
Marco Serra Architekt, Basel
completed in 2012

Sticks "n" 寿司餐厅，哥本哈根
迪纳与迪纳建筑设计事务所，巴塞尔
2011年竣工

Restaurant Sticks 'n' Sushi, Copenhagen
Diener & Diener Architekten, Basel
completed in 2011

察夫塔特安缦度假村，克罗地亚
凯瑞·希尔建筑设计事务所，新加坡
2010年扩初设计

Amanresort Cavtat, Croatia
Kerry Hill Architects, Singapore
design development 2010

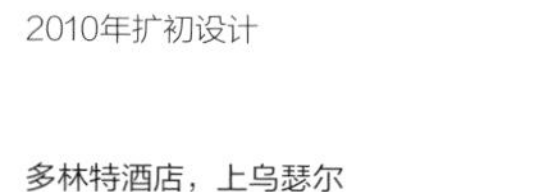

多林特酒店，上乌瑟尔
GHP建筑设计事务所，上乌瑟尔
2010年扩初设计

Dorint Hotel, Oberursel
GHP-Architekten, Oberursel
design development 2010

洲际酒店，布加勒斯特
维吉尔和斯通建筑设计协会，伦敦
2010年竣工

Intercontinental Hotel, Bukarest
Virgile and Stone Associates, London
completed in 2010

圣斯特凡安缦度假村，黑山
丹尼斯顿国际建筑师事务所，吉隆坡
2009年竣工

Amanresort Sveti Stefan, Montenegro
Denniston International Architects, Kuala Lumpur
completed in 2009

茨温格酒店，德累斯顿
福斯特建筑设计事务所，柏林
2008年扩初设计

Hotel near the Zwinger, Dresden
Foster and Partners, Berlin
design development 2008

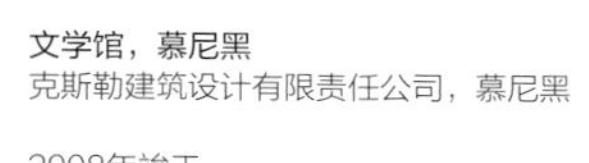

文学馆，慕尼黑
克斯勒建筑设计有限责任公司，慕尼黑
2008年竣工

Literature House, Munich
Kiessler + Partner Architekten GmbH, Munich
completed in 2008

铂尔曼酒吧，科隆
K2建筑设计有限责任公司，亚琛
2008年竣工

Pullmann Bar, Cologne
Architekten K2 GmbH, Aachen
completed in 2008

威斯特法伦省级保险大厦室内咖啡厅区域，明斯特
沙格曼·舒尔特建筑设计有限责任公司，波茨坦
2008年竣工

Westfälische Provinzial Insurance, Münster Catering Parts of the Building
Architekturcontor Schagemann Schulte GmbH, Potsdam
completed in 2008

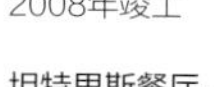

坦特里斯餐厅，慕尼黑
斯蒂芬·布朗费尔斯建筑设计事务所，柏林
2005年竣工

Tantris Restaurant, Munich
Stephan Braunfels Architekten, Berlin
completed in 2005

柏林馆中的汉堡王餐厅，柏林
卡尔费尔特建筑设计有限责任公司，柏林
2005年竣工

Burger King in the Berlin Pavilion, Berlin
Kahlfeldt Architekten GmbH, Berlin
completed in 2005

丽思卡尔顿酒店部分建筑，柏林
希尔默与萨特勒和阿尔布雷希特有限公司，柏林
2004年竣工

The Ritz Carlton Hotel, Berlin Parts of the Building
Hilmer & Sattler und Albrecht GmbH, Berlin
completed in 2004

阿德隆酒店，柏林
Living Architects AB，瑞典于什霍尔姆
1997年竣工

Adlon Hotel, Berlin
Living Architects AB, Djursholm, Sweden
completed in 1997

户外区域

OUTDOOR AREAS

柏林旧博物馆门廊照明，柏林
柏林国家博物馆，普鲁士文化遗产基金会，柏林
2017年竣工

Altes Museum Portico, Berlin
Berlin State Museums, Prussian Cultural Heritage, Berlin
completed in 2017

圣母教堂室外区，柏林
莱文·蒙西尼景观建筑设计有限责任公司，柏林
2017年竣工

Marienkirche Precinct, Berlin
Levin Monsigny Landscape Architects GmbH, Berlin
completed in 2017

多瑙河市场，雷根斯堡
沃尔夫冈·韦泽尔景观建筑设计有限责任公司，英戈尔施塔特
2018年竣工

Donaumarkt, Regensburg
Wolfgang Weinzierl Landschaftsarchitekten GmbH, Ingolstadt
completed in 2018

鹏瑞中心深圳湾壹号，深圳
KPF建筑设计事务所，纽约
AECOM集团，深圳
2018年竣工

Parkland Center, One Shenzhen Bay, Shenzhen
Kohn Pedersen Fox Associates, New York
AECOM, Shenzhen
completed in 2018

工人广场，柏林
莱文·蒙西尼景观建筑设计有限责任公司，柏林
2018年竣工

Platz der Aufbauhelfer, Berlin
Levin Monsigny Landscape Architects GmbH, Berlin
completed in 2018

贸易展览会北入口，法兰克福
英戈·施拉德建筑设计事务所，柏林
2014年竣工

Trade Fair Entrance Gates, Frankfurt am Main, Northern Entrance
Ingo Schrader Architekt, Berlin
completed in 2014

La Lestra会所室外照明，摩纳哥
法布里斯·诺塔里建筑设计事务所，摩纳哥
2014年竣工

La Lestra, Monaco
Architecte Fabrice Notari, Monaco
completed in 2014

华盛顿广场景观照明，柏林
基弗景观建筑设计工作室，柏林
2014年竣工

Grove at Washingtonplatz, Berlin
Büro Kiefer Landschaftsarchitektur, Berlin
completed in 2014

理查德-瓦格纳广场，莱比锡
艾琳·洛豪斯·彼得·卡尔景观建筑设计事务所，汉诺威
2013年竣工

Richard-Wagner-Platz, Leipzig
Irene Lohaus Peter Carl Landschaftsarchitektur, Hannover
completed in 2013

Kurfürstenhöfe选举纪念碑，柏林
伯恩德·德里森建筑设计事务所，科隆
kwp建筑设计事务所，柏林
2013年竣工

Kurfürstenhöfe, Berlin
Bernd Driessen Architekt, Cologne
kwp-Architekten, Berlin
completed in 2013

考古研究展示基地，亚琛
卡达维特费尔德建筑设计有限责任公司，亚琛
2012年竣工

Archeological showcase, Aachen
kadawittfeldarchitektur GmbH, Aachen
completed in 2012

国王大道Galeria Kaufhof购物中心外立面照明，杜塞尔多夫
METRO集团资产管理服务有限公司，杜塞尔多夫
2012年竣工

Facade Illumination Galeria Kaufhof Königsallee, Düsseldorf
METRO Group Asset Management Service GmbH, Düsseldorf
completed in 2012

电视塔翻新项目，柏林
PSP建筑工程事务所，柏林
2011年竣工

Television Tower Redevelopment, Berlin
Architekten Ingenieure PSP, Berlin
completed in 2011

上柯尼希斯大街翻新项目，卡塞尔
海因茨·雅南·普弗吕格城市规划与建筑设计事务所，亚琛
2009年扩初设计

Remodelling of Upper Königsstrasse, Kassel
Heinz Jahnen Pflüger Stadtplaner und Architekten Partnerschaft, Aachen
design development 2009

克莱斯特公园的皇家柱廊，柏林
柏林－滕珀尔霍夫地区管理局

2009年竣工

Königskolonnaden in the Kleistpark, Berlin
District Authority, Berlin-Tempelhof

completed in 2009

查尔大街Galeria Kaufhof购物中心外立面照明，法兰克福
RKW建筑城市规划事务所，杜塞尔多夫

2008年竣工

Facade Illumination, Galeria Kaufhof, Frankfurt-Zeil
RKW Architektur + Städtebau, Düsseldorf

completed in 2008

斯托舍克藏品馆外立面照明，杜塞尔多夫
库恩·马尔维兹建筑设计有限责任公司，柏林

2007年竣工

Facade Illumination Stoschek Collection, Düsseldorf
Kühn Malvezzi Architekten GmbH, Berlin

completed in 2007

华盛顿广场，柏林
加布里埃尔·基弗教授与剑桥大学文学硕士研究生，玛莎·施瓦茨公司，柏林

2007年竣工

Washington Square, Berlin
Martha Schwartz Inc. Cambridge MA with Prof. Gabriele Kiefer, Berlin

completed in 2007

莱茵斯堡的城堡桥梁
莱昂·沃尔哈格·维尔尼克建筑设计事务所，柏林

2007年竣工

Rheinsberg Castle Bridges
Léon Wohlhage Wernik Architekten, Berlin

completed in 2007

亚历山大广场Galeria Kaufhof购物中心外立面照明，柏林
METRO集团资产管理有限责任公司，杜塞尔多夫

2007年竣工

Facade Concept Galeria Kaufhof Alexanderplatz, Berlin
METRO Group Asset Management, Düsseldorf

completed in 2007

Galeria Kaufhof购物中心外立面照明，曼海姆
考夫霍夫仓储股份有限公司，
城市建筑施工管理局，科隆

2007年竣工

Facade Illumination Galeria Kaufhof, Mannheim
Kaufhof Warenhaus AG,
Building Construction Management, Cologne

completed in 2007

瓶子广场，波恩
波恩城市管理局

2007年竣工

Bottler Square, Bonn
City Administration Bonn

completed in 2007

老美术馆部分建筑，慕尼黑
巴伐利亚公共绘画收藏所

2006年竣工

Old Pinakothek, Munich
Parts of the Building
Bavarian Public Collection of Paintings

completed in 2006

"御林广场"道路照明，柏林
柏林参议院

2006年竣工

'Gendarmenmarkt' Streetlighting Survey, Berlin
Senate of Berlin

completed in 2006

尼古拉教堂和教堂广场，莱比锡
莱比锡政府

2005年竣工

Nikolaikirche and Church Square, Leipzig
City of Leipzig

completed in 2005

卡尔施建筑外立面照明，杜塞尔多夫
凯撒教授，METRO集团资产管理有限责任公司，杜塞尔多夫

2005年竣工

Carsch Building Facade Illumination, Düsseldorf
Prof. Cesarz, METRO Group Asset Management GmbH & Co. KG, Düsseldorf

completed in 2005

选帝侯大街临街建筑外立面照明，柏林
卡尔·费尔特建筑设计有限责任公司，柏林

2005年竣工

Kurfürstendamm Facade, Berlin
Kahlfeldt Architekten GmbH, Berlin

completed in 2005

毛奇广场，波恩
schöne aussichten景观建筑设计事务所，卡塞尔

2005年竣工

Moltke Square, Bonn
schöne aussichten landschaftsarchitektur, Kassel

completed in 2005

耶稣之名教堂外立面照明，波恩
波恩住房和城市发展部，波恩

2005年竣工

Facade Illumination of Name of Jesus Church, Bonn
Department of Housing and Urban Development, Bonn

completed in 2005

Galeria Kaufhof购物中心外立面照明，科隆
凯撒教授，METRO集团资产管理有限责任公司，杜塞尔多夫

2004年竣工

Facade Illumination Galeria Kaufhof, Cologne
Prof. Cesarz, METRO Group Asset Management GmbH & Co. KG, Düsseldorf

completed in 2004

特里尔市历史古迹照明项目
特里尔住房和城市发展部，特里尔

2003年竣工

Illumination of the arcaeological historical monuments of the City of Trier
Department of Housing and Urban Development, Trier

completed in 2003

私家住宅

PRIVATE RESIDENCES

百丽宫公寓式酒店新建项目，柏林
Graft建筑设计公司，柏林

2016年竣工

New Construction of Paragon Apartments, Berlin
Graft – Gesellschaft von Architekten mbH, Berlin

completed in 2016

格诺尔特别墅，柏林
施莱尔建筑设计事务所，柏林

2015年竣工

Villa Gnauert, Berlin
Hölzer + Co. Architekten, Berlin

completed in 2015

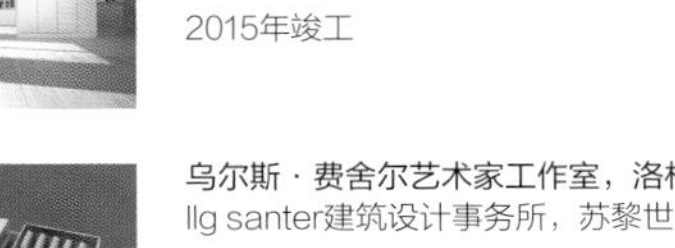

马克格拉芬大街住宅和零售建筑，柏林
马克斯·杜德勒建筑设计事务所，柏林

2015年竣工

Residential and Retail Building Markgrafenstraße, Berlin
Max Dudler, Berlin

completed in 2015

乌尔斯·费舍尔艺术家工作室，洛杉矶
Ilg santer建筑设计事务所，苏黎世

2013年扩初设计

Artist's Workshop Urs Fischer, Los Angeles
Ilg santer gmbh, Zurich

design development 2013

住宅楼，科隆
赫尔施和亨里希建筑设计事务所，科隆

2012年竣工

Residential building, Cologne
Hoersch & Hennrich Architekten, Cologne

completed in 2012

住宅楼，波恩
卡尔－海因茨·肖默建筑设计事务所，波恩

2012年竣工

Residential building, Bonn,
Architekturbüro Karl-Heinz Schommer, Bonn

completed in 2012

厂房住宅楼，柏林
范登瓦伦丁建筑设计事务所，科隆

2012年竣工

Hutfabrik, Berlin
van den Valentyn Architektur, Cologne

completed in 2012

尚菲尔住宅楼，瑞士
迪纳与迪纳建筑设计事务所，巴塞尔

2011年竣工

Residential building Via Suot Chesas, Champfèr
Diener & Diener Architekten, Basel

completed in 2011

蒂尔花园，普福尔茨海姆
莱昂·沃尔哈格·维尔尼克建筑设计事务所，柏林

2011年扩初设计

Tiergarten, Pforzheim
Léon Wohlhage Wernik Architekten, Berlin

design development 2011

住宅楼，柏林
穆勒·莱曼建筑设计有限责任公司，柏林

2011年竣工

Residential building, Berlin
Müller Reimann Architekten GmbH, Berlin

completed in 2011

农庄，弗斯特
库恩建筑设计有限责任公司，达姆施塔特
贝克尔与舍恩建筑设计事务所，诺伊施塔特

2011年竣工

Gutshof, Forst
Kühn Architekten –
Ingenieure modul-AR GmbH, Darmstadt
Becker + Schöne Architekten, Neustadt/Weinstraße

completed in 2011

哈莫尔新建开发项目，布鲁塞尔
迪纳与迪纳建筑设计事务所，巴塞尔

2011年竣工

New Development Hamoir, Brussels
Diener & Diener Architekten, Basel

completed in 2011

马萨霍夫住宅区，柏林
格兰特·恩斯特建筑设计事务所，柏林

2011年竣工

Marthashof, Berlin
Grüntuch Ernst Architekten, Berlin

completed in 2011

沙伊特别墅建设和改造项目，埃森
恩泽瑙尔建筑管理公司，杜塞尔多夫

2009年竣工

Construction and Conversion of Villa Scheidt, Essen
Enzenauer Architekturmanagement, Düsseldorf

completed in 2009

致谢

WITH THANKS TO

安德烈亚斯 · 舒尔茨和 LKL 全体员工，终身荣誉教授简·埃希德

Andreas Schulz, Licht Kunst Licht and Prof. em. Jan Ejhed

Licht Kunst Licht AG
Engineers Designers Architects for Illumination

Bonn
D-53115 Bonn, Jagdweg 16
Telefon + 49 228 91 42 20
Telefax + 49 228 91 12 44
bonn@lichtkunstlicht.com
www.lichtkunstlicht.com

Berlin
D-10997 Berlin, Schlesische Straße 27
Telefon + 49 30 61 79 310
Telefax + 49 30 61 70 083
berlin@lichtkunstlicht.com
www.lichtkunstlicht.com

图片 Photos

施泰德博物馆，法兰克福
STÄDELSCHES KUNSTINSTITUT UND STÄDTISCHE GALERIE, FRANKFURT AM MAIN
诺伯特 · 米古莱茨（Norbert Miguletz）: S. 16 ; 20—28 ; 30—31
安德烈·弗拉克（Andrea Flak）: S. 32—33

蒂森克虏伯新总部园区，埃森
THYSSENKRUPP QUARTER, ESSEN
卢卡斯·罗斯（Lukas Roth）: S. 36 ; 40—47
亚地斯有限公司，维滕（ARDEX GmbH, Witten）: S. 48—51

LWL 艺术与文化博物馆，明斯特
LWL-MUSEUM FÜR KUNST UND KULTUR, MÜNSTER
马库斯·埃本纳（Marcus Ebener）: S. 54 ; 58—73

赛恩钢铁铸件馆，本多夫
SAYN IRON WORKS FOUNDRY, BENDORF
约翰内斯·罗洛夫（Johannes Roloff）: S. 76 ; 80—81; 83
托马斯·奈特（Thomas Naethe）: S. 82

阿伦斯霍普艺术博物馆，阿伦斯霍普
AHRENSHOOP MUSEUM OF ART, AHRENSHOOP
斯特凡·穆勒（Stefan Müller）: S. 86 ; 90—99

理查德 – 瓦格纳广场，莱比锡
RICHARD-WAGNER-PLATZ, LEIPZIG
PUNCTUM / 伯特伦·科伯（Bertram Kober）: S. 102 ; 106—107
迈克尔·莫泽（Michael Moser）: S. 108—109

柏林大道购物中心，柏林
BOULEVARD BERLIN, BERLIN
迪米特里奥斯·卡察马卡斯（Dimitrios Katsamakas）: S. 112 ; 116—123

龙岩山高原餐厅，克尼格斯温特尔
DRACHENFELSPLATEAU, KÖNIGSWINTER
卢卡斯·罗斯（Lukas Roth）: S. 126 ; 130—135

科隆大教堂文化交流中心，科隆
DOMFORUM, COLOGNE
康斯坦丁 · 迈耶（Constantin Meyer）: S. 138 ; 142—147

Abadia Retuerta LeDomaine 酒店和水疗中心，西班牙萨尔东德杜埃罗
HOTEL & SPA ABADIA RETUERTA LEDOMAINE, SARDÓN DE DUERO, SPAIN
马库斯·埃本纳（Marcus Ebener）: S. 150 ; 154—179

深圳湾壹号，深圳
ONE SHENZHEN BAY, SHENZHEN
深圳市鹏瑞地产开发有限公司: S. 182; 186—187; 190—193; 196—197
陈玉生（Sam Chan）: S. 188—189
江怡辰（Yichen Jiang），卢卡斯·金 (Lucas King): S.195

巴伐利亚国王博物馆，霍恩施万高
MUSEUM OF THE BAVARIAN KINGS, HOHENSCHWANGAU
马库斯·埃本纳（Marcus Ebener）：S. 200；204—219

布雷斯劳尔广场地铁站，科隆
BRESLAUER PLATZ SUBWAY STATION, COLOGNE
弗里德·比克力（Frieder Blickle）：S. 222；226－231

巴登－符腾堡州内政部新大楼，斯图加特
BADEN-WÜRTTEMBERG MINISTRY OF THE INTERIOR, STUTTGART
马库斯·埃本纳（Marcus Ebener）：S. 234；238－245

利布弗劳恩圣玛利亚教堂骨灰安置所，多特蒙德
COLUMBARIUM LIEBFRAUENKIRCHE, DORTMUND
卢卡斯·罗斯（Lukas Roth）：S. 248；252—257

威斯滕德校区的高校建筑群，法兰克福
UNIVERSITY BUILDINGS ON CAMPUS WESTEND, FRANKFURT AM MAIN
斯特凡·穆勒（Stefan Müller）：S. 260；264—271；273—279

会议中心，慕尼黑
CONFERENCE CENTRE, MUNICH
diephotodesigner.de：S. 282；286—293

团队 TEAM
HGEsch 图片摄影工作室：安德烈亚斯·舒尔茨（Andreas Schulz）人像照
约翰内斯·罗洛夫（Johannes Roloff）
劳拉·苏德布罗克（Laura Sudbrock）
尼尔斯·冯·李森（Nils von Leesen）
迈克·沙尼亚克（Maik Czarniak）
斯蒂芬妮·格罗斯－布罗克霍夫（Stephanie Grosse-Brockhoff）
丽莎·戈尔克（Lisa Görke）
托马斯·莫瑞兹（Thomas Möritz）

巴伐利亚国王博物馆，霍恩施万高
MUSEUM OF THE BAVARIAN KINGS, HOHENSCHWANGAU
展品的版权属于慕尼黑维特尔斯巴赫补偿基金资产管理部门
The rights of the exhibits belong to the asset management department of Wittelsbacher Ausgleichsfonds, Munich

施泰德博物馆，法兰克福
STÄDELSCHES KUNSTINSTITUT UND STÄDTISCHE GALERIE, FRANKFURT AM MAIN
© VG Bild-Kunst, Bonn 2015
作品版权：
凯瑟琳娜·西弗丁（Katharina Sieverding）
亚历山大·阿奇彭科（Alexander Archipenko）
赫尔曼·格洛克纳（Hermann Glöckner）
让·阿尔普（Jean Arp）
丹尼尔·里希特（Daniel Richter）
伊萨·根茨肯（Isa Genzken）
迪克·施密特（Dierk Schmidt）
伯纳德·舒尔茨（Bernard Schultze）
安东尼乌斯·霍克尔曼（Antonius Höckelmann）
埃米尔·舒马赫（Emil Schumacher）
卡尔·奥托·格茨（Karl Otto Götz）
伊米·诺贝尔（Imi Knoebel）
赫尔曼·尼奇（Hermann Nitsch）
吉里·格奥尔格·多库皮尔（Jiri Georg Dokoupil）
© Elisabeth Nay-Scheibler, Köln / VG Bild-Kunst, Bonn 2015
作品版权：恩斯特·威廉·奈（Ernst Wilhelm Nay）

LWL 艺术与文化博物馆，明斯特
LWL-MUSEUM FÜR KUNST UND KULTUR, MÜNSTER
© VG Bild-Kunst, Bonn 2015
作品版权：
莱纳·鲁滕贝克（Reiner Ruthenbeck）
托马斯·鲁夫（Thomas Ruff）
卡尔·安德烈（Carl Andre）

阿伦斯霍普艺术博物馆，阿伦斯霍普
AHRENSHOOP MUSEUM OF ART，AHRENSHOOP
© VG Bild-Kunst，Bonn 2015
作品版权：
约阿希姆·博彻（Joachim Böttcher）
西奥·鲍登（Theo Balden）
雷内·格雷茨（René Graetz）
维兰德·福斯特（Wieland Förster）
哈拉尔德·梅茨克斯（Harald Metzkes）
曼弗雷德·博彻（Manfred Böttcher）
赫尔曼·巴赫曼（Hermann Bachmann）
威利·西特（Willi Sitte）
沃尔夫冈·马特厄（Wolfgang Mattheuer）
伯恩哈德·海西格（Bernhard Heisig）
埃德蒙·凯斯汀（Edmund Kesting）
汉斯·文特（Hans Vent）
塞萨尔·克莱因（Cesar Klein）
多拉·科赫－斯泰特（Dora Koch-Stetter）
格哈德·马克斯（Gerhard Marcks）
马克斯·考斯（Max Kaus）

2015年9月，LKL团队在米兰普拉达基金会艺术中心　Licht Kunst Licht visiting Fondazione Prada in Milan, September 2015